Transport and Fate of Chemicals in the Environment

This volume collects selected topical entries from the *Encyclopedia of Sustainability Science and Technology* (ESST). ESST addresses the grand challenges for science and engineering today. It provides unprecedented, peer-reviewed coverage of sustainability science and technology with contributions from nearly 1,000 of the world's leading scientists and engineers, who write on more than 600 separate topics in 38 sections. ESST establishes a foundation for the research, engineering, and economics supporting the many sustainability and policy evaluations being performed in institutions worldwide.

John S. Gulliver
Editor

Transport and Fate of Chemicals in the Environment

Selected Entries from the Encyclopedia of Sustainability Science and Technology

Springer

Editor
John S. Gulliver
Department of Civil Engineering
University of Minnesota
Minneapolis, MN 55455, USA

This book consists of selections from the Encyclopedia of Sustainability Science and Technology edited by Robert A. Meyers, originally published by Springer Science+Business Media New York in 2012.

ISBN 978-1-4614-5730-5 ISBN 978-1-4614-5731-2 (eBook)
DOI 10.1007/978-1-4614-5731-2
Springer New York Heidelberg Dordrecht London

Library of Congress Control Number: 2012953699

Printed on acid-free paper

Springer is part of Springer Science+Business Media (www.springer.com)

Contents

Chapter 1
Transport and Fate of Chemicals in the Environment, Introduction

John S. Gulliver

Glossary

Concentration	The quantity of a compound or chemical per unit volume, unit mass, or unit mole, where 1 mole = 6.02×10^{23} molecules of the chemical or compound. In this text, concentration in mass or moles per volume of water, mass per mass of solid, and moles per mole of gas will be discussed, depending upon the media of interest.
Convection	The movement of a constituent with the movement of the fluid.
Density	Total mass per unit volume.
Diffusion	The spreading of fluid constituents through the motion inherent to atoms and molecules.
Diffusion coefficient	A coefficient that describes the tendency of molecules to spread a constituent mass.
Dilution	The mixing of a more concentrated solution with one that is less concentrated. The adage "The solution to pollution is dilution" is still used, sometimes appropriately, for many pollution and mitigation processes.
Kinematic viscosity	The fluid viscosity divided by the fluid density, resulting in units that are similar to a diffusion coefficient, or length squared per time.
Turbulent diffusion	The mixing of chemicals by turbulence, such that a turbulent diffusion coefficient can be defined separately from the temporal mean convection.

This chapter was originally published as part of the Encyclopedia of Sustainability Science and Technology edited by Robert A. Meyers. DOI:10.1007/978-1-4419-0851-3

J.S. Gulliver (✉)
St. Anthony Falls Laboratory, Department of Civil Engineering, University of Minnesota, 500 Pillsbury Drive S.E, 55454 Minneapolis, MN, USA
e-mail: gulli003@umn.edu

J.S. Gulliver (ed.), *Transport and Fate of Chemicals in the Environment: Selected Entries from the Encyclopedia of Sustainability Science and Technology*, DOI 10.1007/978-1-4614-5731-2_1, © Springer Science+Business Media New York 2012

Introduction

Estimating the fate and transport of chemicals released into the environment is an interesting and challenging task. The environment can rarely be approximated as well mixed, and the chemicals in the environment often are not close to equilibrium. Thus, chemical fate and transport in the environment requires a background in the physics of fluid flow and transport, chemical thermodynamics, chemical kinetics, and the biology that interacts with all of these processes. The goal is to follow chemicals as they move, diffuse, and disperse through the environment. These chemicals will inevitably react to form other chemicals, in a manner that approaches, but rarely achieves, a local equilibrium. Many times, these reactions are biologically mediated, with a rate of reaction that more closely relates to an organism being hungry, or not hungry, than to first-order kinetics.

The global environment is large, on the chemical fate and transport scale. Individuals attempt to apply the mathematics of diffusion to the atmosphere, lakes, rivers, groundwater, and the ocean, depending on the system for which the material is most applicable, and to transfer between these systems. Volatilization of a compound from a water body, condensation of a compound from the air, and adsorption of a compound from a fluid onto a solid are all interfacial transport processes. Thus, the fate and transport of chemicals in the environmental media of earth, water, and the atmosphere will be the topic. Contributions in this section will first attempt to formulate fate and transport problems such that they can be solved, regardless of the media or of the transport process, through the mathematics of diffusion (see chapters on Chemicals in the Environment, Diffusive Transport; Chemicals in the Environment, Dispersive Transport; Chemicals in the Environment, Turbulent Transport; Toxic Organic Chemicals; Transport in the Environment). Applications of these principals to specific media and between media will then be described (see chapters on Transport with Jets and Plumes of Chemicals in the Environment; Atmosphere-Water Exchange; Sediment–Water Interfaces, Chemical Flux at; River Fate and Transport; Lake and Reservoir Fate and Transport of Chemicals; Oceanic Fate and Transport of Chemicals; and Subsurface Fate and Transport of Chemicals).

Transport Processes

A transport process, as used herein, is one that moves chemicals and other properties of the fluid through the environment. *Diffusion* of chemicals is one transport process, which is always present. It is a spreading process, which cannot be reversed (without the involvement of another media such as in reverse osmosis). *Convection or advection* is the transport of chemicals from one place to another by fluid flow. The convection and diffusion of a chemical cloud, as represented in Fig. 1.1, is the movement of the cloud and spreading of the cloud over time.

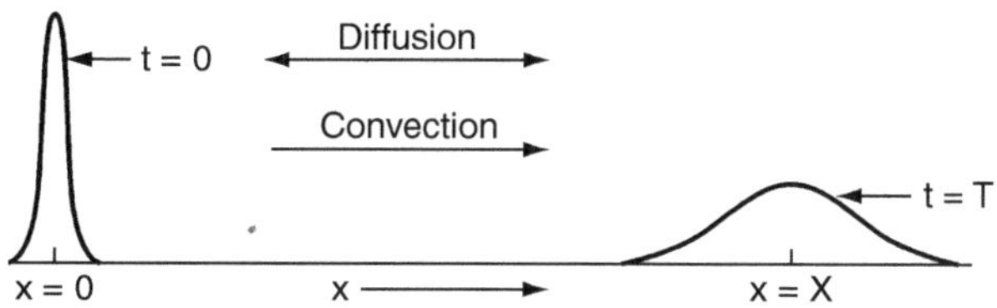

Fig. 1.1 Illustration of convection and diffusion of a chemical cloud along the x-space coordinate (x-axis) [2]

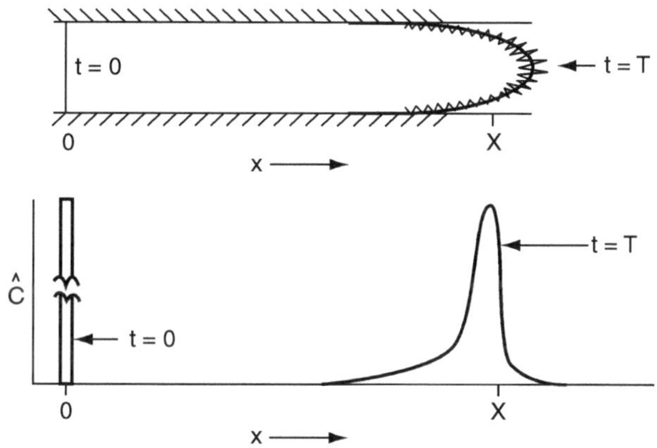

Fig. 1.2 Illustration of longitudinal dispersion of a tracer "plane" at t = 0 to a dispersed "cloud" at $t = T$. is the cross-sectional mean concentration [2]

Turbulent diffusion is actually a form of advection, but the turbulent eddies tend to mix fluid with a random characteristic similar to that of the diffusion process, when viewed from enough distance. The representation given in Fig. 1.1 could also be used to represent convection and turbulent diffusion, except that the pace of turbulent diffusion is normally more than one order of magnitude greater than diffusion. This higher pace of turbulent diffusion means that diffusion and turbulent diffusion do not normally need to be considered together because they can be seen as parallel rate processes, and one has a much different time and distance scale than the other. If two parallel processes occur simultaneously, and one is much faster than the other, the second process can normally be ignored. This is discussed further in this section.

Dispersion is the combination of a nonuniform velocity profile and either diffusion or turbulent diffusion to spread the chemical longitudinally or laterally. Dispersion is something very different than either diffusion or turbulent diffusion because the velocity profile must be nonuniform for dispersion to occur. The longitudinal dispersion of a pipe flow is illustrated in Fig. 1.2. While there is diffusion of the chemical, the nonuniform velocity profile creates a dispersion that is much greater than would occur with diffusion alone. The other important difference is that *dispersion reflects the spreading of a cross-sectional mean concentration*, while diffusion represents the spreading of a local concentration. In some contexts, typically in atmospheric applications, turbulent diffusion is also considered to be a form of dispersion. This is only a semantic difference, and herein, the differentiation will continue to be between turbulent diffusion and the dispersion of a mean concentration.

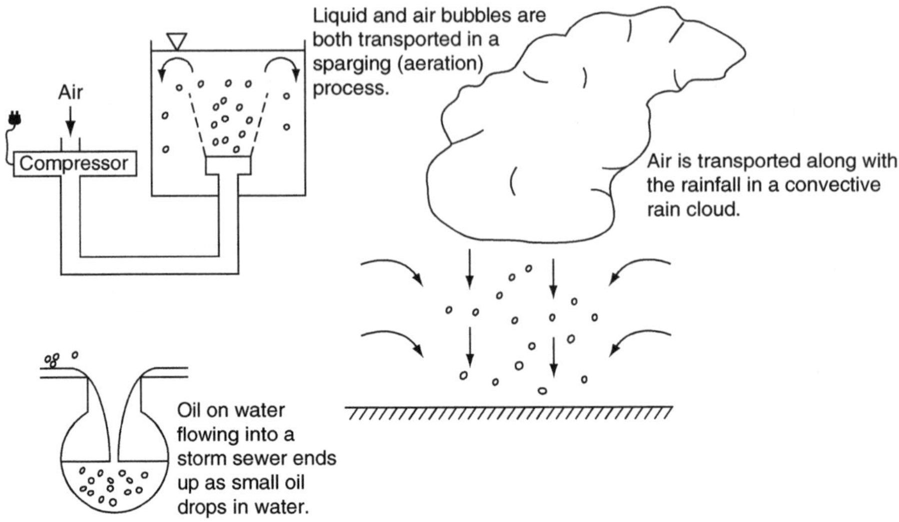

Fig. 1.3 Illustration of sparged multiphase transport. In these two cases, air bubbles create a water flow, and rain drops create an airflow. The oil drops and rain drops create an airflow. The oil drops do not have a significant rise or fall velocity in water and are simply transported [2]

Interfacial transfer is the transport of a chemical across an interface. The most studied form of interfacial transfer is absorption and volatilization, or condensation and evaporation, which is the transport of a chemical across the air–water interface. Another form of interfacial transfer would be adsorption and desorption, generally from water or air to the surface of a particle of soil, sediment, or dust. Illustration of both of these forms of interfacial transfer will be given later in this section.

Finally, there is *multiphase transport*, which is the transport of more than one phase, usually partially mixed in some fashion. The settling of particles in water or air, the fall of drops, and the rise of bubbles in water are all examples of multiphase transport. Figure 1.3 illustrates three flow fields that represent multiphase transport.

Mass transport problems are solved with the *diffusion equation*, often represented as

$$\frac{\partial C}{\partial t}+u\frac{\partial C}{\partial x}+v\frac{\partial C}{\partial y}+w\frac{\partial C}{\partial z}=D\left[\frac{\partial^2 C}{\partial x^2}+\frac{\partial^2 C}{\partial y^2}+\frac{\partial^2 C}{\partial y^2}+\frac{\partial^2 C}{\partial z^2}\right]+S \tag{1.1}$$

1| ← 2 →| |← 3 →| 4

where C is the concentration of a chemical; t is time; u, v, and w represent the temporal mean velocity in the x, y, and z directions, respectively; and D represents a diffusion coefficient. The first term (1) on the far left of Eq. 1.1 represents the rate of accumulation of chemical concentration. The second terms (2) represent the mean convection of the chemical. The third terms (3) to the right of the equal sign represent either diffusion or turbulent diffusion of chemical. The fourth term (4) represents the multitude of reactions that are possible in a fluid in environmental media.

Table 1.1 Examples of short-term and long-term fate

Chemical	Media	Short-term fate	Long-term fate
PCB	Soil and water	Adsorbed to soil	Bioremediated degradation
PCB	Atmosphere	Adsorbed to aerosols	Photocatalyzed degradation
CO_2	Water	Reactions to carbonate and bicarbonate	Photosynthesis to oxygen and biomass
Benzene	Water	Adsorbed to suspended particles	Bioremediated degradation
Ammonia	Soil and water	Reaction to ammonium	Bioremediated degradation to N_2

Chemical Fate

Chemical fate is the eventual short-term or long-term disposition of chemicals, usually to another chemical or storage. Some examples that fit the concept of short-term and long-term fate are given in Table 1.1. If a polychlorinated biphenyl (PCB) compound is in groundwater, the media is soil and water. The "short-term" fate will be that the PCB will primarily adsorb to the soil. The "long-term" fate is that the chemical will desorb, when the PCB-laden water has left, and eventually be bioremediated by microbacteria looking for carbon sources. If this PCB is in the atmosphere, it will be adsorbed primarily to aerosols and particles in the short-term, while its long-term fate will probably be photocatalyzed degradation.

There are as many examples of short-term and long-term fate as there are chemical–media combinations. An important consideration for this topic is whether one is interested in short-term or long-term fate. This is often a question to be answered by toxicologists because it is the most toxic forms of a chemical that are of most interest.

The Importance of Mixing

Mixing is a rate-related parameter in that most rates of reaction or transport are dependent upon mixing in environmental systems. When mixing is dominant (the slowest process), the first-order rate equation can be described as

$$\text{Rate of process} = \text{Mixing parameter} \times \text{Difference from equilibrium} \tag{1.2}$$

Thus, two items are needed to compute the rate of the process: the equilibrium concentrations for all species involved and the mixing rate parameter. A common example would be dissolved oxygen concentration in aquatic ecosystems.

One of the most common chemicals of concern in water bodies is oxygen. Without sufficient oxygen, the biota would be changed because the "desirable" organisms in the water body require oxygen to live. The rate of oxygen transfer between the atmosphere and a water body is therefore important to the health of the aquatic biota. For air-water oxygen transfer, Eq. 1.2 can be formulated as

$$\frac{dM}{dt} = K_L A\left(\frac{C_a}{H} - C\right) \tag{1.3}$$

where dM/dt is the rate of mass transfer into the water, K_L is a bulk oxygen transfer coefficient, A is the surface area for transfer, C_a is the concentration of oxygen in the air, H is a coefficient that partitions oxygen between the air and water at equilibrium (called Henry's Law constant for liquids and gas equilibrium), and C is the concentration of oxygen in the water. Air is 20.8% oxygen, so the concentration of oxygen in the atmosphere is determined primarily by atmospheric pressure. Henry's Law constant for oxygen is a function of pressure as well as temperature. Thus, the equilibrium concentration of oxygen is influenced by the thermodynamic variables pressure and temperature. The rate parameter is K_L A, which has units of volume/second. The difference from equilibrium partitioning is represented by $C_a/H - C$. It is C that is typically needed to bring as close to equilibrium with the atmosphere as possible, and the means to do it is by having a large dM/dt. This usually means a large K_LA because it would be difficult to alter either C_a or H in the atmosphere. While the surface area is often established by the boundary conditions, K_L is determined by the turbulence and diffusion coefficient (i.e., mixing) close to the water surface and represents the rate of mixing per unit surface area. Thus, the primary variable that can be changed in order to increase dM/dt is the mixing parameter represented by K_L. Some further examples of mixing rate and equilibrium parameters in environmental processes are given in Table 1.2.

Resistance to Transport

An important concept for environmental transport is *resistance*. That is, the inverse of a rate parameter is a resistance to chemical transport, or in equation form:

$$\frac{1}{\text{Rate parameter}} = \text{Resistance to chemical transport} = R \tag{1.4}$$

Figure 1.4 gives an example of the adsorption of a compound to suspended sediment, modeled as two resistances in series. At first, the compound is dissolved in water. For successful adsorption, the compound must be transported to the sorption sites on the surface of the sediment. The inverse of this transport rate can also be considered as a resistance to transport, R_I. Then, the compound, upon reaching the surface of the suspended sediment, must find a sorption site for adsorption. This second rate parameter is more related to surface chemistry than

Table 1.2 Examples of important mixing rate and equilibrium parameters in environmental processes

Process	Mixing-rate-related parameter	Equilibrium parameter
1. Treatment processes		
Coagulation/flocculation	Size of coagulation and flocculation basins and proper mixing (residence time)	Dose of coagulants (alum)
Softening	Design of softening tank to increase mixing	Dose of softening agent (lime)
Settling	Design of settling basin to reduce mixing	$\rightarrow 0$
Chlorination	Design of chlorination and dechlorination chambers for proper mixing and res. time	Dose of chlorine
Filtration	Size of filter bed	Length of time before backflushing
2. Surface waters		
Oxygen transfer	Diffusion and turbulent mixing	Atmospheric O_2 conc. Henry's Law constant
Volatilization of pollutants	Diffusion and turbulent mixing	$\rightarrow 0$
Toxic spills	Diffusion and turbulent mixing	Spill–water equilibrium
Internal loading of nutrients	Hypolimnetic mixing	O_2 concentration in hypolimnion
Sorption onto suspended sediments	Turbulent mixing exposes chemicals to sediment	Sediment–water partitioning
3. Atmosphere		
Greenhouse gases (CO_2, CH_4)	Turbulent mixing	Atmospheric concentrations
Volatilization of spills	Turbulent mixing and dispersion	$\rightarrow 0$
Aerosols	Turbulent mixing, dispersion, and settling	None
4. Groundwater and sediments		
Spills	Advection and dispersion	$\rightarrow 0$
O_2	Diffusion and advection	Atmospheric conc. of O_2
Sorption of soil	Advection and diffusion	Equilibrium soil–water partitioning

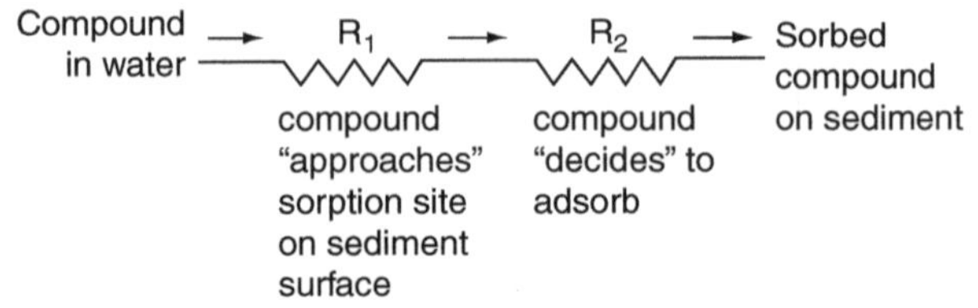

Fig. 1.4 Adsorption analogy to two resistors in a series: adsorption of an organic compound to sediment [2]

to diffusive transport, and is considered as a second resistance, R_2, that acts in series to the first resistance. The second resistance cannot occur without crossing the first resistance of transport to the sorption site, so they must occur in series.

Now, if R_1 is much greater than R_2, it can be assumed that R_2 is zero without compromising the accuracy of the rate calculation. In electric circuits, two resistances applied in series are simply added together in calculating the line resistance. The same is true for resistance to chemical transport. If R_1 is 1,000 resistance units and R_2 is 1 resistance unit, R_2 can be ignored and still be within 99.9% of the correct answer.

Another example is the air–water transfer of a compound, illustrated in Fig. 1.5. This example will be used to explain volatile and nonvolatile compounds. There is resistance to transport on both sides of the interface, regardless of whether the compound is classified as volatile or nonvolatile. The resistance to transport in the liquid phase is given as $R_L = 1/K_L$. If chemical transfer is being described through an equation like Eq. 1.3, the resistance to transfer in the gas phase is given as $R_G = 1/(HK_G)$. The equilibrium constant is in the R_G equation because the equivalent waterside concentrations are being used to represent the concentration difference from equilibrium, and the gas phase resistance needs to be a resistance to an equivalent water concentration.

The gas phase and the liquid phase resistances are applied in series. In general, gas film coefficients are roughly two orders of magnitude greater than liquid film coefficients. It is also true that Henry's Law constant, H, varies over many orders of magnitude as the transported compounds are varied. Nitrogen gas, for example, has a Henry's Law constant of approximately 15, using mass concentrations. The herbicide, atrazine, has a Henry's Law constant of 3×10^{-6}. Thus, the ratio R_G/R_L would vary by seven orders of magnitude between nitrogen gas and atrazine.

If these orders of magnitude are put into a series resistance equation,

$$R = R_L + R_G = \frac{1}{K_L} + \frac{1}{HK_G} \tag{1.5}$$

Due to the Henry's Law constants, it can be seen that for nitrogen gas, $R \cong R_L$, and for atrazine, $R \cong R_G$. If a typical ratio of $K_G/K_L \sim 100$ is applied, $R_G = R_L$ when $H = 0.01$.

Now, the mass transfer between phases is given as

$$\frac{dM}{dt} = \frac{A}{R}\left(\frac{C_a}{H} - C\right) \tag{1.6}$$

Fig. 1.5 Air–water transfer analogy to two resistors in a series [2]

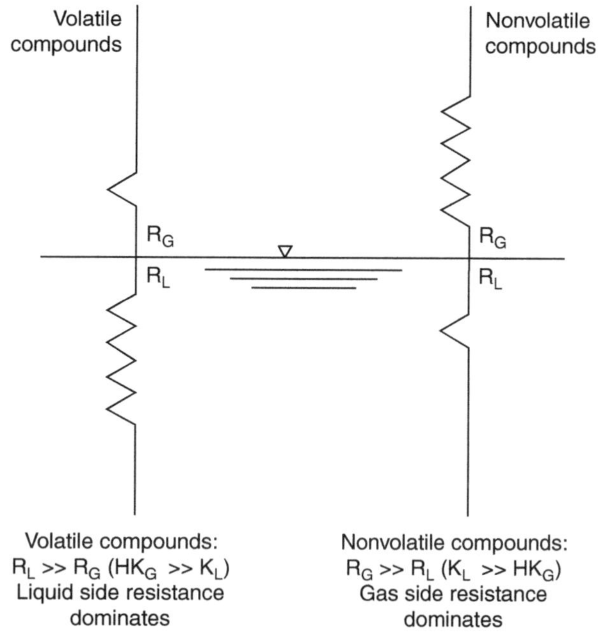

or

$$\frac{dM}{dt} = \frac{A}{1/K_L + 1/HK_G}\left(\frac{C_a}{H} - C\right) \tag{1.7}$$

Nitrogen gas would be a volatile compound because the equilibrium is strongly to the gas phase, and there is little gas phase resistance to its transfer, i.e., $1/K_L \gg 1/(HK_G)$. For that reason, N_2 is generally called a gas, as are many other volatile compounds such as methane, oxygen, and propane.

Atrazine, on the other hand, would be a nonvolatile compound, $1/(HK_G) \gg 1/K_L$, because equilibrium is strongly to the liquid phase due to the small Henry's Law constant. There is also a strong gas phase resistance to the transfer. Atrazine was manufactured to remain in the liquid phase, where it will act as a herbicide, rather than in the gas phase, where the farm personnel will be breathing this toxic chemical. If you were going to pick a nonionic compound that is not made by humans from the list of those that are gas or liquid in our environment, a good guess is that it would be a volatile or semivolatile compound. There are only a few environmental compounds that are nonvolatile. Remarkably, one of them is water. While the atmosphere may be as much as 3% water, the water bodies in the world are very close to 100% water. The equilibrium is strongly to the liquid side, due to a large equilibrium-partitioning constant.

One theme of this discussion can now be stated as follows: when transport processes occur in series, it is the slower transport processes that are important

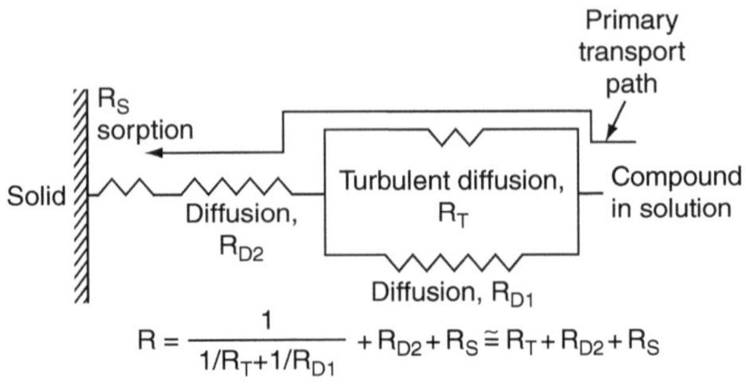

Fig. 1.6 Transport to a sorption site and the resistor analogy [2]

for chemical transport calculations because the *resistance* to transport is large, just as the large resistors of a series in an electronic circuit are the most important.

Now is the time for the second theme: when transport processes *occur in parallel*, the fast transport process with the low resistance dominates. The result is the opposite of resistances in series. Figure 1.6 illustrates this concept with the transport of a compound from a water body to a sorption site on a solid. In the bulk solution, there is diffusion and turbulent diffusion occurring simultaneously. Transport can occur due to either process, so there are two different paths that may be followed, without the need of the other path. These transport processes are operating in parallel, and the faster transport process will transport most of the compound. The analogy to electronic circuits applies in this case as well. Beginning with a compound in solution in Fig. 1.6, there are two parallel transport paths, each with a resistance to transfer. Most of the compound will be transported through the path with the least resistance. Many times, the path with the greater resistance can be ignored because the quantity of compound transported through this path is very small. When the compound comes close to the solid, however, the turbulent diffusion dissipates because eddies become so small that they are dissipated by the viscous action of the water. Now, one is back to one transport path, with the act of sorption and diffusion acting in series. Thus, the slowest transport path once again becomes the important process.

The overall resistance to the sorption process illustrated in Fig. 1.6 can be written as follows:

$$R = \frac{1}{1/R_T + 1/R_{D1}} + R_{D2} + R_S \cong R_T + R_{D2} + R_S \qquad (1.8)$$

where R_T, R_{D1}, R_{D2}, and R_S are the resistances to turbulent transport, diffusive transport in the bulk of the fluid, diffusive transport near the solid surface, and adsorption, respectively. It can be seen that, in Fig. 1.6 and Eq. 1.8, the resistance due to diffusion in the bulk of the fluid can be neglected because turbulent diffusion is a parallel path. The resistance due to diffusion only needs to be considered when there is no parallel path for turbulent diffusion, such as very near the surface of the solid. Thus, R_{D1} can be ignored, but not R_{D2}.

Conclusion

In this chapter, some of the topics that will be covered and applied in the other chapters of the section have been introduced, where the physics of mass transport are essential. These and similar engineering concepts will be revisited throughout the first half of this volume, in an attempt to develop models in the environmental fate and transport of chemicals that are close to realistic, but can be solved, even if that solution is approximate.

Future Directions

Global climate change is a topic for the present and future because society has only begun to assess and tackle the causes, implications, and remediation of the new Anthropocene [1]. There are global warming gases (CO_2 and CH_4) and global cooling gases (refrigerants) that are transported between the oceans and the atmosphere and could be transported between the atmosphere and the earth (carbon sequestration). These global transport rates are not quick, so we humans will be dealing with transport to a greater extent in the future. On a more local scale, urbanization is creating nonsource pollution of water and the atmosphere that is increasing. Roads, parking lots, and rooftops are often directly connected to lakes and rivers through storm sewers, and the pollution of urban activities is not insubstantial. This will be a bigger concern as discovery about this source of pollution occurs. Finally, the nitrogen of various forms released from agricultural activities continues to create hypoxic (dead) zones when the rivers have transported this nutrient to the oceans. The nitrification–denitrification processes as organic nitrogen is decomposed to ammonia, utilized by aerobic bacteria to make nitrites and nitrates, and finally converted by anaerobic bacteria to nitrogen gas occurring in streams and rivers add an interesting complexity to the fate and transport calculations.

Bibliography

1. Ellis E (2011) A man-made world. The Economist London, UK, 26 May 2011
2. Gulliver JS (2007) Introduction to chemical transport in the environment. Cambridge University Press, Cambridge, UK

Chapter 2
Chemicals in the Environment, Diffusive Transport

Edward Cussler

Glossary

Convection	Mass transfer effected by flow due to applied forces like pressure (forced convection) or to density differences (free convection).
Diffusion	Mixing caused by molecular motion.
Diffusion coefficient	The negative of the flux per concentration gradient.
Diffusivity	Another name for the diffusion coefficient.
Dispersion	Mixing caused by diffusion and simultaneous flow.
Flux	Mass or moles transferred per area per time.
Mass transfer	Diffusion and dispersion, especially across interfaces.
Mass transfer coefficient	The flux per concentration difference, especially near an interface.
Overall mass transfer coefficient	The flux per virtual concentration difference from one phase across an interface into a second phase.

This chapter was originally published as part of the Encyclopedia of Sustainability Science and Technology edited by Robert A. Meyers. DOI:10.1007/978-1-4419-0851-3

E. Cussler (✉)
Department of Chemical Engineering, University of Minnesota, 55455 Minneapolis, MN, USA
(612)-625-1596(612)-626-7246
e-mail: cussl001@umn.edu; cussler@cems.umn.edu

J.S. Gulliver (ed.), *Transport and Fate of Chemicals in the Environment: Selected Entries from the Encyclopedia of Sustainability Science and Technology*,
DOI 10.1007/978-1-4614-5731-2_2, © Springer Science+Business Media New York 2012

Definition of the Subject

Diffusion is mixing without stirring. It is mixing caused by Brownian motion, that is, by thermally induced random motion of molecules or small particles. Because diffusion is often slow, it frequently limits the overall rate of the process. Diffusion has the reputation of being a difficult subject, which it can be; however, the difficulty most often comes from complicated units, from interfaces, or from the combination of diffusion and convection. By itself, diffusion is not hard. It is easier than viscous flow and much easier than ideas like entropy or chemical potential.

Two other phenomena, closely related to diffusion, are also reviewed in this entry. Dispersion is mixing caused by the interaction of flow and diffusion. Often, it is described using mathematics similar to those which describe diffusion. In environmental problems, these two phenomena are sometimes treated without distinction and without penalty. Mass transfer, an alternative description of diffusion, assumes that all concentration changes occur near interfaces. While it is used largely to describe chemical processing, it has considerable value in environmental problems.

Introduction

This entry is organized as four sections. The first section gives the mathematical description of diffusion itself. The second reviews dispersion, a different phenomenon that is mathematically similar to diffusion but which is caused by different physical effects. The third section, "Diffusion Coefficients," reviews values of the diffusion coefficients themselves; and the fourth section "Diffusion Across Interfaces," explains mass transfer across interfaces, especially air–water interfaces. Finally, the section "Important Special Cases" reports some common situations which have important features.

Basic Diffusion

The basic mathematical description of diffusion is Fick's Law, suggested by Adolph Fick (1829–1910) when he was just 26 years old. For a dilute solution, Fick's Law is:

$$j_1 = -D\frac{dc_1}{dz} \tag{2.1}$$

where j_i is the flux of a solute "1," and dc_1/dz is the concentration gradient, that is, the change of the solute concentration with position. The diffusion coefficient D is a proportionality constant which is nearly constant in almost all situations. The flux is the amount of solute moving per cross-sectional area per time and so has the

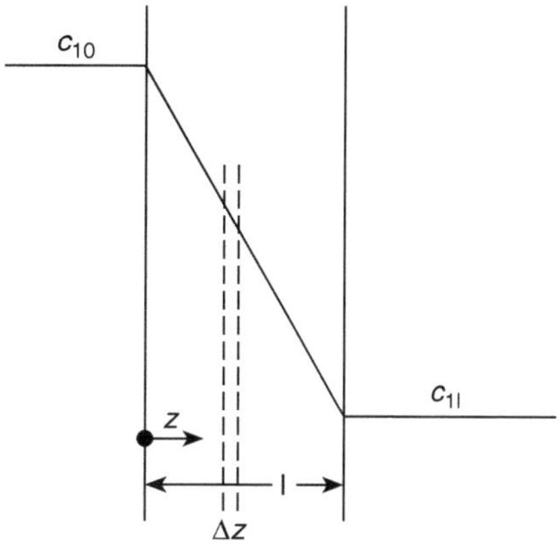

Fig. 2.1 Concentration profile across a thin film

dimensions of mass (M) per area (length squared L^2) per time (t). The concentration gradient has dimensions of concentration (M/L^3) per distance (L). Thus the diffusion coefficient has dimensions of (L^2/t).

The form of Fick's Law in Eq. 2.1 has some hidden implications. First, it is a one-dimensional equation of what is actually a more general vector relation. Because at least four out of five diffusion problems are one-dimensional, this is often not a major issue. Second, the concentration can be expressed in different units. If it were expressed in moles per volume, then the flux would be in moles per area per time. If it were expressed as a mole fraction or a mass fraction or a partial pressure, then unit conversions would be necessary and annoying, but not difficult. Third, the minus sign in Eq. 2.1 is arbitrary, stuck in to make the diffusion coefficient positive. The only difficult implication of Eq. 2.1 is the restriction to dilute solutions, explored in more detail at the end of this section. The restriction is rarely important because solutions in the environment are so often dilute. For example, liquid water contains 55 mol/l, so almost every aqueous solution is dilute.

The most important case of Fick's law is diffusion across a thin film, described next. Other important cases and concentrated solutions are reviewed later.

Diffusion Across a Thin Film

The simplest case, steady diffusion across a film, is also the most important. Imagine a thin film separating two well-stirred solutions, as shown in Fig. 2.1. On the left, the solution has a concentration c_{10}; on the right, the concentration is $c_{1\ell}$. The key parts of this case are the variation of concentration across the film $c_1(z)$ and the flux j_1. Finding these requires a mass balance on a differential volume Δz thick and located at an arbitrary position z within the film:

$$(mass\ accumulation) = (mass\ diffusing\ in - out)$$

$$\frac{\partial}{\partial t}(c_1 A \Delta z) = (j_1 A)_z - (j_1 A)_{z+\Delta z} \tag{2.2}$$

where A is the constant cross-sectional area of the film. Because diffusion is steady, the concentration does not change with time, the left-hand side of Eq. 2.2 is zero, and

$$0 = \frac{j_1|_z - j_1|_{z+\Delta z}}{(z + \Delta z) - z}$$

$$0 = -\frac{dj_1}{dz} \tag{2.3}$$

This restates the assumption of steady-state diffusion, independent of time. Combining this relation with Fick's Law (Eq. 2.1) yields:

$$0 = D\frac{d^2 c_1}{dz^2} \tag{2.4}$$

This is subject to two boundary conditions:

$$z = 0 \quad c_1 = c_{10} \tag{2.5}$$

$$z = \ell \quad c_1 = c_{1\ell} \tag{2.6}$$

This is enough to solve this important problem.
Equation 2.4 may be integrated once to find:

$$\frac{dc_1}{dz} = A \tag{2.7}$$

where A is an integration constant. Integrating a second time gives:

$$c_1 = Az + B \tag{2.8}$$

where B is a second integration constant. Evaluating A and B from Eqs. 2.5 and 2.6 gives:

$$\frac{c_1 - c_{10}}{c_{1\ell} - c_{10}} = \frac{z}{\ell} \tag{2.9}$$

The flux can now be found by combining this result with Fick's Law:

$$\begin{aligned} j_1 &= -D\frac{dc_1}{dz} \\ &= \frac{D}{\ell}(c_{10} - c_{1\ell}) \end{aligned} \tag{2.10}$$

If the concentration difference across the film is doubled, the flux doubles. If the diffusion coefficient is twice as big, the flux will be twice as big, too. If the film thickness increases two times, the flux will be cut in half.

This important example is so simple mathematically that many novices tend to skip over it. This is a mistake. Its nuances are explored by the following questions:

1. *How does the flux change if the film is chemically different than the adjacent solutions?*

In this case, the boundary conditions in Eqs. 2.5 and 2.6 change to:

$$z = 0 \quad c_1 = Hc_{10} \tag{2.11}$$

$$z = \ell \quad c_1 = Hc_{1\ell} \tag{2.12}$$

where H is a partition coefficient, the ratio at equilibrium of the concentration inside the film to that in the adjacent solution. Paralleling the arguments above,

$$j_1 = \frac{(DH)}{\ell}(c_{10} - c_{1\ell}) \tag{2.13}$$

The diffusion coefficient D in Eq. 2.10 is replaced with the product (DH), which is called the permeability. (The term (DH/ℓ) is called the permeance.) As will be shown later, diffusion coefficients in gases and liquids do not vary much, but partition coefficients vary a lot. Thus partition is often the key to permeability.

2. *How is the flux changed by a fast reversible reaction giving an immobile product?*

This case occurs surprisingly frequently in, for example, adsorption in soil or dyeing of wool. The answer is that at steady state, the flux does not change. Every point crossing the film has a different concentration which is in equilibrium with a different absorbed amount. Still, at steady state, the reaction is at local equilibrium and does not affect the flux. This is not the case for unsteady state, for irreversible reactions, or for mobile reaction products.

Other Important Cases

Many diffusion problems are not thin films. Surprisingly, many do behave as if they were thin films. For example, for a sphere of radius R slowly dissolving in a stagnant fluid with a concentration of $c_{1\infty}$, the flux is:

$$j_1 = \frac{D}{R}(c_{10} - c_{1\infty}) \tag{2.14}$$

Fig. 2.2 Concentration profile into a semi-infinite slab

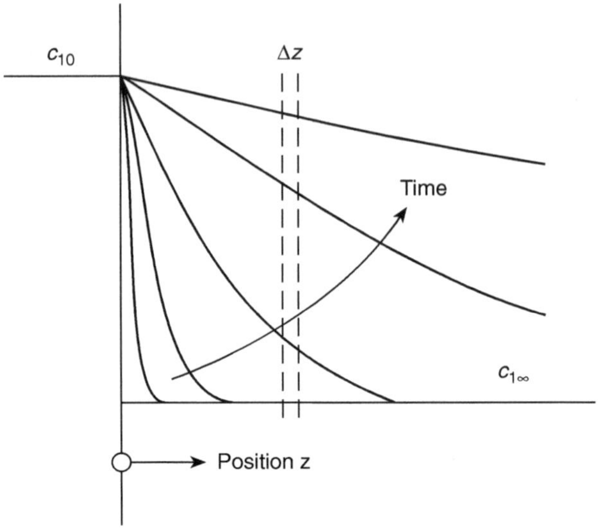

where c_{10} is the concentration in solution at the surface of the sphere. For diffusion from a solution of c_{10} through a very thin, impermeable film with a cylindrical orifice of radius R, and into a solution at $c_{1\infty}$, the flux is:

$$j_1 = \frac{D}{\frac{\pi}{2}R}(c_{10} - c_{1\infty}) \tag{2.15}$$

The fluxes in these cases are strong mathematical parallels to that in Eq. 2.10. The mathematics in these cases is different, but the final result is remarkably similar. The thin film limit is a good guide about 80% of the time.

For unsteady state diffusion, this is not true. Fluxes and concentration profiles for a wide variety of unsteady cases have been calculated and are tabulated in a few clear texts. One of these cases, useful in perhaps 10% of all cases, is unsteady diffusion into a semi-infinite slab, shown in Fig. 2.2. In this case, a mass balance gives:

$$\frac{\partial c_1}{\partial t} = D\frac{\partial^2 c_1}{\partial z^2} \tag{2.16}$$

where t is the time. For this semi-infinite slab, the initial and boundary conditions are:

$$t = 0 \quad \text{all } z \quad c_1 = c_{1\infty} \tag{2.17}$$

$$t = 0 \quad z = 0 \quad c_1 = c_{10} \tag{2.18}$$

$$t = 0 \quad z = \infty \quad c_1 = c_{1\infty} \tag{2.19}$$

Table 2.1 Flux across a film or into a slab. These two cases are important because they bracket almost all diffusion problems. In this table, K is the equilibrium constant of the rapid chemical reaction

	Thin film	Semi-infinite slab
Concentration difference	Δc_1	Δc_1
Diffusion coefficient	D	$\sqrt{D}$
Thickness	ℓ^{-1}	–
Time	–	t^{-1}
Flux without reaction	$j_1 = \frac{D}{\ell}\Delta c_1$	$j_1 = \sqrt{\frac{D}{\pi t}}\Delta c_1$
Flux with fast reversible reaction	$j_1 = \frac{D}{\ell}\Delta c_1$	$j_1 = \sqrt{\frac{D(1+K)}{\pi t}}\Delta c_1$

The concentration profile in this case is:

$$\frac{c_1 - c_{1\infty}}{c_{10} - c_{1\infty}} = 1 - erf\frac{z}{\sqrt{4Dt}} \tag{2.20}$$

where *erf* is the error function. The flux at the edge of the slab, that is, at $z = 0$, is:

$$j_1 = \left(\sqrt{\frac{D}{\pi t}}\right)(c_1 - c_{1\infty}) \tag{2.21}$$

This is the key result for this case.

The cases of a thin film and a semi-infinite slab are especially important because they bracket observed behavior. More specifically, the result in Eq. 2.21 for a semi-infinite slab is compared with the flux across a thin film in Table 2.1. In both cases, the flux will double if the concentration difference doubles. If the diffusion coefficient doubles, the flux across a thin film doubles, but the flux into the semi-infinite slab increases $\sqrt{2}$ times. If the film's thickness doubles, the flux drops two times for the film, but is unchanged for the slab. If diffusion occurs for twice as long, the flux for the thin film keeps its steady value, but that for the slab drops by a factor of $1/\sqrt{2}$. These limits will usually bracket all diffusion behavior because all shapes will be between the film and the slab. These two cases are key to understanding the mathematics of diffusion.

A third special case is especially important for environmental engineering. This is the decay of a pulse. In this case, a large amount of solute is released at a particular plane at $z = 0$. Solute diffuses away from this position in only one dimension. The solute concentration as a function of position z and time t gives the details of any environmental impact.

The mathematics follows the same route as the slab: a mass balance is subject to initial and boundary conditions, which are combined with Fick's Law and solved to give the concentration profile. The mass balance is:

$$\frac{\partial c_1}{\partial t} = D\frac{\partial^2 c_1}{\partial z^2} \tag{2.22}$$

This mass balance, identical with the mass balance for the slab (Eq. 2.16), occurs so frequently that some just call it "the diffusion equation." The initial and boundary conditions are different from Eqs. 2.17–2.19 for the slab:

$$t = 0 \quad \text{all } z \qquad c_1 = \frac{M}{A}\delta(z) \tag{2.23}$$

$$t = 0 \quad z = 0 \quad \frac{dc_1}{dz} = 0 \tag{2.24}$$

$$z = \infty \quad c_1 = 0 \tag{2.25}$$

In these conditions, M is the total solute injected, A is the cross-sectional area, and $\delta(z)$ is the Dirac function, equal to zero everywhere except at $z = 0$, where it is infinity. While the mathematical solution of Eqs. 2.22–2.25 is tricky, the answer is simple:

$$c_1 = \left[\frac{M/A}{\sqrt{4\pi Dt}}\right] e^{-\frac{z^2}{4Dt}} \tag{2.26}$$

In this Gaussian concentration profile, the quantity in square brackets is the maximum concentration, which drops as time grows. The mathematical form of this important result is also observed for other environmental problems which do not depend only on diffusion, as described in the section "Dispersion."

Concentrated Diffusion

One reason that diffusion has the reputation of being difficult comes from the major complexities of concentrated solutions. These complexities are rarely important in environmental engineering and so should be ignored unless there are good experimental reasons not to do so. These complexities are mentioned here only to illustrate when they are important.

To explore this, imagine putting a pot of room temperature water of 25°C on a stove that is turned off. The flux of any evaporating water will be given by Eq. 2.10. If the air above the stove were dry, $c_{1\ell}$ would be zero. Because the water at 25°C has a vapor pressure of about 25 mmHg, c_{10} is about:

$$\begin{aligned} c_{10} &= \left[\frac{25 \text{ mm Hg}}{750 \text{ mm Hg}}\right] \frac{1 \text{ mol}}{22.4 \times 10^{-3}\,\text{m}^3} \\ &= 1.5\frac{\text{mol}}{\text{m}^3} \end{aligned} \tag{2.27}$$

If the liquid water in the pot is 0.1 m below the rim and the diffusion coefficient of water vapor in air is about 2.8×10^{-5} m^2/s, the flux is:

$$\begin{aligned} j_1 &= \frac{D}{\ell}(c_{10} - c_{1\ell}) \\ &= \frac{2.8 \times 10^{-5}\text{m}^2/\text{s}}{0.1\text{m}}\left(1.5\frac{\text{mol}}{\text{m}^3} - 0\right) \\ &= 4 \times 10^{-4}\frac{\text{mol}}{\text{m}^2\text{s}} \end{aligned} \tag{2.28}$$

Now imagine heating the liquid water in the pot to boiling. If the heat flux q is 20 kJ/m^2 s, then the molar flux caused by boiling is:

$$\begin{aligned} n_1 &= \frac{q}{\Delta\tilde{H}_{vap}} \\ &= \frac{20 \text{ kJ /m}^2\text{s}}{48 \text{ kJ/mol}} \\ &= 0.4\frac{\text{mol}}{\text{m}^2\text{s}} \end{aligned} \tag{2.29}$$

The boiling flux n_1 is 1,000 times greater than the diffusion flux at room temperature, but it is not a function of D. Thus slow dilute evaporation is a function of the diffusion coefficient; but boiling depends not on diffusion but on heating rate.

But what about intermediate cases? For example, how fast will evaporation take place at 50°C?

Answering this question requires a more complete form of Fick's Law. Unfortunately, there is no single way to do this. One choice is the following:

$$[\text{Total flux}] = [\text{Diffusion flux}] + [\text{Convective flux}]$$

$$\begin{aligned} n_1 &= c_1(v_1 - v) + c_1 v = j_1 + c_1 v \\ &= -D\frac{dc_1}{dz} + c_1 v \end{aligned} \tag{2.30}$$

where v is most often a volume average velocity. The most common alternative form of Fick's law, strongly advocated by a few zealots, may be approximated as:

$$\nabla c_1 = \frac{c_1 c_2}{cD}(v_2 - v_1) \tag{2.31}$$

where c_2 and v_2 are the concentration and velocity of the solvent. This form avoids choosing a convective velocity, but it can cloud the physical significance of the problem.

Fortunately, the result for a problem like the water evaporation given above is the same for both forms of Fick's Law given in Eqs. 2.30 and 2.31. If water at 50°C is evaporating from the pot above into dry stagnant air, the total flux is:

$$\begin{aligned} n_1 &= -\frac{Dc}{\ell}\ln\left(1-\frac{c_{10}}{c}\right) \\ &= \frac{-2.8\times10^{-5}\mathrm{m}^2/\mathrm{s}\left(\frac{1\ \mathrm{mol}}{22.4\times10^{-3}\mathrm{m}^3}\right)}{0.1\ \mathrm{m}}\ln\left(1-\frac{92.5}{760}\right) \\ &= 1.62\times10^{-3}\mathrm{mol/m^2s} \end{aligned} \tag{2.32}$$

This is about 7% greater than the result would be if calculated from Eq. 2.10 for a dilute solution at 50°C. The moral is clear: the effects of concentrated diffusion will only rarely be important in the atmospheric and aquatic environments.

This completes our basic description of the three cases key to understanding diffusion. These are the thin film (80% of the cases), the semi-infinite slab (10% of the cases), and the decay of a pulse (5% of the cases). Other cases with different boundary conditions do occur, and solutions for these are tabulated in the literature. However, these other cases are not as common in practice. Similarly, diffusion in concentrated solutions is complicated but infrequently important. The simple form of Fick's Law in these three cases is the best way to get started.

Dispersion

We now turn to an environmentally important problem mathematically similar to diffusion but with a different physical origin. To make this problem specific, imagine dealing with the spill of a single toxin on the ground. Imagine the concentration of the toxin spreads in one dimension with time and groundwater flow, producing a roughly Gaussian concentration profile. We want to know how the spread of this toxin varies with the diffusion coefficient of the toxin.

The answer is surprising: if the diffusion coefficient increases, the spread of the toxin may be bigger, smaller, or unchanged. The toxin's concentration profile is:

$$c_1 = \frac{M/A}{\sqrt{4\pi Et}}e^{-\frac{(z-vt)^2}{4Et}} \tag{2.33}$$

where M is the total amount of toxin, A is the cross-sectional area across which the dispersion occurs, and v is the velocity of any flow through the soil. The dispersion coefficient E has the same units as the diffusion coefficient but will often be much larger. This result, a complete analogue to Eq. 2.26, can be derived from parallels to

Eqs. 2.22–2.25 by replacing the diffusion coefficient D with the dispersion coefficient E. However, while this mathematical parallel is complete, it does not explain the physics responsible for dispersion.

To explore the physics involved, imagine the toxin is injected as a pulse into a small tube of diameter d. The toxin's dispersion will be a strong function of how much flow is in the tube. If there is absolutely no flow, then the dispersion coefficient equals the diffusion coefficient:

$$E = D \tag{2.34}$$

Increasing diffusion increases dispersion. If there is a small, laminar flow of velocity v, then:

$$E = \frac{v^2 d^2}{192D} \tag{2.35}$$

Increasing diffusion decreases dispersion. The velocity where Eq. 2.35 becomes dominant is when:

$$\frac{v^2 d^2}{192D^2} \gg 1 \tag{2.36}$$

For non-absorbing soil with the equivalent of 500 μm particles and diffusion in water liquid, D is about 10^{-9} m^2/s, so v must be much greater than 30 μm/s or 2 m/day for Eq. 2.35 to swamp Eq. 2.34. If there is a large, turbulent flow (dv/v > 2,000), then:

$$E = \frac{dv}{2} \tag{2.37}$$

Dispersion, now independent of diffusion, is due to the coupled turbulent fluctuations of concentration and velocity. The physical basis of Eqs. 2.33–2.37 is associated with G.I. Taylor.

Those with a more practical bent may correctly be skeptical of modeling flow through a soil as occurring in a straight tube. Others sharing this skepticism have extended this analysis to flow in packed beds. The key results involve two new quantities:

$$\bar{t} = t(1 + k') \tag{2.38}$$

$$k' = \left[\frac{\text{Soil concentration}}{\text{Solution concentration}}\right]\left(\frac{1-\varepsilon}{\varepsilon}\right) \tag{2.39}$$

where ε is the void fraction available for flow. In physical terms, $\bar{t}$ is the time corrected for any absorption by the soil, including material that diffuses into the soil's pores. The quantity k' is a type of equilibrium constant between the soil and the solution, lumping together adsorption and absorption. In this case, a pulse of toxin may still be dispersed to give the Gaussian concentration profile in Eq. 2.24, but with time t replaced by $\bar{t}$. The dispersion coefficient E is now given by:

$$E = D(1+k') + \frac{d^2v^2}{192D}\left(\frac{1+6k'+11(k')^2}{1+k'}\right) + \frac{\delta^2 v^2}{3D'}\left(\frac{k'}{1+k'}\right) \tag{2.40}$$

where δ is an equivalent thickness of an absorbent and D' is the diffusion coefficient in the absorbent, not in the solution. The first term on the right-hand side of Eq. 2.40, the parallel of Eq. 2.34, is due to diffusion in the direction of flow. The second term, the analogue of Eq. 2.37, comes from Taylor dispersion and is often the most important. The third term is new, the result of the rate of absorption.

This overview of dispersion is intended as a caution and a starting point. The caution is that many Gaussian concentration profiles are due to diffusion coupled with other phenomena. The starting point in understanding these profiles is recognizing that their spread can depend inversely on diffusion. In other words, slow diffusion may result in wide dispersion.

Diffusion Coefficients

Diffusion is an important process because it is slow, and diffusion coefficients thus often control the overall rate of processes involving diffusion, flow, and chemical reaction. Typical values of diffusion coefficients, shown in Table 2.2, are chosen from the wide number of references in the literature but corrected to a temperature of 25°C. This wide literature is much less extensive than studies of other physical properties like viscosity or Young's modulus, because diffusion coefficients are relatively difficult to measure.

The values in Table 2.2 show diffusion in gases is about 10,000 times faster than diffusion in liquids, which is in turn over a billion times faster than diffusion in solids. Diffusion coefficients in gases fall around 10^{-5} m^2/s. Diffusion coefficients in liquids fall around 10^{-9} m^2/s. Diffusion coefficients in solids are much more variable, but are so slow that most of the mass transport occurs in fluid-filled gaps and pores within the solid. For example, in a bed of sand, most transport occurs in the spaces between sand grains and relatively little within the bulk of the grains themselves. In environmental problems, diffusion in gases and liquids is more important.

Table 2.2 Diffusion coefficients. Values given are in m^2/s and at 298 K and 1 atm

Gases	
Gas pair	Diffusion coefficient
Air-H_2O	2.6×10^{-5}
CO_2-O_2	1.6×10^{-5}
H_2-N_2	7.8×10^{-5}
H_2-O_2	8.9×10^{-5}
N_2-O_2	2.2×10^{-5}
N_2-H_2O	2.9×10^{-5}
O_2-H_2O	2.8×10^{-5}
O_2-octane	0.7×10^{-5}
Solids	
	Diffusion coefficient
C in Fe (BCC)	6×10^{-25}
Fe in Fe (BCC)	3×10^{-52}
B in Si	7×10^{-33}
He in SiO_2	4×10^{-14}
Na^+ in NaCl	1×10^{-36}
Ag^+ in AgCl	1×10^{-19}
Liquids	
Solute-solvent	Diffusion coefficient
O_2-H_2O	2.10×10^{-9}
CO_2-H_2O	1.92×10^{-9}
H_2S-H_2O	1.41×10^{-9}
HCl-H_2O	3.33×10^{-9}
NaCl-H_2O	1.61×10^{-9}
CaCl-H_2O	1.33×10^{-9}
NH_3-H_2O	1.64×10^{-9}
Urea-H_2O	1.38×10^{-9}
Sucrose-H_2O	0.52×10^{-9}
Albumin-H_2O	0.08×10^{-9}
H_2O-C_2H_5OH	1.24×10^{-9}
Benzene-butanol	0.99×10^{-9}
Hexane-heptane	4.21×10^{-9}

The diffusion coefficients, given in Table 2.2 for 1 atm and 25°C, do change with process variables, as outlined in Table 2.3. The variation with temperature in gases and liquids is small. For example, the temperature must be increased from 25°C (= 298° K) to 200°C (= 473° K) to double the diffusion coefficient in gases. Because the viscosity of a liquid drops as the temperature rises, the diffusion coefficient in a liquid changes faster with temperature, but the change is still modest. In contrast, diffusion coefficients in solids usually change more rapidly with temperature, doubling every 10°C or so.

Other process variables also have relatively small effects. The diffusion coefficient in gases does vary inversely with pressure; but the gas concentration varies

Table 2.3 Variations of diffusion coefficients

	Gases	Liquids	Solids
Typical value, m^2/s	10^{-5}	10^{-9}	Much smaller
vs. T	$T^{3/2}$	T	Large
vs. p	p^{-1}	–	–
vs. solute diameter	size^{-2}	size^{-1}	size^{2}
vs. viscosity μ	μ^{1}	μ^{-1}	–

directly with pressure; so the flux, related to the diffusion coefficient times the concentration, may remain more constant. The diffusion coefficient does vary inversely with the size of the diffusing species. While these effects are usually modest for gases and liquids, they can be much larger for solids. These generalizations are justified by the approximate physical arguments given next.

Gases. The diffusion coefficients in gases can be predicted with reasonable accuracy from kinetic theory. This theory assumes that a gas contains individual molecules moving with thermal motion and colliding with each other only as pairs. Under these cases, the diffusion coefficient is given by:

$$D = \frac{1}{3}\lambda v \tag{2.41}$$

where λ is the distance between collisions and v is the molecular velocity. For a monatomic gas, this velocity is kinetic, related to the thermal energy:

$$\frac{1}{2}mv^2 = k_\mathrm{B}T \tag{2.42}$$

where m is the molecular mass and k_B is Boltzmann's constant. Keep in mind that here v is a molecular velocity. It is the sonic velocity; it is much greater than the average velocity v used in Eqs. 2.30 and 2.31.

We must now estimate the distance between collisions λ. There are two cases. First, for the bulk gas, λ is found from the volume occupied by one molecule:

$$\begin{aligned}&[\text{Volume of one molecule}] =\\&[\text{Distance between collisions}] \times\\&[\text{Area swept out between collisions}]\end{aligned}$$

$$\frac{k_\mathrm{B}T}{p} = \lambda\left[\frac{\pi}{4}\sigma^2\right] \tag{2.43}$$

where σ is the molecular diameter. Equations 2.41–2.43 can be combined to find:

$$D = \left(\frac{4\sqrt{2}}{3\pi}\right)\frac{(k_\mathrm{B}T)^{\frac{3}{2}}}{p\sigma^2\sqrt{m}} \tag{2.44}$$

This approximate relation is close to that found from more complex theories: D does vary inversely with p, σ^2, and $\sqrt{m}$; it does vary with T to a power greater than one and less than two.

The second important limit of Eq. 2.41 occurs only for a gas diffusing in small pores of diameter d. In this case, the diffusing species is much more likely to collide with the pore walls than with other molecules. Thus the combination of Eqs. 2.41 and 2.34 becomes:

$$D = \frac{\sqrt{2}}{3} d \sqrt{\frac{k_B T}{m}} \tag{2.45}$$

This case, called Knudsen diffusion, has a diffusion coefficient which depends on the pore diameter d, but not on the molecular diameter σ. Now, the diffusion coefficient is independent of pressure, though it does vary inversely with the square root of solute mass. Under ambient temperature and pressure, Knudsen diffusion is important when the pores are much less than 0.1 μm.

Liquids. Diffusion in liquids is normally not described by a kinetic theory but as the motion of a rigid, spherical solute diffusing in a continuum of solvent. Despite the major approximations obviously made by this simple model, it gives remarkably good results. It is the standard against which new predictions are always judged.

The model begins by describing the friction on a solute sphere:

$$\text{Force} = [\text{Coefficient of friction } f] \times \text{Velocity } v_1 \tag{2.46}$$

The velocity v_1 now is the average and not the sonic value v used for gases. The coefficient of friction f is given by Stokes Law:

$$f = 6\pi\mu R \tag{2.47}$$

where μ is the solvent velocity and R is the solute radius. The force was suggested by Einstein to be the negative of the gradient of the chemical potential μ_1. Thus:

$$\begin{aligned} -\frac{d\mu_1}{dz} &= [6\pi\mu R] v_1 \\ &= -\nabla\left(k_B T \ln c_1\right) = -\frac{k_B T}{c_1}\frac{dc_1}{dz} \end{aligned} \tag{2.48}$$

Rearranging:

$$-c_1 v_1 = \left(\frac{k_B T}{6\pi\mu R}\right)\frac{dc_1}{dz} \tag{2.49}$$

But $(c_1 v_1)$ is the total flux n_1, equal in dilute solution to the diffusion flux j_1. Comparing this with Eq. 2.33 gives:

$$D = \frac{k_B T}{6\pi\mu R} \tag{2.50}$$

The diffusion coefficient in liquids varies inversely with solute size and with solvent velocity. This simple relation is called the Stokes–Einstein equation.

Equation 2.50 often gives good estimates of diffusion in liquids. Its simplicity is an invitation to attempt improvements. These include assuming the sphere is not solid but gas, replacing the sphere with an ellipsoid, allowing the sphere to spin, and putting it in a small pore. Other attempts at improvement allow for nonideal solutions, assigning different friction coefficients to solute and solvent, and considering changes close to the spinoidal. While none of these efforts is definitive, each can clarify the perspective of a particular chemical system. Still, the simple Stokes–Einstein equation is the best place to start for understanding diffusion in liquids.

Solids. As explained above, the diffusion in solids is so slow that most transport usually occurs in any fluid-filled flows and voids within the solid. Some solid processes are certainly dramatically affected by diffusion – metallic welds and doped semiconductors are two good examples – but the diffusion of chemicals in the environment is usually through fluids.

The relative unimportance of diffusion in solids is fortunate, because diffusion coefficients in solids scatter. These coefficients do not cluster around a single value, and they depend strongly on crystal structures. For example, the diffusion of carbon in body-centered cubic iron is 10^{10} times faster than the diffusion of carbon in face-centered cubic iron. Sometimes, an anomalously high coefficient reflects different types of vacancies in the solid crystals. For example, silver ion diffuses 10^{17} times faster in AgCl than sodium ion diffuses in NaCl.

Estimates of diffusion in solids, which normally begin with a face-centered cubic lattice, assume a coefficient given by:

$$D = R^2 N\omega \tag{2.51}$$

where R is now the distance between atoms or ions in the crystal; N is the dimensionless fraction of vacant sites; and ω is the jump frequency, the number of atomic or ionic movements per time. The size of R is estimated from crystal structure, and the fraction N from the free energy of mixing. The jump frequency ω is often felt to have an Arrhenius temperature dependence. Arguments like this are not predictions but are rationales to organize data.

Diffusion Across Interfaces

Diffusion from one phase to another is an important and complex limit, a source of confusion for many. In this case, there are two limits that are close parallels to the cases of a thin film and a semi-infinite slab discussed above in the sections "Diffusion Across a Thin Film" and "Other Important Cases." One case is exemplified by the so-called infinite couple, when two alloy bars of different but homogeneous composition are closely joined together. In this case, each atomic species can diffuse between the two bars, giving concentration profiles that are known. This limit is rarely important in environmental science and engineering.

The second, much more important limit occurs when solutes diffuse from one relatively well-mixed phase across a phase boundary to a second relatively well-mixed phase. This limit approximates what happens when sulfur dioxide in the air diffuses into a lake. In this case, bulk air is often well-mixed, and the bulk water in the lake is, too. However, this good mixing does not extend all the way to the air–water interface. About the last one millimeter of air and about the last one-tenth millimeter of the water are not well-mixed. Diffusion across these two films, one in air and the other in the water, is what governs the rate of sulfur dioxide dissolution in the lake.

We develop these ideas below. "The Mathematics of Mass Transfer" derives the mathematical framework. "Concentration Units" details transport across interfaces. "The Meaning of $c_1{}^*$" uses this framework to calculate the mass transfer in several environmentally relevant situations.

The Mathematics of Mass Transfer

To begin our study of mass transfer, imagine a small volume of air containing hydrogen sulfide at concentration c_{10} that is suddenly contacted with a large volume of water. The sulfide dissolves in the water so that its concentration c_1 drops with time. Predicting this concentration change with the diffusion equations given above is possible, but difficult. Often, an easier prediction is to use an alternative tool, a mass transfer analysis, which is more suitable for engineering applications.

This mass transfer analysis begins by writing a mass balance on the H_2S in the air:

$$\begin{aligned}&[\text{Accumulation in the air}] = \\ &[\text{Amount dissolved in water}]\end{aligned}$$

$$V\frac{dc_1}{dt} = -AK(c_1 - c_1{}^*) \tag{2.52}$$

where V is the air volume, A is the interfacial area between air and water, and $c_1{}^*$ is proportional to the concentration of the H_2S in the water. When there is a lot of pure, well-mixed water present, this concentration is zero. The rate constant K in Eq. 2.52 is an overall mass transfer coefficient, a function of H_2S diffusion in both the water and the air. It has the units of velocity, that is, of length L per time t. This mass balance is subject to the initial condition:

$$t = 0 \quad c_1 = c_{10} \tag{2.53}$$

Integrating, Eq. 2.52 becomes:

$$\frac{c_1}{c_{10}} = e^{-K\left(\frac{A}{V}\right)t} \tag{2.54}$$

The H_2S concentration in the air decays exponentially with time, as if it were undergoing a first-order chemical reaction. The rate constant of this reaction (KA/V) has units of reciprocal time. However, the concentration in air is not dropping because of a chemical reaction but because of diffusion of H_2S from the air into the water.

As a second example, imagine absorbing carbon dioxide from flue gas. The flue gas is steadily flowing upward in a small absorption tower. Excess strong base is steadily flowing downward through the tower. A mass balance on the carbon dioxide in a small differential volume dV in the tower results in:

$$\begin{aligned}[\text{Accumulation in } dV] = {} & [\text{CO}_2 \text{ Flow in} - \text{out}] \\ & + [\text{CO}_2 \text{ Absorbed by base}]\end{aligned}$$

$$0 = Q\frac{dc_1}{dV} - Ka(c_1 - c_1{}^*) \tag{2.55}$$

where Q is the volumetric flow rate of flue gas, a is the interfacial area per volume in the tower, and $c_1{}^*$ is about zero because the base is strong and there is a lot of it. As before, K is an overall mass transfer coefficient describing the rate of reaction. This mass balance is subject to a boundary condition:

$$V = 0 \quad c_1 = c_{10} \tag{2.56}$$

Integration gives:

$$\frac{c_1}{c_{10}} = e^{-Ka\left(\frac{V}{Q}\right)} \tag{2.57}$$

The CO_2 concentration exiting the absorption column decreases exponentially as the column volume V is increased or as the column flow Q is decreased. Note that Eqs. 2.54 and 2.57 are complete mathematical parallels, even though the former describes unsteady dissolution without flow, and the latter describes steady absorption with flow.

Interfacial mass transfer is not hard. It is just an alternative description of diffusion which complements that given by Fick's Law. The three features do make interfacial mass transfer complicated. These three complications are the units of concentration, the detailed meaning of $c_1{}^*$, and the values of the mass transfer coefficient K. Details of these features follow.

Concentration Units

The first issue, concentration units, results because the units used for clearly explaining the ideas are not always those easiest to use in practice. The concentration units implied in this essay are of the amount per volume, for example, moles per liter or grams per cubic meter. The concentration units used in practice are different. In gases, the units are sometimes partial pressures; in liquids, the units are often mole fractions.

Expressing concentrations as partial pressures or mole fractions leads to different definitions of mass transfer coefficients. In particular, the total flux across the interface N_1 from one dilute gaseous solution into another dilute liquid solution may be defined as:

$$\begin{aligned} N_1 &= n_1|_{\text{interface}} = j_1|_{\text{interface}} \\ &= K(c_1 - c_1{}^*) \end{aligned} \tag{2.58}$$

where c_1 is the concentration of species "1" in the gas. The restriction to dilute solution is not a major constraint. Alternatively, the interfacial flux can be defined as:

$$N_1 = K_p(p_1 - p_1{}^*) \tag{2.59}$$

where p_1 is the partial pressure of solute "1" in the gas, and K_p is a new, different overall mass transfer coefficient. But from the ideal gas law:

$$p_1 = \frac{n_1 RT}{V} = c_1 RT \tag{2.60}$$

Comparing the two equations shows:

$$K_p = \frac{K}{RT} \tag{2.61}$$

If K has units of meters per second, then K_p may have units of moles per square meter per second per pascal.

Similarly, for mass transfer from a liquid into a gas, an alternative definition is:

$$N_1 = K(c_1 - c_1{}^*) \tag{2.62}$$

where c_1 is now the concentration of species "1"in the liquid, and K is an overall mass transfer coefficient different from that in Eq. 2.58. Alternatively,

$$N_1 = K_x(x_1 - x_1{}^*) \tag{2.63}$$

where x_1 is the mole fraction of species "1" in the liquid, and K_x is still another overall mass transfer coefficient. Because

$$c_1 = cx_1 \tag{2.64}$$

where c is the total concentration in the liquid, the two overall coefficients are related:

$$K_x = cK \tag{2.65}$$

For example, if K is in meters per second, and c is in moles per meter cubed, then K_x will have units of moles per square meters per second. Other definitions of coefficients are also possible, but are no harder.

The Meaning of $\mathbf{c_1}^*$

The meaning of the concentration $c_1{}^*$ appearing in Eqs. 2.52, 2.55, 2.58, 2.59, and 2.62 is the hardest step in this description. These flux equations all assert that the flux is proportional to a concentration difference. The flux will be zero when the concentration is zero. Thus, $c_1{}^*$ must be the hypothetical gaseous concentration of species "1" that is in equilibrium with species "1" dissolved in the liquid. This is harder than interfacial heat transfer: there, the heat flux is proportional to the temperature on one side of the interface minus that on the other side. Here, the mass flux is proportional to a concentration difference which equals to one real concentration that does exist minus a second one which is hypothetical.

To be more specific, imagine the case in Eq. 2.58, where c_1 is the actual concentration of species "1" in the well-mixed, bulk gas on one side of the interface. The concentration $c_1{}^*$ is equal to the concentration in the well-mixed, bulk liquid times some type of Henry's Law constant, which describes equilibrium between gas and liquid. Sometimes, those studying this point for the first time can be helped by silently chanting

▼ $c_1{}^*$ is the concentration that would be in the gas if it were in equilibrium with the liquid (which it isn't).

Remembering this chant may help mastering this difficult point.

To try to make this point clearer, imagine in calculating the flux of oxygen in air into wastewater with a concentration of 1×10^{-4} mol/ℓ. At equilibrium,

$$c_1(\text{gas}) = 30c_1(\text{liquid}) \tag{2.66}$$

Thus,

$$c_1{}^*(\text{gas}) = 30 \times 10^{-4}\ \text{mol}/\ell \tag{2.67}$$

As a result,

$$\begin{aligned} c_1 - c_1{}^* &= \frac{0.21\ \ \text{mol}}{22.4\ell} - 30 \times 10^{-4}\frac{\text{mol}}{\ell} \\ &= 64 \times 10^{-4}\ \text{mol}/\ell \end{aligned} \tag{2.68}$$

Understanding problems like these is often helped by always checking what happens when the system is at equilibrium.

Values of Mass Transfer Coefficients

We now turn to the variations of the overall mass transfer coefficient with quantities like the diffusion coefficient in the adjacent phases. The most common case is that of transfer from a gas into a liquid. The concentration in the gas is expressed as a partial pressure, and the concentration in the liquid is expressed as a mole fraction. The flux N_1 across the interface is then:

$$\begin{aligned} N_1 &= K_p(p_1 - p_1{}^*) \\ &= k_p(p_1 - p_{1i}) \\ &= k_x(x_{1i} - x_1) \end{aligned} \tag{2.69}$$

where p_1 and x_1 are the average concentrations in the gas and liquid, respectively; and p_{1i} and x_{1i} are the corresponding but unknown gas and liquid concentrations at the interface. The mass transfer coefficients k_p and k_x describe transport in the gas and in the liquid. Sensibly, the individual mass transfer coefficient k_p is a function of diffusion in the gas, but not of diffusion in the liquid; and the individual mass transfer coefficient k_x is the reverse.

The concentrations across the interface will normally be in equilibrium, so that:

$$p_{1i} = Hx_{1i} \tag{2.70}$$

where H is a Henry's Law constant. Combining this constraint with Eq. 2.69 gives:

$$N_1 = \left[\frac{1}{\frac{1}{k_p} + \frac{H}{k_x}}\right](p_1 - Hx_1) \tag{2.71}$$

By comparing this with the overall mass transfer coefficient K_p yields:

$$K_p = \frac{1}{\frac{1}{k_p} + \frac{H}{k_x}} \tag{2.72}$$

$$p_1{}^* = Hx_1 \tag{2.73}$$

These are the results sought. Obviously, similar equations are possible for other concentrations and other equilibria analogous to Henry's Law.

The way in which k_p and k_x vary with the diffusion coefficients can be estimated either from experiments or from theories. The experiments are summarized as correlations, most often in terms of dimensionless numbers. For example, for mass transfer into a liquid flowing through a packed tower with packing of size d, the most widely accepted correlation is:

$$k_x = k(\text{liquid})c(\text{liquid}) \tag{2.74}$$

$$k(\text{liquid})\left(\frac{1}{\nu g}\right)^{\frac{1}{3}} = 0.0051\left(\frac{v}{a\nu}\right)^{0.67}\left(\frac{D}{\nu}\right)^{0.50}(ad)^{0.4} \tag{2.75}$$

where ν is the kinematic viscosity of the liquid, g is the acceleration due to gravity, v is the superficial liquid velocity, a is the surface area per volume of the packing, and D is the diffusion coefficient in the liquid. The quantity ($v/a\nu$) is one form of the dimensionless Reynolds number; the quotient (ν/D) is the dimensionless Schmidt number. Correlations like this, which are based on extensive experiments, should be used for estimates whenever possible.

In many cases, however, appropriate correlations may not be reliable, or may not be available at all. In these cases, estimates for liquids can be made by assuming that:

$$k(\text{liquid}) = \frac{D}{\ell} = \frac{D}{10^{-4}\text{m}} \tag{2.76}$$

where D is the diffusion coefficient in the liquid and ℓ is often called the film thickness or the boundary layer. This casual description can be confusing, because these terms are more specifically defined in theories of mass transfer. A corresponding estimate for gases is:

$$k(\text{gas}) = \frac{D}{\ell} = \frac{D}{10^{-3}\text{m}} \tag{2.77}$$

where D is now the diffusion coefficient in gases, typically 10^4 times larger than that in liquids. Equations 2.76 and 2.77 are major approximations to be used only in desperation.

Important Special Cases

The sections above describe the mathematics of diffusion and dispersion. They have summarized characteristics of diffusion coefficients and listed some typical values. They have discussed mass transfer coefficients as an alternative description of interfacial diffusion frequently valuable in environmental engineering. None of the ideas presented are especially difficult to understand.

However, actually putting these ideas into practice can be complicated, largely because of difficult units and subtle definitions. This final section considers specific chemical examples that illustrate the ideas involved. These examples are approximate but can serve as a warning of where trouble can occur.

Overall Mass Transfer Coefficient of Oxygen (Case #1)

Imagine wanting to estimate the mass transfer coefficient K_p of oxygen from air into water. From Table 2.2 and Eq. 2.77,

$$k(\text{gas}) = \frac{D}{\ell} = \frac{2 \times 10^{-4}\ \text{m}^2/\text{s}}{10^{-3}\text{m}} = 0.02\ \text{m/s} \tag{2.78}$$

The coefficient k_p is found from this by a unit conversion:

$$k_p = \frac{k(\text{gas})}{RT} = \frac{0.02\ \text{m/s}}{\left(8.2 \times 10^{-6} \frac{\text{m}^2\ \text{atm}}{\text{mol} \circ \text{K}}\right) 298^\circ\text{K}} = 8 \frac{\text{mol}}{\text{m}^2\text{s atm}} \tag{2.79}$$

Similarly, from Table 2.3 and Eq. 2.76,

$$k(\text{liquid}) = \frac{2 \times 10^{-9}\ \text{m}^2/\text{s}}{10^{-4}\ \text{m}} = 2 \times 10^{-5}\ \text{m/s} \tag{2.80}$$

The coefficient k_x has a different conversion:

$$\begin{aligned} k_x &= k(\text{liquid})c = \left(2 \times 10^{-5}\ \text{m/s}\right)\left(\frac{1\ \text{mol}}{18 \times 10^{-6}\ \text{m}^3}\right) \\ &= 1.1 \frac{\text{mol}}{\text{m}^2\text{s}} \end{aligned} \tag{2.81}$$

Henry's Law for this system is (cf. Eq. 2.70):

$$p_1 = \left(4.3 \times 10^4\ \text{atm}\right)x_1 \tag{2.82}$$

Thus from Eq. 2.72,

$$K_p = \frac{1}{\frac{\text{m}^2\text{s atm}}{8 \text{ mol}} + \frac{4.3\times10^4 \text{ atm m}^2\text{s}}{1.1 \text{ mol}}} = 2.6 \times 10^{-5} \frac{\text{mol}}{\text{m}^2\text{s atm}} \tag{2.83}$$

This coefficient describes oxygen transport between air and water when the concentration difference is expressed as partial pressures of oxygen.

Overall Mass Transfer Coefficient of Oxygen (Case #2)

The example above is straightforward because it matches the detailed equations given earlier. The same problem can be solved in different units. Thus, the flux N_1 is:

$$N_1 = K_c({c_1}^* - c_1) \tag{2.84}$$

where c_1 is the actual oxygen concentration in water, ${c_1}^*$ is the oxygen concentration in water that is in equilibrium with air, K_c is a different overall mass transfer coefficient given by:

$$K_c = \frac{1}{\frac{m}{k(\text{gas})} + \frac{1}{k(\text{liquid})}} \tag{2.85}$$

and m is a different form of Henry's Law constant, defined by the equilibrium:

$$c_1(\text{liquid}) = mc_1(\text{gas}) \tag{2.86}$$

Comparing Eqs. 2.81 and 2.82 gives:

$$\begin{aligned} m &= \frac{c(\text{liquid})RT}{H} \\ &= \left(\frac{\text{mol}}{18 \times 10^{-6}\text{m}^3}\right)\left(8.2 \times 10^{-6}\frac{\text{m}^3 \text{ atm}}{\text{mol}\circ \text{K}}\right) \\ &\quad \left(\frac{298°\text{K}}{4.3 \times 10^{-4}\text{atm}}\right) \\ &= 0.03 \end{aligned} \tag{2.87}$$

The numerical value of Henry's Law constant is completely different. The combination of Eqs. 2.78, 2.80, 2.85, and 2.87 gives:

$$K_c = \frac{-1}{\frac{0.03}{0.02\ \mathrm{m/s}} + \frac{3}{2\times10^{-5}\ \mathrm{m/s}}} \\ = 2 \times 10^{-5}\ \mathrm{m/s} \tag{2.88}$$

In both this formulation and that in Eq. 2.83, diffusion in the liquid dominates the mass transfer. This is often taken as a consequence of the slower diffusion in the liquid. This is not completely true, as the next example shows.

Overall Mass Transfer Coefficient of Ammonia from Air into Water

This example illustrates how the rate at which ammonia is dissolved in water can be estimated. The individual mass transfer coefficients of ammonia are easily found:

$$k(\mathrm{gas}) = \frac{D(\mathrm{gas})}{\ell} = \frac{2.3 \times 10^{-5}\mathrm{m}^2/\mathrm{s}}{10^{-3}\mathrm{m}} = 0.023\ \mathrm{m/s} \tag{2.89}$$

and

$$k(\mathrm{liquid}) = \frac{D(\mathrm{liquid})}{\ell} = \frac{1.6 \times 10^{-9}\mathrm{m}^2/\mathrm{s}}{10^{-4}\mathrm{m}} \\ = 1.6 \times 10^{-4}\mathrm{m/s} \tag{2.90}$$

One Henry's Law constant for dilute acid is given in the literature as:

$$p_1(\mathrm{atm}) = \left(7000 \frac{\mathrm{atm}\ \ell}{\mathrm{mol}}\right) c_1(\mathrm{liquid,\ molar}) \tag{2.91}$$

Converting the pressure into a molar concentration:

$$c_1(\mathrm{gas,\ molar}) = \left(7000 \frac{\mathrm{atm}\ \ell}{\mathrm{mol}}\right)\left(\frac{\mathrm{mol}\circ\ \mathrm{K}}{0.082\ \mathrm{atm}\ \ell}\right) \\ \left[\frac{c_1(\mathrm{liquid,\ molar})}{298^\circ\mathrm{K}}\right] \\ = 290 c_1(\mathrm{liquid,\ molar}) \tag{2.92}$$

Combining these results with Eq. 2.85 gives:

$$K_c = \frac{1}{\frac{290}{0.023\ \mathrm{m/s}} + \frac{1}{1.6\times10^{-5}\ \mathrm{m/s}}} \\ = 1.3 \times 10^{-4}\ \mathrm{m/s} \tag{2.93}$$

The much higher solubility of ammonia in dilute acid means the mass transfer is now more affected by diffusion in air.

Toxin Diffusion in a Biofilm

The final example imagines a dilute toxin dissolved in water and metabolized irreversibly by microorganisms immobilized in a biofilm. This example, which is not as chemically specific as the first three, also assumes that the concentration of dissolved oxygen is much greater than the concentration of the toxin. Thus, the rate per biofilm area N_1 is given by the overall rate of diffusion of the toxin to the biofilm, followed by the diffusion and reaction of the toxin within the biofilm.

This overall rate is mathematically equivalent to mass transfer across an interface, where the solute diffused through the gas to reach the interface, quickly crossed the interface, and then diffused into the liquid. In fact, the biofilm case is often easier because most biofilms are largely water and hence their partition coefficient m is one. Thus,

$$N_1 = \frac{c_1}{\frac{1}{k(\text{liquid})} + \frac{1}{k(\text{biofilm})}} \tag{2.94}$$

The coefficient k (liquid) can often be found from mass transfer correlations; Equation 2.76 provides a first guess. The value in the biofilm depends on the details of the reaction. However, delightfully, most theories give the same result:

$$k(\text{biofilm}) = \left[\frac{D(\text{biofilm})}{\tau}\right]^{\frac{1}{2}} \tag{2.95}$$

where τ is the half-life of the reaction. While beyond the scope of this entry, this result is carefully derived in most books on diffusion and reaction. This result underscores the value of the simple ideas of diffusion and reaction presented here.

Future Directions

After 150 years of concentrated effort, diffusion is an established subject. Active research does continue on, for example, semiconductors and polymer membranes, but this does not have major environmental application.

Diffusion is an important tool for describing environmentally significant mass transfer. In many cases, this transfer can be described in terms of diffusion coefficients. In many air pollution problems, mass transfer can be described in terms of dispersion, which is mathematically similar to diffusion but due to coupled diffusion and flow.

The underused description of mass transfer, especially across interfaces, is in terms of mass transfer coefficients. These are functions of diffusion coefficients and of other parameters, like velocity and viscosity. Exploiting this topic offers potential gain for environmental engineering.

Bibliography

Books and Reviews

Aris R (1999) Mathematical modeling. Academic, San Diego. ISBN 0-12-604585-1

Astarita G, Savage DW, Bisio A (1983) Gas treating with chemical solvents. Wiley, New York. ISBN 0-471-057681

Berg HC (1993) Random walks in biology. Princeton University Press, Princeton. ISBN 0-691-000646

Bird RB, Stewart WS, Lightfoot EN (2002) Transport phenomena, 2nd edn. Wiley, New York. ISBN 0-471-41077-2

Carslaw HS, Jaeger JC (1976) Conduction of heat in solids. Oxford University Press, New York. ISBN 0-19-853303-9

Crank J (1980) The mathematics of diffusion. Oxford University Press, New York. ISBN 0-19-853411-6

Cussler EL (2009) Diffusion: mass transfer in fluid systems, 3rd edn. Cambridge University Press, Cambridge. ISBN 0-521-56477-8

Dutta BK (2009) Mass transfer and separation processes. PHI Learning, New Delhi. ISBN 978-81-203-2990-4

Seader JD, Henley EJ (2006) Separation processes principles, 2nd edn. Wiley, New York. ISBN 0-471-46480-5

Seinfeld JH, Pandis S (2006) Atmospheric chemistry and physics, 2nd edn. Wiley, New York. ISBN 0-471-72018-6

Sherwood TK, Pigford RL, Wilke CR (1975) Mass transfer. McGraw Hill, New York. ISBN 0-070-566929

Taylor GI (1953) Dispersion of soluble matter in solvent flowing slowly through a tube. Proc R Soc London Ser A 219:186–203

Taylor GI (1954) The dispersion of matter in turbulent flow through a pipe. Proc R Soc Lond Ser A 223:446–468

Taylor R, Krishna R (1993) Multicomponent mass transfer. Wiley, New York. ISBN 0-47-157417-1

Treybal RE (1980) Mass transfer operations, 3rd edn. McGraw-Hill, New York. ISBN 0-070-651760

Chapter 3
Toxic Organic Chemicals

Simanga Gama, Jon A. Arnot, and Don Mackay

Glossary

Bioaccumulation	The phenomenon similar to bioconcentration but including uptake from food as well as uptake from the ambient environment. This is expressed as a bioaccumulation factor (BAF) and generally applies to organisms in the environment. The BAF is the steady-state ratio of the chemical concentration in the organism to that in the environment.
Bioconcentration	The phenomenon by which an organism, such as a fish, absorbs chemical from its ambient environment of water or air by respiratory uptake and/or dermal absorption. The steady-state ratio of the chemical concentration in the organism to the chemical concentration in its ambient environment of water or air is the bioconcentration factor, (BCF) and the BCF is usually measured in a laboratory test.
Biomagnification	The ratio of the chemical concentration in the predator to that of the prey.

This chapter was originally published as part of the Encyclopedia of Sustainability Science and Technology edited by Robert A. Meyers. DOI:10.1007/978-1-4419-0851-3

S. Gama (✉) • D. Mackay
Canadian Centre for Environmental Modelling and Chemistry, Trent University, 1600 West Bank Drive, Peterborough, ON K9J 7B8, Canada
e-mail: SimangaGama@trentu.ca; dmackay@trentu.ca

J.A. Arnot
Department of Physical and Environmental Sciences, University of Toronto Scarborough, 1265 Military Trail, Toronto, ON M1C 1A4, Canada
e-mail: jon.arnot@utoronto.ca

J.S. Gulliver (ed.), *Transport and Fate of Chemicals in the Environment: Selected Entries from the Encyclopedia of Sustainability Science and Technology*, DOI 10.1007/978-1-4614-5731-2_3, © Springer Science+Business Media New York 2012

Hazard — The inherent toxic potency of a chemical, usually expressed as the quantity or concentration of the substance necessary to elicit a defined adverse effect in the organism.

Mass Balance Model — A mathematical description of the fate and transport of a chemical in the environment, usually in the form of a computer program. The model provides a complete accounting of all processes experienced by the chemical and is used to estimate environmental concentrations, persistence, and exposures. Models may also be used to forecast future changes in concentrations as a result of actions to reduce contamination.

Persistence — The average time that a discharged chemical survives in the environment before it is degraded into another substance or substances. It may be expressed as a "half-life" by analogy to radioactive substances or as a residence time.

Risk — The likelihood that there will be an adverse effect as a result of exposure to the chemical. Risk thus depends both on toxic potency (hazard) and the prevailing exposure.

Toxicity — The phenomenon by which a chemical substance elicits an adverse effect on an exposed organism. The effect may be death (lethality) or a less severe effect such as a failure to reproduce, an increased vulnerability to predation or significant behavioral changes.

Definition of the Subject

Organic chemicals play an invaluable role in the modern lifestyle. They include pharmaceuticals, pesticides, plastics, fuels, solvents, explosives, surface coatings, adhesives, disinfectants, and fire retardants. From the perspective of conservation and sustainability, the preferred strategy is to use chemicals such that they perform their desired function, cause no unintended adverse effects, and hence leave no legacy of contamination. This is a fundamental component in the move toward "green chemistry" which also strives to reduce resource depletion, energy use, ozone depletion, and interference with natural biogeochemical cycles. Of the over 50 million chemicals that have been characterized, most are organic compounds. Some 100,000 are commercially produced in quantities large enough to raise concerns that they may become present in the environment in sufficient quantities and at sufficient concentrations to cause risks to the well-being of humans or other organisms. As a response to this concern, most regulatory agencies have listed chemicals of national concern, a typical number being in the hundreds to thousands. These "toxic organic chemicals" merit regulatory scrutiny and possible controls over synthesis and use.

At the international level, the Stockholm Convention has listed over 20 substances as worthy of regulation or even bans. Most of these are synthetic chlorinated organic substances, but some naturally occurring and inadvertently produced substances are also of concern. In this entry, the chemical attributes or properties of these substances are described and criteria that dictate the level of concern about hazard and risk to environmental and human health are outlined. Selected classes of organic chemicals are then discussed in more detail.

Introduction

A consensus has emerged among the scientific and regulatory communities that it is a combination of properties of chemical substances that dictates their designation as "toxic organic substances" and thus the need for their regulation. The result has been the identification of classes of chemicals as persistent organic pollutants (POPs) and persistent bioaccumulative and toxics (PBTs) and those that undergo long-range transport (LRT) on a global scale. Although numerical criteria have been developed for certain key chemical properties such as toxicity and persistence, it is apparent that there is a continuum of properties, thus there is no clear demarcation between for example, PBTs and non-PBTs. There are varying degrees of PBT-like character among "toxic organic chemicals." It can be justifiably argued that all chemical substances are toxic if the administered dose or exposure is sufficient, that is, as stated over 500 years ago by Paracelsus, "it is the dose that makes the poison." Accordingly, there is no clear demarcation between toxic and nontoxic substances: the demarcation must be on the basis of both potency as a toxic agent and the exposure experienced in the environment. The regulatory focus must be on ensuring the principles of sustainability and conservation by reducing exposure.

It is important to discriminate between the terms hazard and risk, since these terms are often wrongly used interchangeably. Hazard reflects the inherent properties of the chemical such as its toxic potency, regardless of the actual exposure. Risk reflects the probability of adverse effects as a function of both hazard and the actual exposure. A highly hazardous substance may pose a low risk and vice versa.

Properties and Characteristics of Toxic Organic Chemicals

Organic chemicals tend to fall into distinct classes based on their molecular structure, an example being the alcohols of increasing carbon chain length, methanol, ethanol, propanol, butanol, etc. There is often a systematic change in properties such as boiling point or vapor pressure in such homologous series. This feature is invaluable in that interpolation and modest extrapolation of properties is often possible. This is formalized in linear free energy relationships (LFER) and is exploited in the development of quantitative structure activity (or property) relationships (QSARs and QSPRs) that are widely used for property estimation purposes. As quantum chemical molecular modeling computation methods become more reliable and accessible, it is clear that the key properties of chemical substances are amenable to a priori estimation from molecular structure.

A key LFER arises from a systematic increase in halogenation of a parent organic substance. For example, the series benzene, chlorobenzene, dichlorobenzene to hexachlorobenzene displays consistent changes in properties such as a decrease in vapor pressure and aqueous solubilities. This also occurs with bromine and to an extent with fluorine substitution and with increased methylation, for example, benzene, toluene, xylene, etc. These systematic property changes are exploited when later reviewing the properties of groups of chemicals such as the polychlorinated biphenyls (PCBs). The key properties as discussed by Mackay et al. [1] are addressed below.

Molecular structure and molar mass (g/mol) are obvious fundamental properties that reflect the constituent elements (molar mass) and their location in the chemical (molecular structure). It is necessary to discriminate between isomers – chemicals comprised of the same elements but with different structural or spatial patterns. For example, the four stereoisomers of hexachlorocyclohexane (HCH) are comprised of the exact same elements (six carbon, six chlorine, and six hydrogen atoms) contained in the same general structure (a cyclic ring of six carbon atoms, each with a bond to a chlorine and a hydrogen atom), but with different three-dimensional spatial patterns of these bonds. The seemingly minor differences in HCH isomer structure result in significant differences in toxicity.

Melting point and boiling point dictate the state of the pure chemical at atmospheric temperatures and pressure as either solids, liquids, or gases.

Chemical solubility and partitioning: the tendency of a substance to partition from its pure state into the atmosphere, water, and octanol, is expressed, respectively, as the saturation vapor pressure, aqueous solubility, and solubility in octanol. These properties are useful but are not always measurable. For example, the solubility of ethanol in water is not measurable because of miscibility. Octanol is widely accepted as a surrogate phase for partitioning into lipids (fats) in biota and for natural organic matter (OM) and organic carbon (OC) in soils and sediments. Since saturation conditions for organic chemicals rarely exist in the environment, it is more convenient to use ratios of these three solubilities, that is, the three partition coefficients or ratios of concentrations, K_{AW} (air–water), K_{OW} (octanol–water), and K_{OA} (octanol–air). Clearly only two of these are independent since K_{OA} is K_{OW}/K_{AW}. The air–water partition coefficient K_{AW} is $H/S^S{}_W$ or $(P^S/RT)/S^S{}_W$ where P^S is the saturation vapor pressure (Pa), R is the gas constant (8.314 Pa m^3/mol K), T is absolute temperature (K), $S^S{}_W$ is the solubility in water (mol/m^3), H is the Henry's law constant (Pa m^3/mol) and is $P^S/S^S{}_W$. For most environmental purposes, K_{OW} is assumed to be equivalent to the lipid–water partition coefficient. K_{OW} is also used to estimate partitioning to organic carbon from water (K_{OC}) as approximately 0.35 K_{OW} plus or minus a factor of three [2].

The K_{AW}, K_{OW}, K_{OA} properties thus largely dictate the equilibrium partitioning characteristics of the chemicals between air, water, soils, and sediments that contain organic carbon as well as biota in terms of their lipid content. A chemical space diagram provides a convenient depiction of these properties as plots of either log K_{AW} versus log K_{OW}, (in which case the constant log K_{OA} values lie on a diagonal), or as log K_{AW} versus log K_{OA}, with log K_{OW} values lying on a diagonal. Figure 3.1 is such a plot and shows the points corresponding to the series of chemicals representing particular chemical classes. For example, points for 1,1′-(2,2,2-trichloroethylidene)bis 4-chloro-benzene (DDT), benzo(a)pyrene, and PCBs are plotted showing the wide variation in partitioning properties which translate into differences in environmental fate. On this plot are lines corresponding to various estimated mass percentage partitioning between air, water, and octanol phases for an assumed volumetric proportion of these phases.

These simple relationships apply only if the molecule does not dissociate or ionize as applies to organic acids such as phenols at high pH and organic bases such as amines at low pH. For these substances, knowledge of the dissociation constant pK_A and the prevailing pH is essential.

Reactivity, expressed as a rate constant or half-life, is an essential property because it determines the persistence, that is, the residence time of the chemical in the environment. Persistence is important because to a first approximation the quantity of the chemical residing in the environment (kg) and hence the concentrations that exert toxicity (kg/m^3) are proportional to the discharge rate of chemical to the environment (kg/h) and to the chemical's residence time (h). Chemicals usually react following second-order kinetics

$$\text{Rate} = k_2 C_C C_R = k_1 C_C = 0.693 C_C / t_{1/2}$$

where k_2 is a second-order rate constant, C_C is chemical concentration, and C_R is the concentration of a reacting species such as hydroxyl radicals in the atmosphere [3]. It is often convenient to lump k_2 and C_R as a first-order rate constant k_1 (h^{-1}) but this assumes that C_R and k_2 are constant. In reality, both vary spatially and temporally, thus the half-life $t_{1/2}$ is also expected to vary. Despite this variability, persistence is such an important property that it is necessary to provide estimates of half-lives of the chemical for all relevant media to which the chemical partitions. Some literature and databases exist on the half-lives of well-studied priority chemicals in air, water, soils, sediments, and in biota by metabolic conversions or biotransformation [1, 4–6]. There is, however, a paucity of data for other substances including "emerging" contaminants. The most common mechanisms are reaction with oxidizing species such as hydroxyl radicals, hydrolysis, photolysis (direct and indirect), biodegradation by microorganisms, and biotransformation in animals and plants.

Regulatory agencies have set half-life criteria for specific environmental media (for example, 60 days in water and 2 days in the atmosphere) [7]; however, in addition to medium-specific half-lives, the overall or average persistence of

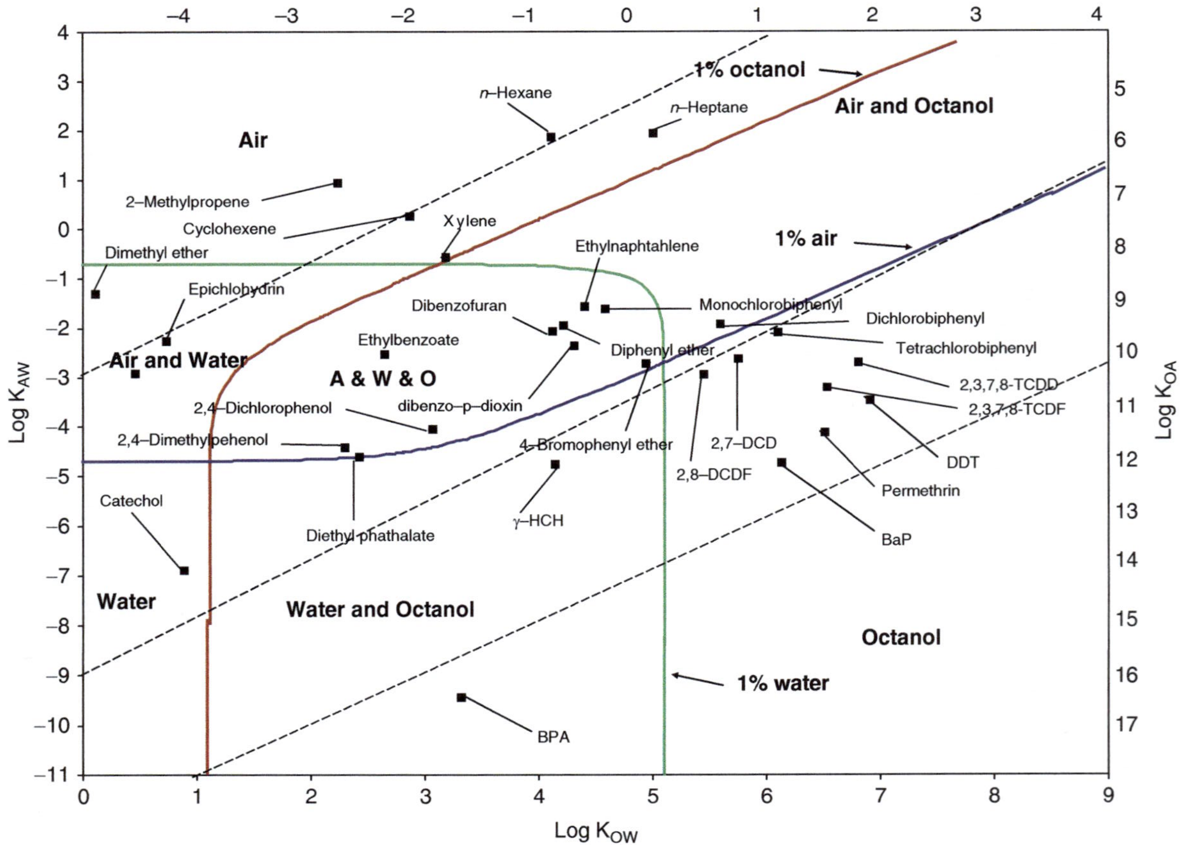

Fig. 3.1 Chemical space plot of the air–water partition coefficient (log K_{AW}) versus the octanol–water partition coefficient (log K_{OW}) showing positions of chemical classes on the three media of air, water, and octanol. Octanol serves as a surrogate for organic matter in soils and sediments. Constant log K_{OA} values form the dashed diagonals on this plot of 1, 6, and 10 (from top to bottom) since log K_{OA} is log K_{OW}−log K_{AW}. The heavy lines represent 1% partitioning into each medium

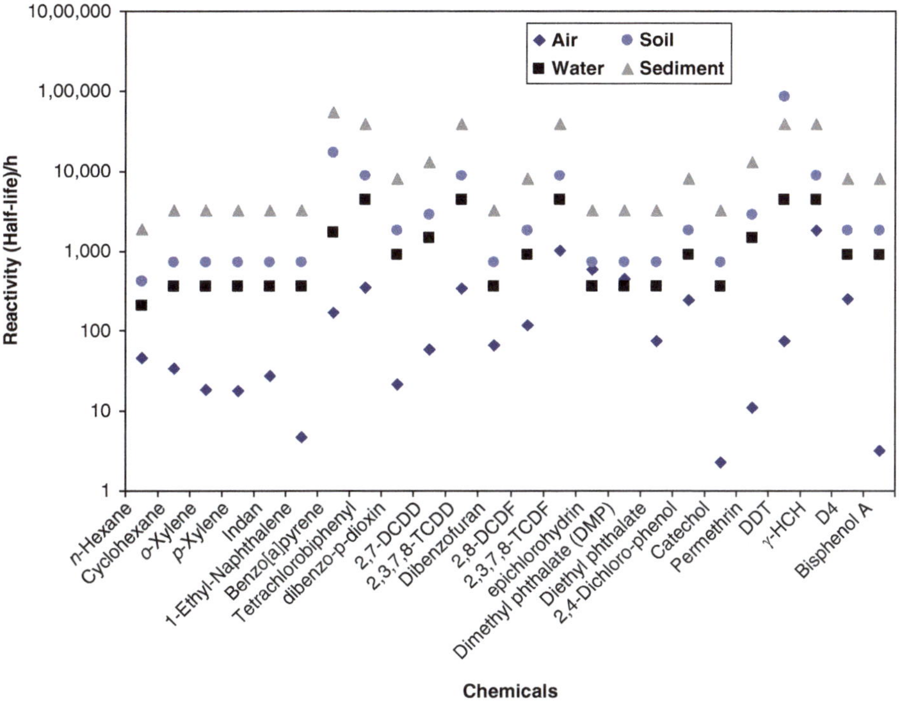

Fig. 3.2 Estimated half-lives (hours) of selected chemicals in air, water, soil, and sediment [6]

a chemical in the environment is also important and is controlled by the relative quantities of chemical in each medium [8]. The relative quantities in each environmental medium, often referred to as the environmental fate and distribution, are largely a function of the partitioning properties of the chemical and the composition of the environment. In principle, the overall reaction rate constant (k_{OV}) is the weighted mean of the rate constants, the weighting being done according to the proportions in each medium. Computer programs such as the OECD Tool [9] are used to perform such calculations and yield an "overall persistence." This quantity is essential in any assessment of sustainability because the year-to-year carryover of a chemical is controlled by overall persistence. In principle, an overall persistence ($P_{OV} = 0.693/k_{OV}$) of say 2 years implies that an appreciable fraction of the chemical discharged in year 1 will remain after 5 or more years. Indeed, this is the most important single property of an organic chemical, hence their designation as persistent organic pollutants (POPs) [10]. See also the discussion by Gouin et al. [11], Mackay [12] and Scheringer [13]. Figure 3.2 illustrates the variation in reactivity of a selection of chemicals in each medium.

Potential for long-range transport (LRT) in air or oceans is increasingly recognized as a cause for concern because the adverse effects may be experienced at locations remote from the source [13]. Obvious regions of concern are the Arctic and

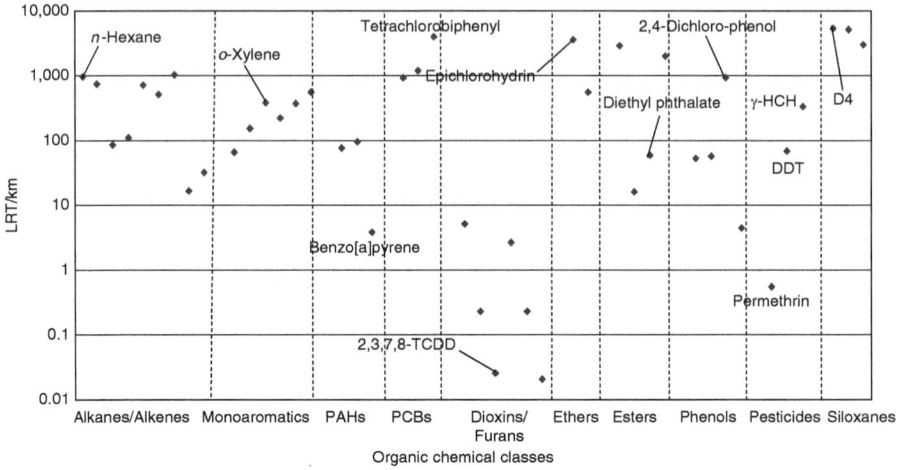

Fig. 3.3 Long-range transport (LRT; km) in air estimated by Level II fate calculations using the RAIDAR 1.00 model [16]. The chemicals are categorized into general classes

Antarctic where local biota may be unacceptably contaminated and used as food by mammals including humans. Multimedia mass balance models have been developed that predict the influence of a chemical's environmental phase distribution on its ability to be transported over long distances [14]. Persistence, and hence reactivity, of the chemical in air influences its potential to be transported over long distances in air. Multimedia models currently used to estimate LRT parameter include Globo-POP [15], RAIDAR [16], and the OECD Tool [9]. The models simulate LRT by incorporating the chemical partitioning properties discussed above. More complex global atmospheric and oceanic circulation models have also been used to simulate and forecast the global cycling of chemicals, for example, Guglielmo et al. [17].

Figure 3.3 shows the variation in long-range transport potential as calculated by the RAIDAR model for a selection of substances.

Bioaccumulation/Bioconcentration and Toxicity

Bioaccumulation refers to the tendency for an organic chemical to partition into biota from their surrounding environment. The bioconcentration factor (BCF) is a metric of bioaccumulation and is the steady-state ratio of the chemical concentration in an organism, such as a fish, to the chemical concentration in an environmental medium, such as water. The BCF is usually measured in a laboratory test and does not include exposures to chemical from dietary sources. Bioaccumulation includes chemical

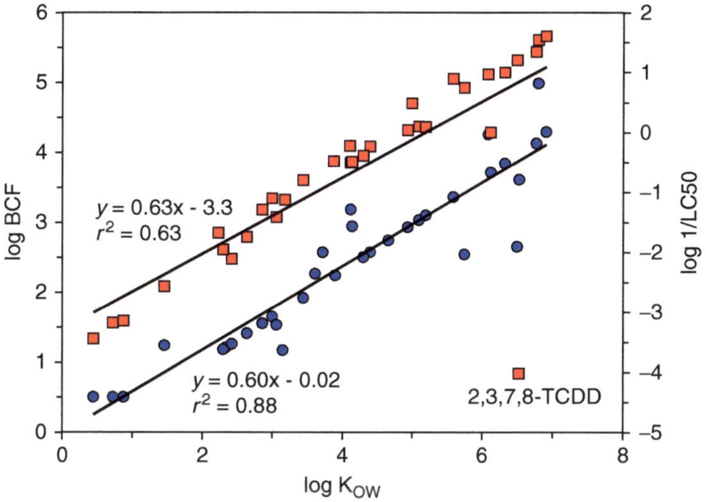

Fig. 3.4 Plot of bioconcentration factors (BCF; L/kg, *blue circles*) in fish and 14-day aquatic toxicity estimates for fish (1/LC50; L/mg, *red squares*) of selected organic chemicals as a function of chemical hydrophobicity (K_{OW}) [6]

uptake from the environment, that is, bioconcentration, and from food. Biomagnification refers to the increase in chemical concentration from prey to predator [18].

Chemical hydrophobicity, as characterized by K_{OW}, plays a large role in determining the BCF and the toxicity of the chemical as measured in aquatic tests for fish. Hydrophobic organic chemicals bioconcentrate in aquatic organisms as a result of the chemical partitioning from the water into the lipids of the organism. Prolonged bioconcentration of the chemical may induce toxicity by, for example, disrupting the integrity of the lipid bilayer of cell membranes, that is, baseline narcotic mode-of-toxic action. Indeed, the bioaccumulation process results in the transport of chemical from external media, where there are generally no sites for toxic action, into the organism, where there are sites for toxic action [19]. Figure 3.4 illustrates the fundamental role of partitioning (or bioconcentration) on aquatic toxicity when the parameters for the BCF (L/kg) and toxicity in fish (LC50; mg/L) are plotted as functions of K_{OW}. Bioconcentration and aquatic toxicity (expressed as 1/LC50) both increase with K_{OW} and the slopes of the two relationships are almost identical. Chemicals with log K_{OW} greater than 5 include polycyclic aromatic hydrocarbons (PAHs), many pesticides, chlorinated furans and dioxins, and polychlorinated biphenyls (PCBs) which are implicated as being both toxic and persistent. Chemicals that are highly toxic, such as 2,3,7,8-tetrachlorodibenzo-*p*-dioxin (2,3,7,8-TCDD), and that have modes of action that are more specific than baseline narcosis will typically fall well below the 1/LC50 regression line.

The Stockholm Convention on Persistent Organic Pollutants (POPs)

The Stockholm Convention is an international agreement designed to limit the use of chemicals that are persistent, bioaccumulative, and subject to long-range transport on a continental or global scale such that they cause significant adverse effects (toxicity) in regions distant from where they are used or manufactured [10]. The initial group comprised of 12 substances or classes was colloquially designated the "dirty dozen." In recent years, other substances have been added. Table 3.1 lists these substances. It is

Table 3.1 Chemicals listed under the Stockholm Convention or under review as of 2010; these chemicals are categorized under Annex A (Elimination), Annex B (Restriction), or Annex C (Unintentional production)

Initial 12 chemicals (i.e., "Dirty Dozen")	
Aldrin (Annex A)	Pesticide
Chlordane (Annex A)	Pesticide
DDT (Annex B)	Pesticide
Dieldrin (Annex A)	Pesticide
Endrin (Annex A)	Pesticide
Heptachlor (Annex A)	Pesticide
Hexachlorobenzene (Annex A, C)	Industrial chemical
Mirex (Annex A)	Pesticide
Toxaphene (Annex A)	Pesticide
Polychlorinated biphenyls (Annex A, C)	Industrial chemical and flame retardant
Polychlorinated dibenzo-*p*-dioxins (Annex C)	By-product
Polychlorinated dibenzofurans (Annex C)	By-product
Recently listed chemicals	
α-hexachlorocyclohexane (Annex A)	Pesticide, by-product
β-hexachlorocyclohexane (Annex A)	Pesticide, by-product
γ-hexachlorocyclohexane (Annex A)	Pesticide
Chlordecone (Annex A)	Pesticide
Hexabromobiphenyl (Annex A)	Industrial chemical and flame retardant
Pentachlorobenzene (Annex A, C)	Pesticide, industrial chemical
Perfluorooctane sulfonic acid, its salts and perfluorooctane sulfonyl fluoride (Annex B)	
Polybrominated diphenyl ethers (Annex A)	Industrial chemical
Tetra, pentabromodiphenyl ether (Commercial penta)	
hexa- and heptabromodiphenyl ether (Commercial octa)	
Chemicals under review as of 2010	
Endosulfan (Annex F)	Insecticide
Hexabromocyclododecane (Annex E)	Flame retardant
Short-chained chlorinated paraffins (Annex E)	Used in pressure lubricants in metal industry. Also used in sealers, glue coatings in building industry, and in leather and rubber treatments.

noteworthy that many are chlorinated hydrocarbons, the reason being that chlorination impacts stability and hence persistence to the molecule.

Chemical Classes

A brief description of some chemical classes is given and provides references for further reading. Molecular structures for selected chemicals in each class are also provided.

Alkanes and Alkenes

The alkanes are present in fuels and generally exhibit low solubilities in water, which when coupled with their relatively high vapor pressures results in relatively high air–water partition coefficients (K_{AW}). Values of K_{OW} are high and increase with carbon number, for example, *n*-hexane has a log K_{OW} value of 4.11 while the higher molecular weight *n*-heptane has a log K_{OW} value of 5.0 [1]. Half-lives in air and water are relatively short (approximately 40 and 200 h in air and water) [1, 3]. The corresponding unsaturated hydrocarbons or alkenes are more soluble in water, have lower K_{AW} and K_{OW} values and are more reactive, especially in air in which half-lives may be only a few hours. They are thus implicated in the formation of photochemical smog due to the presence of the double bond which renders them susceptible to oxidation by hydroxyl radical. This class of chemicals predominantly partitions to air as indicated in Fig. 3.1.

The halogenated (primarily chlorinated) alkanes are less flammable and are thus valuable solvents but they have longer half-lives and are more environmentally persistent, especially in soils and ground water. They are more hydrophobic than the parent alkanes as a result of their lower solubility in water. Of particular environmental concern are the chlorinated C_2 alkanes and alkenes such as trichloroethylene that are widely used as solvents and have become frequent groundwater contaminants. The chlorinated long- and short-chain paraffin compounds have many uses, notably as cutting oils.

The cycloalkanes are similar in properties to the alkanes. The fully chlorinated cyclohexanes are of concern because of their persistence, volatility, and potential for long-range transport. The pesticide lindane (γ-hexachlorocyclohexane or γ-HCH) and its isomers α-HCH and β-HCH have been distributed globally by transport in air and ocean currents. Uses of these chemicals are now limited. Brominated cyclohexanes that are used as fire retardants are also persistent and bioaccumulative.

n-hexane

hexene

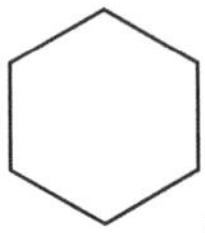

cyclohexane

cyclohexene

Mono-aromatics

The mono-aromatics, for example, benzene, toluene, and the isomers of xylene, are significant components of fuels and are important chemical intermediates. They have higher solubilities in water than the corresponding cycloalkanes and they are relatively volatile and reactive in air contributing to photochemical smog [3]. Mono-aromatics such as xylene partition mostly to air as indicated in Fig. 3.1, while higher molecular weight aromatics such as polyaromatic hydrocarbons (e.g., 1-ethyl naphthalene) have lower volatilities and water solubilities. Chlorination results in increased persistence and hydrophobicity (K_{OW}). Hexachlorobenzene is very persistent and has become globally distributed by atmospheric transport and is banned under the Stockholm Convention.

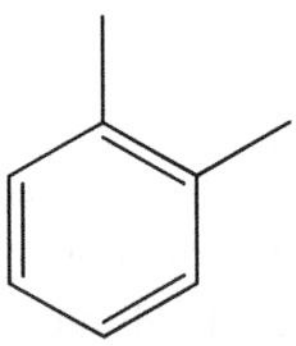

o-xylene

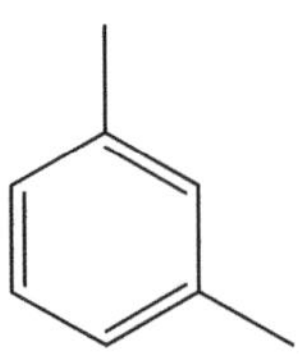

m-xylene

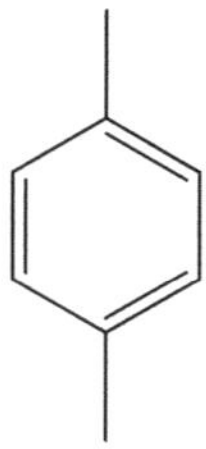

p-xylene

Polyaromatic Hydrocarbons (PAHs)

Fusion of a number of benzene rings results in an important series of PAHs such as the 2-ring naphthalene, 3-ring phenanthrene, 4-ring pyrene, and 5-ring benzo-a-pyrene (BaP). These substances are formed along with smoke and soot during incomplete combustion. They are hydrophobic, relatively involatile, and are regarded as particularly toxic. BaP is a well-studied human carcinogen. The primary environmental concern is human inhalation in the vicinity of combustion sources. Chlorinated naphthalenes are used as PCB substitutes and are hydrophobic and persistent. Chlorination of PAHs increases their hydrophobicity and stability, and decreases their water solubility and volatility. For further reading, see Haritash and Kaushik [20], Mastral and Callen [21], and Wilson and Jones [22].

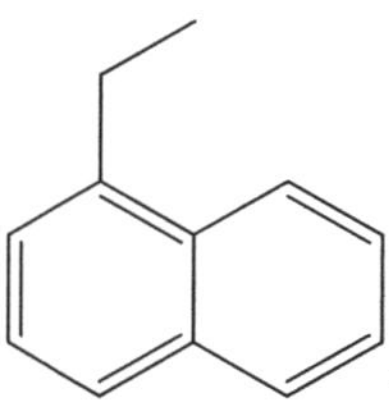

1-ethyl-naphthalene

Polychlorinated Biphenyls (PCBs)

Polychlorinated biphenyls (PCBs) are a class of stable synthetic organic chemicals consisting of 209 possible congeners. PCBs are used industrially as dielectric fluids in capacitors and transformers but also as flame retardants, inks, sealants, and plasticizers. The stability of the C–Cl covalent bond makes these chemicals more persistent in the environment. Most biodegradation reactions involve oxidation or reduction of substrate chemicals. PCBs are also resistant to oxidation, making them resistant to biodegradation. As one of the most notorious toxic organics they have been banned under the Stockholm Convention, but large quantities remain in use and they are widespread in distribution. PCBs are associated with adverse health effects in many fish, marine mammal, and avian species. For instance, levels of PCBs above 20 mg/kg in the blubber of whales and seals have been linked to thyroid, immune, and reproductive disruption [23, 24]. PCBs are predicted to partition to octanol more than air and water, but they are sufficiently volatile to be subject to atmospheric transport. For further reading, see also Fox et al. [25], Hoffman et al. [26], Borga et al. [27], Aken et al. [28], Van den Berg et al. [29], and Domingo and Bocio [30].

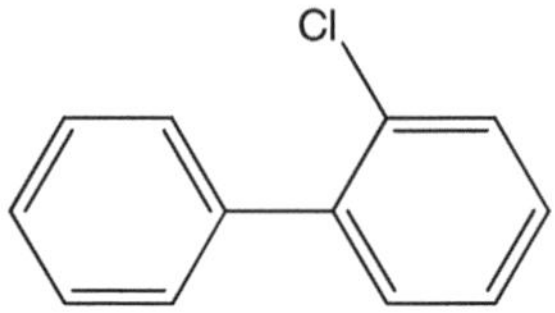

2-chlorobiphenyl

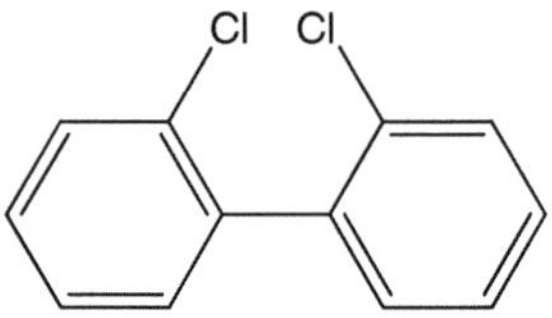

2,2'-dichlorobiphenyl

Dioxins and Furans

Non-chlorinated dioxins and furans such as dibenzo-dioxin and dibenzo-furan, respectively, are relatively water soluble and volatile as indicated in Fig. 3.1 and they can partition to all three media. Chlorinated dioxins and furans are by-products of industrial processes and naturally occurring, and are categorized as being unintentionally produced under Annex C of the Stockholm Convention [10]. Chlorination of dioxins and furans results in increased hydrophobicity, lower water solubilities, and lower volatilities, as demonstrated in Fig. 3.1 where chloro-furans and chloro-dioxins such as 2,8-dichlorodibenzo-furan (2,8-DCDF), 2,7-dichlorodibenzo-dioxin (2,7-DCDD), respectively, are regarded as partitioning largely to octanol, that is, organic phases. Increased chlorination further increases their hydrophobicity as indicated with 2,3,7,8-TCDD and 2,3,7,8-TCDF. A notable congener is 2,3,7,8-tetrachlorodibenzo-p-dioxin (2,3,7,8-TCDD), which is a very potent and persistent toxicant, indeed one of the most toxic organic chemicals known. See also Lohmann and Jones [31], Aberg et al. [32], Van den Berg et al. [29], and Domingo and Bocio [30].

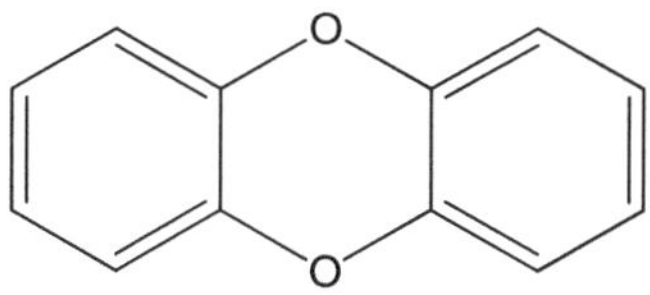

dibenzo-*p*-dioxin

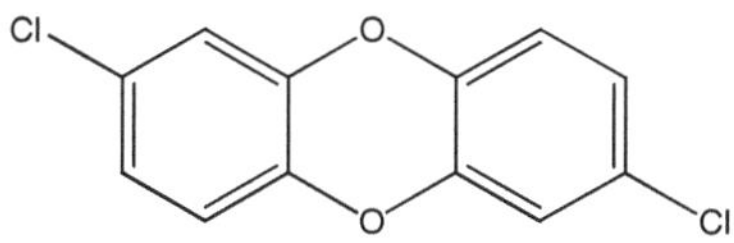

2,7-dichlorodibenzo-*p*-dioxin

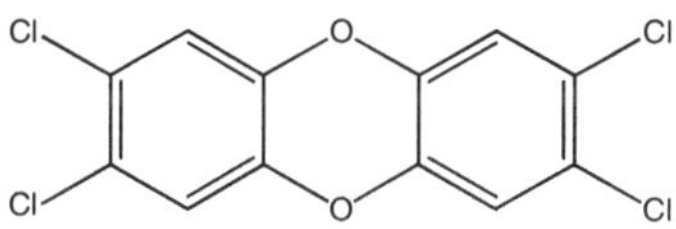

2,3,7,8-tetrachlorodibenzo-*p*-dioxin

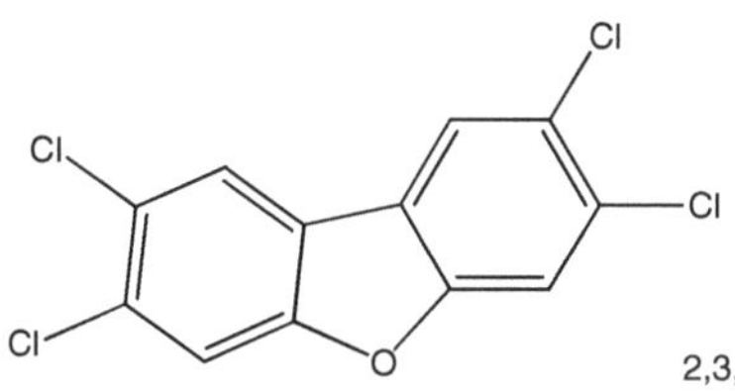

2,3,7,8-tetrachlorodibenzofuran

Ethers

Of more recent environmental concern are certain halogenated ethers such as polybrominated diphenyl ethers (PBDEs). PBDEs are used as fire retardants and are added to a wide variety of commercial and household products. For example, PBDEs are added to polyurethane foam and used in upholstered furniture to reduce their flammability [33]. Commercial pentabromodiphenyl ether, which is mainly composed of tetrabromodiphenyl ether and pentabromodiphenyl ether, is categorized as a persistent organic pollutant and has recently been banned under the Stockholm Convention [10]. Because of their low water solubility, relatively high log K_{OW}, low log K_{AW}, and low vapor pressure, in the environment they partition primarily to sediments and soils. Their persistence in soil and sediments can result in accumulation in the environment and in food webs. Short non-brominated ethers such as dimethyl ether (a common laboratory solvent) are highly soluble, have a high vapor pressure and low log K_{OW} (see Fig. 3.1), and are less persistent. See also Vonderheide et al. [34], Wang et al. [35], and de Wit et al. [36].

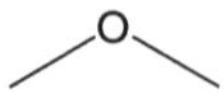

dimethyl ether

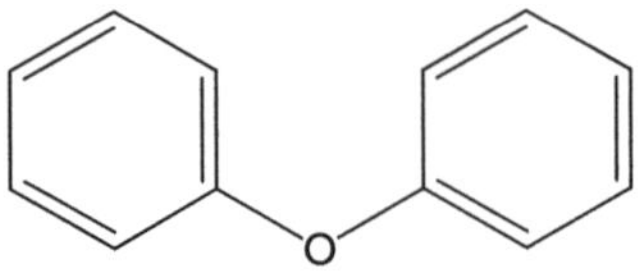

diphenyl ether

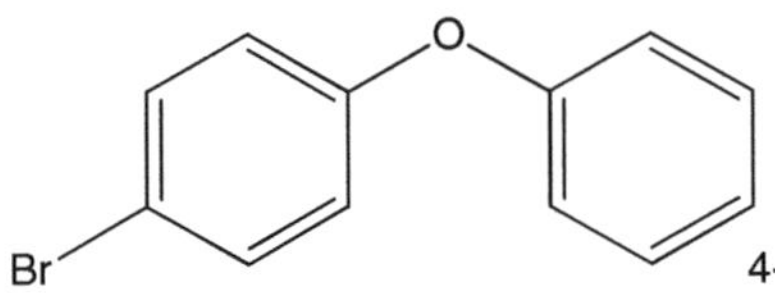

4-bromodiphenyl ether

Esters

Esters are a product of a condensation reaction between alcohols and carboxylic acids. Esters of low molecular weight are relatively volatile and are components of fragrances. Notably among the esters of environmental concern are the phthalate acid esters (PAEs). These commercial chemicals have been used in a wide range of products namely; household electronics, building materials, insecticides and pharmaceutical products. These compounds are found worldwide in oceans, groundwater, sediment, and the atmosphere [37, 38]. Examples of PAEs include dimethyl phthalate (DMP) and diethyl phthalate (DEP).

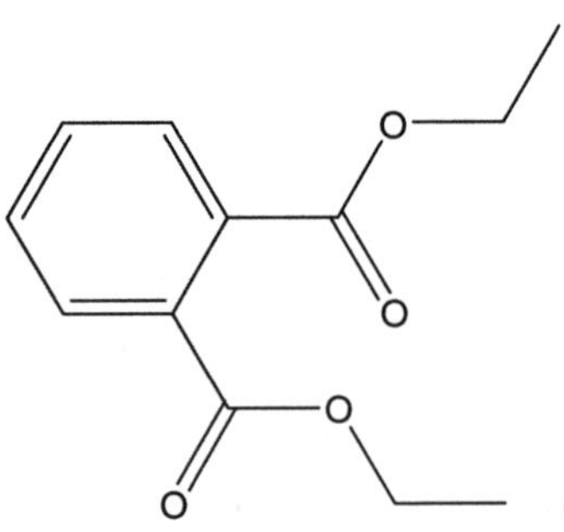

diethyl phthalate

Phenols

Phenols are derivatives of benzene possessing a hydroxyl (−OH) group. They are used as reagents in the manufacture of dyes, drugs, fungicides, and pesticides and in chemical and paper industries. Sources of emissions of phenolic waste to the environment are from petroleum refineries, industrial plants, mine discharges, and through general uses in products containing resins and paints. The hydroxyl group can dissociate into a phenolate and hydrogen ions imparting acidity and relatively high solubility in water [39]. Substituted phenols such as chlorinated and methylated (2, 4-dichlorophenol, pentachlorophenol (PCP) and 2, 4-dimethylphenol) are more susceptible to dissociation and are more acidic and toxic. PCP has been widely used as a wood preservative, while phenol was one of the first medical disinfectants. Catechol is a unique phenol with two hydroxyl groups and is implicated in affecting the human nerve center system, inhibits DNA replication, and leads to chromosomal aberration [40]. Derivatives of catechol include the toxic agents in poison ivy and poison oak.

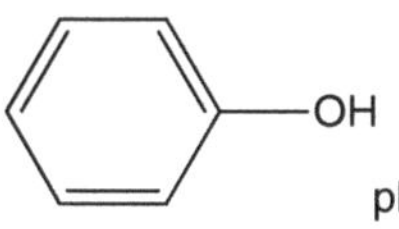

phenol

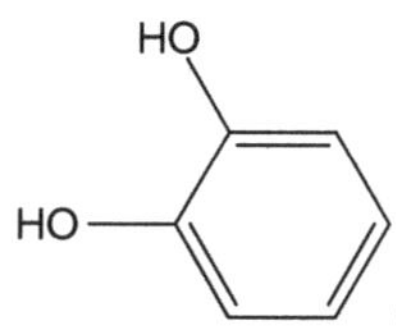

catechol

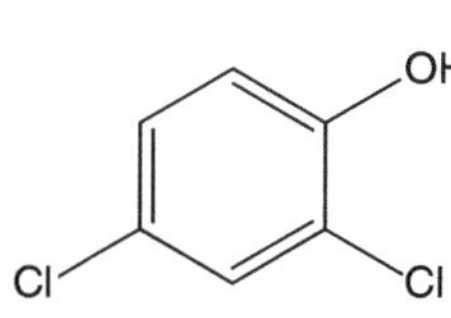

2,4-dichlorophenol

Pesticides

Pesticides are designed to have properties that enable them to reach and impact the target organism or pest including insects, rodents, and vegetation. Effectiveness is increased by persistence, but this can result in nonselective toxicity to other non-targeted organisms including birds and humans. They may have a greater potential to contaminate a wide range of environmental media. Many of the original pesticides were organochlorine (OC) compounds which are now either banned or highly restricted under the Stockholm Convention [10]. Many OCs are relatively volatile; thus, evaporation is an important loss process for pesticides from the areas where they are applied initially, they then move to the atmosphere and eventually get transported through air to remote areas. Notable among the OCs are DDT and its related compounds DDE and DDD, and lindane (γ-HCH). Permethrin, a synthetic pyrethroid insecticide used worldwide to control a broad range of insect pests such as mosquitoes, is also highly hydrophobic and partitions in the environment in a manner similar to DDT. OC pesticides have been largely replaced by organophosphates, pyrethroids, and other substances that are less persistent, but often more potent toxicants. For further reading on pesticides, especially on their global distribution, see Weber et al. [41], Hoferkamp et al. [42], Li and Macdonald [43], and Muir and de Wit [44].

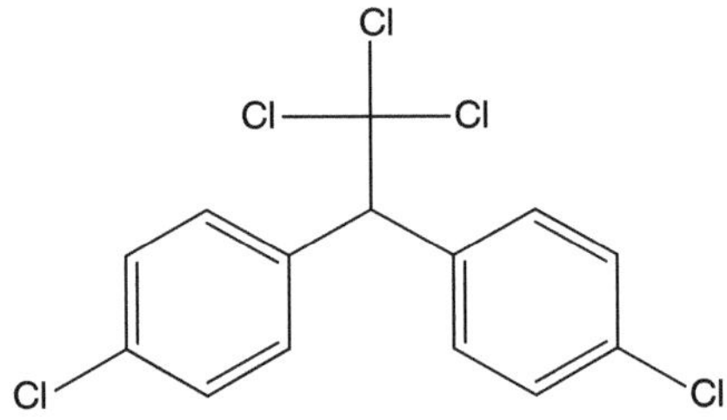

DDT

Personal Care, Indoor, and Household Products

Most organic substances in personal care and household products can impact humans directly in the home and they may enter the environment through sewage wastes (down the drain releases) or by direct releases to air. Included in this group are detergents, fragrances, flame retardants, additives to plastics, and indoor pesticides and solvents. An example is the class of cyclic siloxanes whose backbone is formed of repeated oxygen and silicon atoms with methyl groups attached. These compounds are components of toiletries such as antiperspirants, soaps, and shampoos and are used in many other "household" products including textiles, paints, sealants, lubricants, and non-medicinal ingredients in pharmaceuticals. They are relatively nontoxic but they can be quite persistent [45]. They are highly volatile with large K_{AW} values and hence predominantly partition to air as illustrated in Fig. 3.1. Bisphenol A (BPA) (2,2-bis(4-hydroxyphenyl) propane) has been used often in the production of hard plastic (polycarbonate) bottles and metal-based food and beverage can liners in the past few decades. BPA is regarded as an endocrine-disrupting chemical (EDC) [46]. Regulatory agencies have recently expressed serious concerns about the potential effects of BPA on the brain and prostate gland in fetuses, infants, and young children and some have banned the sale and importation of baby bottles containing BPA [47]. A variety of other organic substances are emitted from various indoor sources, including office equipment such as printers [48].

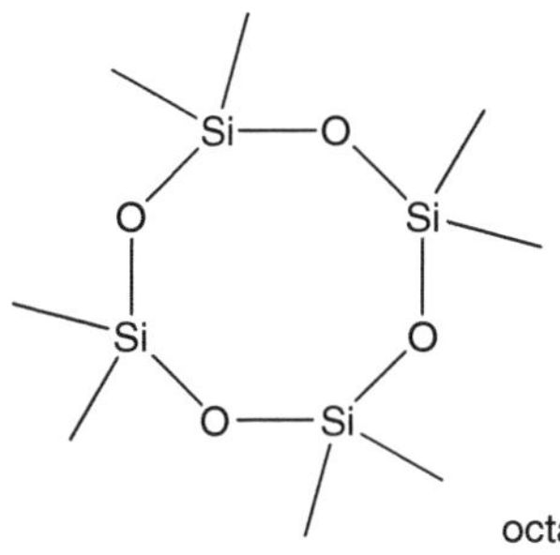

octamethylcyclotetrasiloxane (D4)

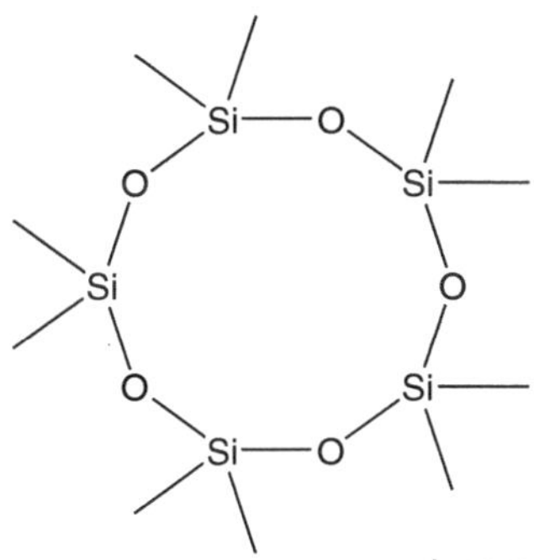

decamethylcyclopentasiloxane (D5)

Conclusions

There is a wide range of organic chemicals that have the potential to be toxic with a multitude of uses and with properties that vary greatly resulting in a diversity of environmental behavior characteristics and exposure routes. Management of these substances presents a considerable challenge to governments and industries. Mistakes have been made in the past by synthesizing and dispersing substances such as DDT and PCBs that have proven to be persistent, bioaccumulative, toxic and have become globally distributed. There is optimism that such mistakes can be avoided in the future if there is an adequate understanding of the factors that dictate environmental partitioning, fate, exposure, and toxicity. Effort is now being devoted to searching for new toxic organics by advance knowledge of their structural and property characteristics [49–51]. It is increasingly possible to estimate the critical chemical properties from molecular structure and assess environmental fate and transport quantitatively using mass balance models that can be validated using monitoring data. If this information can be exploited in advance of commercial distribution, it should be possible to enjoy the many benefits of synthetic organic chemicals without the adverse effects that have been, and are being experienced, from uses that do not satisfy the principles of conservation and sustainability.

Future Directions

The effective future management of toxic organic chemicals in the environment can be improved by a number of scientific and regulatory actions. There is a need for better information on the production, uses and rates of discharge of chemicals, both nationally and internationally. Improved methods of estimating chemical properties from molecular structure are desirable including partitioning, degradability, and toxicity. These improvements could enhance the assessment of chemical fate and possible effects by industry, thus avoiding repetition of past mistakes. Undoubtedly, there are as yet unidentified toxic organic substances, thus there is an incentive to improve techniques and sensitivity of chemical analyses and monitoring of environmental media. As a complement to these monitoring efforts there is a need to further develop, apply, and validate the use of computer models that predict environmental fate and effects. Finally there is a strong incentive to increase the effectiveness of regulatory programs, both nationally and internationally by incorporating the emerging scientific advances and providing a holistic and transparent evaluation of risk to the public and the ecosystem. Ultimately it is an informed public that is the best guarantee that society will maintain and enhance actions to improve sustainability and conservation.

Bibliography

1. Mackay D, Shiu WY, Ma KC (2000) Physical-chemical Properties and Environmental Fate Handbook. Chapman and Hall CRCnetBASE, CRC Press LLC, Boca Raton
2. Seth R, Mackay D, Muncke J (1999) Estimating the organic carbon partition coefficient and its variability for hydrophobic chemicals. Environ Sci Technol 33(14):2390–2394
3. Atkinson R, Arey J, Aschmann SM (2008) Atmospheric chemistry of alkanes: review and recent developments. Atmos Environ 42(23):5859–5871
4. Howard PH, Boethling RS, Jarvis WF, Meylan WM, Michalenko EM (1991) Handbook of Environmental Degradation Rates. Lewis Publishers, Chelsea
5. Arnot JA, Mackay D, Parkerton TF, Bonnell M (2008) A database of fish biotransformation rates for organic chemicals. Environ Toxicol Chem 27(11):2263–2270
6. U.S. EPA (2009) Exposure assessment tools and models, estimation programs interface (EPI) suite, ver. 4.0. U.S. Environmental Protection Agency, Exposure Assessment Branch, Washington, DC
7. Government of Canada, Persistence and Bioaccumulation Regulations. Canada Gazette Part II March 29, 2000

8. Webster E, Mackay D, Wania F (1998) Evaluating environmental persistence. Environ Toxicol Chem 17:2148–2158
9. Wegmann F, Cavin L, MacLeod M, Scheringer M, Hungerbühler K (2009) The OECD software tool for screening chemicals for persistence and long-range transport potential. Environ Modell Softw 24(2):228–237
10. UNEP (2001) Final Act of the Conference of Plenipotentiaries on the Stockholm Convention on Persistent Organic Pollutants. In: United Nations Environment Program, Geneva, Switzerland, p 44
11. Gouin T, Mackay D, Webster E, Wania F (2000) Screening chemicals for persistence in the environment. Environ Sci Technol 34(5):881–884
12. Mackay D, McCarthy LS, MacLeod M (2001) On the validity of classifying chemicals for persistence, bioaccumulation, toxicity and potential for long-range transport. Environ Toxicol Chem 20(7):1491–1498
13. Scheringer M (2002) Persistence and Spatial Range of Environmental Chemicals. Wiley-VCH, Weinheim
14. Wania F (2003) Assessing the potential of persistent organic chemicals for long-range transport and accumulation in polar regions. Environ Sci Technol 37(7):1344–1351
15. Wania F, Mackay D (1995) A global distribution model for persistent organic-chemicals. Sci Total Environ 161:211–232
16. Arnot JA, Mackay D, Webster E, Southwood JM (2006) Screening level risk assessment model for chemical fate and effects in the environment. Environ Sci Technol 40(7):2316–2323
17. Guglielmo F, Lammel G, Maier-Reimer E (2009) Global environmental cycling of γ-HCH and DDT in the 1980s – a study using a coupled atmosphere and ocean general circulation model. Chemosphere 76(11):1509–1517
18. Gobas FAPC, Morrison HA (2000) Bioconcentration and Biomagnification in the Aquatic Environment. In: Boethling RS, Mackay D (eds) Handbook of Property Estimation Methods for Chemicals: Environmental and Health Sciences. CRC Press, Boca Raton
19. McCarty LS, Mackay D (1993) Enhancing ecotoxicological modeling and assessment: body residues and modes of toxic action. Environ Sci Technol 27(9):1719–1728
20. Haritash AK, Kaushik CP (2009) Biodegradation aspects of polycyclic aromatic hydrocarbons (PAHs): a review. J Hazard Mater 169(1–3):1–15
21. Mastral AM, Callen MS (2000) A review on polycyclic aromatic hydrocarbon (PAH) emissions from energy generation. Environ Sci Technol 34(15):3051–3057
22. Wilson SC, Jones KC (1993) Bioremediation of soil contaminated with polynuclear aromatic hydrocarbons (PAHs): a review. Environ Pollut 81(3):229–249
23. Kannan K, Blankenship AL, Jones PD, Giesy JP (2000) Toxicity reference values for the toxic effects of polychlorinated biphenyls to aquatic mammals. Hum Ecol Risk Assess 6:181–201
24. Hornbuckle K, Robertson L (2010) Polychlorinated Biphenyls (PCBs): sources, exposures, toxicities. Environ Sci Technol 44(8):2749–2751
25. Fox GA, Weseloh DV, Kubiak TJ, Erdman TC (1991) Reproductive outcomes in colonial fish-eating birds – a biomarker for developmental toxicants in Great-Lakes food-chains .1. Historical and ecotoxicological perspectives. J Great Lakes Res 17(2):153–157
26. Hoffman DJ, Rice CP, Kubiak TJ (1996) PCBs and dioxins in birds. In: Beyer WN, Heinz GH, Redmon-Norwood AW (eds) Environmental Contaminants in Wildlife: Interpreting Tissue Concentrations. Lewis Publishers, Boca Raton, pp 165–207
27. Borga K, Fisk AT, Hargrave B, Hoekstra PF, Swackhamer D, Muir DCG (2005) Bioaccumulation factors for PCBs revisited. Environ Sci Technol 39(12):4523–4532
28. Aken BV, Correa PA, Schnoor JL (2009) Phytoremediation of polychlorinated biphenyls: new trends and promises. Environ Sci Technol 44(8):2767–2776

29. Van den Berg M, Birnbaum LS, Bosveld ATC, Brunstrom B, Cook P, Feeley M, Giesy JP, Hanberg A, Hasegawa R, Kennedy SW, Kubiak T, Larsen JC, van Leeuwen FXR, Liem AKD, Nolt C, Peterson RE, Poellinger L, Safe S, Schrenk D, Tillitt DE, Tysklind M, Younes M, Waern F, Zacharewski T (1998) Toxic equivalency factors (TEFs) for PCBs, PCDDs PCDFs for humans and wildlife. Environ Health Perspect 106(12):775–792
30. Domingo JL, Bocio A (2007) Levels of PCDD/PCDFs and PCBs in edible marine species and human intake: a literature review. Environ Int 33(3):397–405
31. Lohmann R, Jones KC (1998) Dioxins and furans in air and deposition – a review of levels, behavior and process. Sci Total Environ 219:53–81
32. Aberg A, MacLeod M, Wiberg K (2008) Physical-chemical property data for dibenzo-*p*-dioxin (DD), dibenzofuran (DF), and chlorinated DD/Fs: a critical review and recommended values. J Phys Chem Ref Data 37(4):1997–2008
33. Hites RA (2004) Polybrominated diphenyl ethers in the environment and in people: a meta-analysis of concentrations. Environ Sci Technol 38(4):945–956
34. Vonderheide AP, Mueller KE, Meija J, Welsh GL (2008) Polybrominated diphenyl ethers: causes for concern and knowledge gaps regarding environmental distribution, fate and toxicity. Sci Total Environ 400(1–3):425–436
35. Wang Y, Jiang G, Lam PKS, Li A (2007) Polybrominated diphenyl ethers in the East Asian environment: a critical review. Environ Int 33(7):963–973
36. de Wit CA, Herzke D, Vorkamp K (2010) Brominated flame retardants in the Arctic environment – trends and new candidates. Sci Total Environ 408(15):2885–2918
37. Staples CA, Peterson DR, Parkerton TF, Adams WJ (1997) The environmental fate of phthalate esters: a literature review. Chemosphere 35(4):667–749
38. Peijenburg WJGM, Struis J (2006) Occurrence of phthalate esters in the environment of the Netherlands. Ecotoxicol Environ Saf 63:204–215
39. Shiu W, Ma K-C, Varhanickova D, Mackay D (1994) Chlorophenols and alkylphenols: a review and correlation of environmentally relevant properties and fate in an evaluative environment. Chemosphere 29(6):1155–1224
40. Topping DC, Bernard LG, O'Donoghue JL, English JC (2007) Hydroquinone: acute and subchronic toxicity studies with emphasis on neurobehavioral and nephrotoxic effects. Food Chem Toxicol 45:70–78
41. Weber J, Halsall CJ, Muir D, Teixeira C, Small J, Solomon K, Hermanson M, Bidleman T (2010) Endosulfan, a global pesticide: a review of its fate in the environment and occurrence in the Arctic. Sci Total Environ 408:2966–2984
42. Hoferkamp L, Hermanson MH, Muir DCG (2010) Current use pesticides in Arctic media; 2000–2007. Sci Total Environ 408(15):2985–2994
43. Li YF, Macdonald RW (2005) Sources and pathways of selected organochlorine pesticides to the Arctic and the effect of pathway divergence on trends in biota: a review. Sci Total Environ 342(1–3):87–106
44. Muir DCG, de Wit CA (2010) Trends of legacy and new persistent organic pollutants in the circumpolar arctic: overview, conclusions, and recommendations. Sci Total Environ 408:3044–3051
45. Graiver D, Farminer KW, Narayan R (2003) A review of the fate and effects of silicones in the environment. J Polym Environ 11(4):129–136
46. Hiroi T, Okada K, Imaoka S, Osada M, Funae Y (2006) Bisphenol A binds to protein disulfide isomerase and inhibits its enzymatic and hormone-binding activities. Endocrinology 147:2773–2780
47. Meeker JD, Calafat AM, Hauser R (2010) Urinary bisphenol A concentrations in relation to serum thyroid and reproductive hormone levels in men from an infertility clinic. Environ Sci Technol 44(4):1458–1463
48. Destaillats H, Maddalena RL, Singer BC, Hodgson AT, McKone TE (2008) Indoor pollutants emitted by office equipment: a review of reported data and information needs. Atmos Environ 42(7):1371–1388

49. Brown TN, Wania F (2008) Screening chemicals for the potential to be persistent organic pollutants: a case study of arctic contaminants. Environ Sci Technol 42(14):5202–5209
50. Howard PH, Muir DCG (2010) Identifying new persistent and bioaccumulative organics among chemicals in commerce. Environ Sci Technol 44(7):2277–2285
51. Muir DCG, Howard PH (2006) Are there other persistent organic pollutants? A challenge for environmental chemists. Environ Sci Technol 40(23):7157–7166

Chapter 4
Transport in the Environment

John S. Gulliver

Glossary

Adsorption	The process of dissolved chemicals sticking to a solid.
Convection	The movement of a constituent with movement of the fluid.
Desorption	The detachment of a chemical from a solid.
Diffusion	The spreading of fluid constituents through the motion inherent to atoms and molecules.
Diffusion coefficient	A coefficient that describes the tendency of molecules to spread a constituent mass.
Dirac delta	An impulse of a given quantity (mass) that occurs over an infinitely short time or space.
Kinematic viscosity	The fluid viscosity divided by the fluid density, resulting in units that are similar to a diffusion coefficient, or length squared per time.
Laminar flow	Flow that has no turbulent eddies, where the fluid flows in laminas and diffusion creates the mixing of the fluid.
Retardation factor	A divisor that indicates the slowing of chemical movement through a media due to adsorption.
Reynolds number	The ratio of inertial to viscous forces, resulting in a meaningful velocity times a meaningful distance divided by kinematic viscosity.

This chapter was originally published as part of the Encyclopedia of Sustainability Science and Technology edited by Robert A. Meyers. DOI:10.1007/978-1-4419-0851-3

J.S. Gulliver (✉)
St. Anthony Falls Laboratory, Department of Civil Engineering, University of Minnesota, 500 Pillsbury Drive S.E., Minneapolis, MN 55454, USA
e-mail: gulli003@umn.edu

J.S. Gulliver (ed.), *Transport and Fate of Chemicals in the Environment: Selected Entries from the Encyclopedia of Sustainability Science and Technology*, DOI 10.1007/978-1-4614-5731-2_4, © Springer Science+Business Media New York 2012

Definition of Transport in the Environment

In this section various solution techniques for the convection-diffusion equation are reviewed, which is generally defined as the mass transport equation with diffusive terms. These techniques will be applied to chemical transport solutions in sediments. There are also a number of applications to chemical transport in biofilms. There are many other applications of the convection-diffusion equation, but they require more background with regard to the physics of mixing processes, which will be addressed in later sections of the volume.

Introduction

What is *mass (or chemical) transport*? It is the transport of a solute (the dissolved chemical) in a solvent (everything else). The solute is the dissolvee and the solvent is the dissolver. There are liquids that are generally classified as solvents because they typically play that role in industry. Some examples would be degreasing and dry-cleaning solvents, such as trichloroethylene (TCE). In environmental applications, these "solvents" are the solutes, and water or air is usually the solvent. In fact, when neither water nor air are the solvents, a general term "nonaqueous phase liquid," or NAPL, is applied. NAPL is defined as a liquid that is not water, which could be composed of any number of compounds.

The substance being transported can either be dissolved (part of the same phase as the solvent) or particulate substances. The diffusion equation will also be discussed by considering mass conservation in a fixed control volume. The mass conservation equation can be written as:

$$\begin{gathered} Flux\ rate\ IN - Flux\ rate\ OUT + Rate\ of \\ (Sources - Sinks) = Rate\ of\ Accumulation \end{gathered} \tag{4.1}$$

Now that there is our mass conservation equation, it must be decided which control volume would be the most convenient for our applications. The control volumes used most for this type of mass balance are given in Fig. 4.1. The general control volume, given in Fig. 4.1a, is used for descriptive purposes, to maintain generality. It is rare that one works with something that approximates such a contorted control volume. The control volumes that are used in practice are given in Fig. 4.1b, c, and d. For the environmental applications of chemical transport, the rectangular control volume, Fig. 4.1b, has proven to be the most useful. The cylindrical control volume, Fig. 4.1c, is used to make pipe or tube flow problems easier to solve, and the spherical control volume, Fig. 4.1d, is often helpful when dealing with transport in and around particles

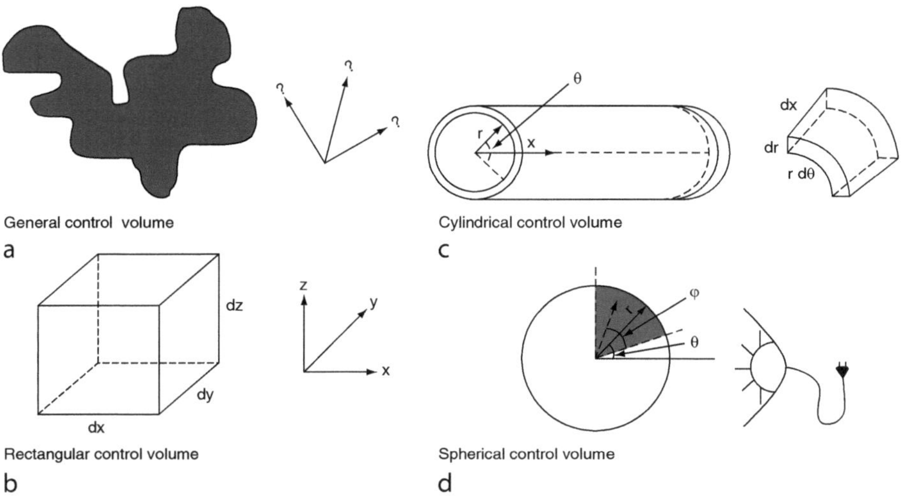

Fig. 4.1 Common control volumes found in engineering texts and (for the latter three) used in solving the diffusion equation. (From [1])

or drops. For this control volume, it is convenient to imagine a light being shined along the axis, which casts a shadow of the vector on to a plane normal to the light. The φ angle is measured from the reference axis to the shadow in this plane.

A rectangular control volume will be used for the development of our mass conservation (diffusion) equation.

Development of the Diffusion Equation

The diffusion equation will be developed by considering each term in Eq. 4.1 separately. In addition, the flux terms will be divided into diffusive and convective flux rates.

Diffusive Flux Rate

The molecules of a fluid "at rest" are still moving because of their internal energy. They are vibrating. In a solid, the molecules are held in a lattice. In a gas or liquid, they are not, so they move around because of this vibration. Since the molecules are vibrating in all directions, the movement appears to be random. Diffusive fluxes are

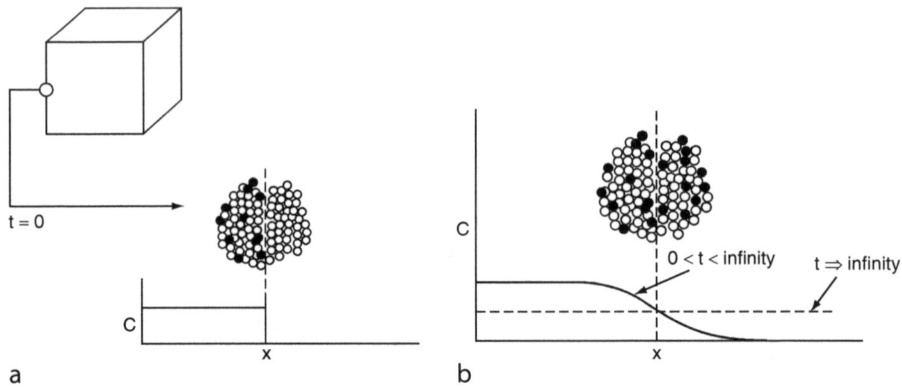

Fig. 4.2 Illustration of net diffusive flux through one side of the rectangular control volume. (From [1])

described by Fick's law [9], given in the section by Dr. Cussler on diffusion. For this purpose, let us consider one side of our control volume, normal to the x-axis, with an area A_x, shown in Fig. 4.2. Fick's law describes the diffusive flux rate as:

$$\text{Diffusive flux rate (g/s)} = -\mathrm{D}\,(\mathrm{m}^2/\mathrm{s})\,\frac{\partial C}{\partial x}\,(\mathrm{g/m}^4)\,A_x(m^2) \tag{4.2}$$

where C is concentration of the solute (tracer), D is the diffusion coefficient of the solute in the solvent (water), which relates to how fast how and far the tracer molecules are moving to and fro, and $\partial C/\partial x$ is the gradient of concentration with respect to x, or the slope of C with x, as shown in Fig. 4.2. Thus, the diffusive flux rate depends upon the diffusion coefficient and the *gradient* of concentration with distance.

Convective Flux

The convective flux rate into our control volume is simply the chemical mass carried in by convection. If the same box of Fig. 4.2 is considered, except with a velocity component u in the x-direction, the convective flux rate into the box from the left-hand side is:

$$\text{Convective flux rate (g/s)} = \text{Velocity component normal to surface (m/s)} \times \text{Surface area (m}^2\text{)} \times \text{Concentration (g/m}^3\text{)} \quad (4.3)$$

or

$$\text{Convective flux rate} = uA_xC \quad (4.4)$$

where u is the component of velocity in the x-direction and A_x is the surface area normal to the x-axis on that side of the box. All six sides of our box would have a convective flux rate through them, just as they would have a diffusive flux.

Rate of Accumulation

The rate of accumulation is the change of chemical mass per unit time, or:

$$\text{Rate of accumulation (g/s)} = \forall(\text{m}^3)\frac{\partial C}{\partial t}(\text{g/m}^3\text{/s}) \quad (4.5)$$

where $\forall$ is the volume of our box.

Source and Sink Rates

The solute chemical can appear or disappear through chemical reaction. In addition, interfacial transfer is often integrated over the control volume and considered as a source or sink throughout the control volume. This type of pseudo-reaction can be of significant help in solving chemical transport problems when averages over a larger control volume, such as cross-sectional mean concentrations, are being computed. For both cases (chemical reactions and pseudo-reactions), the source and sink rates are given as:

$$\text{Source} - \text{sink rate (g/s)} = S\,(\text{g/m}^3\text{/s})\forall(\text{m}^3) \quad (4.6)$$

where S is the net source/sink rate per unit volume. The particular reactions that a given chemical is likely to undergo will determine the form of S used in Eq. 4.6. These are listed in Table 4.1. The source/sink term could be a combination of two or more of these reactions. For convenience in determining analytical solutions to

Table 4.1 Common source and sink terms used in the convection-diffusion equation

Source/sink name	Equation	Units of constant
Zero order	$S = k_o$	k_o – g/m^3-s
First order	$S = k_1 C$	$k_1 = S^{-1}$
Second order	$S = k_2 C^2$	k_2-m^3/g-s
Independent variable	$S = k_{1i}\, P^a$	k_{1i} – s^{-1}
	$S = k_{2i}\, PC^b$	k_{2i} – m^3/g-s
Monod kinetics[c]	$S = \dfrac{\mu_m C}{k_c + C} P$	μ_m = maximum growth rate (s^{-1}) k_c = half-saturation coefficient (g/m^3)

[a]If P is nearly constant, then k_{1i} can be provided as a zero-order term
[b]Often called second order
[c]Common for biologically mediated reactions

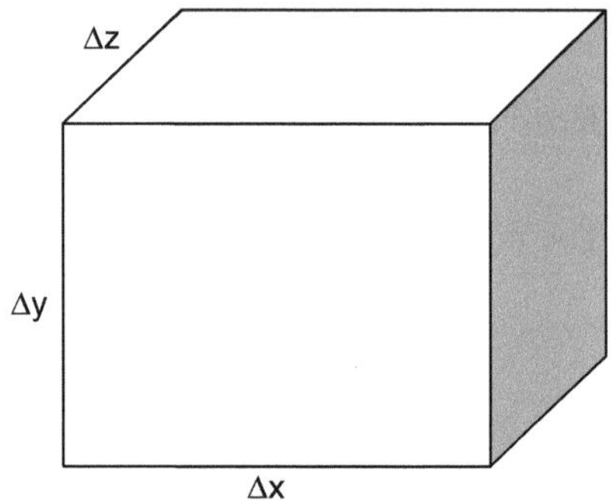

Fig. 4.3 Dimension of the rectangular control volume

the diffusion equation, most source/sink terms are approximated as either a first-order or zero-order reaction.

Mass Balance on Control Volume

A mass balance on one compound in our box is based upon the principle that whatever comes in must do one of three things: be accumulated in the box, flux out of another side, or react in the source/sink terms. If it seems simple, it is.

We will begin by assigning lengths to the sides of our box of dx, dy, and dz, as shown in Fig. 4.3. Then, for simplicity in this mass balance, we will arbitrarily designate the flux as positive in the $+x$-direction, $+y$-direction, and $+z$-direction. The x-direction flux, so designated, is illustrated in Fig. 4.4. Then, the two flux terms in Eq. 4.1 become:

$$\text{Flux rate in} + \text{Difference in flux rate} = \text{Flux rate out} \tag{4.7}$$

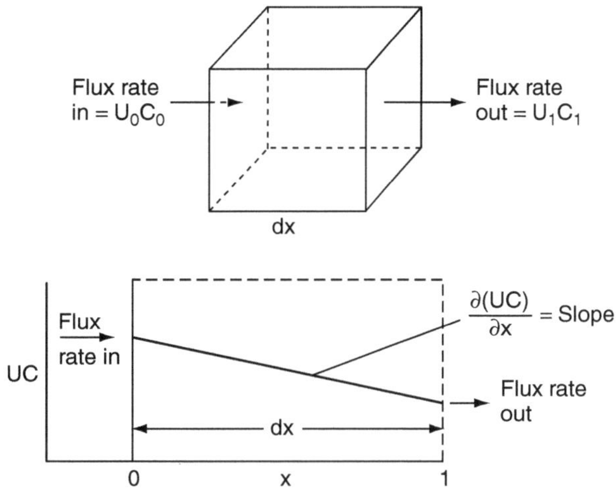

Fig. 4.4 Illustration of the x-component of mass flux rate into and out of the rectangular control volume. (From [1])

or, because a *difference* can be equated to a *gradient times the distance over which the gradient is applied*:

$$\text{Flux rate out} - \text{Flux rate in} = \text{Gradient in flux rate} \times \text{Distance} \tag{4.8}$$

Equation 4.8 can thus be applied along each spatial component as:

$$\text{Flux rate (out} - \text{in})_x = \frac{\partial}{\partial x}(\text{flux rate})dx \tag{4.9a}$$

$$\text{Flux rate (out} - \text{in})_y = \frac{\partial}{\partial y}(\text{flux rate})dy \tag{4.9b}$$

$$\text{Flux rate (out} - \text{in})_z = \frac{\partial}{\partial z}(\text{flux rate})dz \tag{4.9c}$$

Convective flux rates. The convective and diffusive flux rates are dealt with separately. They will eventually be separated in the final diffusion equation, and it is convenient to make that break now. The x-component of the convective flux rate is equal to the x-component of velocity times the concentration times the area of our box normal to the x-axis. Therefore, in terms of convective flux rates, Eq. 4.9a becomes:

$$\begin{aligned}\text{Convective flux rate(out} - \text{in})_x &= \frac{\partial}{\partial x}(uCA_x)dx \\ &= \frac{\partial}{\partial x}(uC)dx\,dy\,dz\end{aligned} \tag{4.10a}$$

Because the normal area, $Ax = dy\,dz$, of our box does not change with x, it can be pulled out of the partial with respect to x. This is done in the second part of Eq. 4.10a. The same can be done with the y- and z-components of the convective flux rate:

$$\begin{aligned}\text{Convective flux rate(out} - \text{in)}_{\text{y}} &= \frac{\partial}{\partial y}(vCA_y)dy \\ &= \frac{\partial}{\partial y}(vC)\,dx\,dy\,dz\end{aligned} \tag{4.10b}$$

$$\begin{aligned}\text{Convective flux rate (out} - \text{in) z} &= \frac{\partial}{\partial y}(w\,C\,A_z)dz \\ &= \frac{\partial}{\partial z}(w\;C)dx\,dy\,dz\end{aligned} \tag{4.10c}$$

Finally, adding Eqs. 4.10a, 4.10b, and 4.10c results in the total net convective flux rate.

$$\begin{aligned}&\text{Net convective flux rate} \\ &= \left[\frac{\partial}{\partial x}(u\,C) + \frac{\partial}{\partial y}(vC) + \frac{\partial}{\partial z}(wC)\right]dx\,dy\,dz\end{aligned} \tag{4.11}$$

Diffusive flux rates. For net diffusive flux rate in the x-direction, Eq. 4.9a becomes:

$$\begin{aligned}&\text{Diffusive flux rate(out} - \text{in)}_{\text{X}} \\ &= \frac{\partial}{\partial x}\left(-D\frac{\partial C}{\partial x}A_x\right)dx \\ &= \frac{\partial}{\partial x}\left(-D\frac{\partial C}{\partial x}\right)dx\,dy\,dz\end{aligned} \tag{4.12a}$$

The y- and z-directions give a result similar to Eq. 4.12a:

$$\begin{aligned}&\text{Diffusive flux rate (out} - \text{in)}_{\text{y}} \\ &= \frac{\partial}{\partial y}\left(-D\frac{\partial C}{\partial y}A_y\right)dy \\ &= \frac{\partial}{\partial y}\left(-D\frac{\partial C}{\partial y}\right)dx\;dy\;dz\end{aligned} \tag{4.12b}$$

$$\begin{aligned}&\text{Diffusive flux rate (out} - \text{in)}_z \\ &= \frac{\partial}{\partial z}\left(-D\frac{\partial C}{\partial z}A_z\right)dz \\ &= \frac{\partial}{\partial z}\left(-D\frac{\partial C}{\partial z}\right)dx\;dy\;dz\end{aligned} \tag{4.12c}$$

Finally, Eqs. 4.12a, 4.12b, and 4.12c can be added to write an equation describing the net diffusive flux rate (out–in) out of the control volume:

$$\text{Net diffusive flux rate} = -\left[\frac{\partial}{\partial x}\left(D\frac{\partial C}{\partial x}\right)+\frac{\partial}{\partial y}\left(D\frac{\partial C}{\partial y}\right)+\frac{\partial}{\partial z}\left(D\frac{\partial C}{\partial z}\right)\right] dx\, dy\, dz \tag{4.13}$$

The diffusion coefficient is often not a function of distance, such that Eq. 4.13 can be further simplified by putting the constant value diffusion coefficient in front of the partial derivative. However, we will also be substituting turbulent diffusion and dispersion coefficients for D when appropriate to certain applications, and they are not always constant in all directions. We will therefore leave the diffusion coefficient inside the brackets for now.

Control volume mass balance. Now Eqs. 4.1, 4.5, 4.6, 4.11, and 4.13 can be combined into a mass balance on our box for Cartesian coordinates. After dividing by $\forall = dx\, dy\, dz$ and moving the diffusive flux terms to the right-hand side, this mass balance is:

$$\frac{\partial C}{\partial t}+\frac{\partial}{\partial x}(uC)+\frac{\partial}{\partial y}(vC)+\frac{\partial}{\partial z}(wC) = \left[\frac{\partial}{\partial x}\left(D\frac{\partial C}{\partial x}\right)+\frac{\partial}{\partial y}\left(D\frac{\partial C}{\partial y}\right)+\frac{\partial}{\partial z}\left(D\frac{\partial C}{\partial z}\right)\right]+S \tag{4.14}$$

When working with a computational transport code, there is little reason to further simplify Eq. 4.14. One primary objective of this section, however, is to develop approximate analytical solutions to environmental transport problems, and we will normally be assuming that diffusivity is not a function of position, or x, y, and z. The convective transport terms can be expanded with the chain rule of partial differentiation:

$$\frac{\partial}{\partial x}(uC) = u\frac{\partial C}{\partial x}+C\frac{\partial u}{\partial x} \tag{4.15a}$$

$$\frac{\partial}{\partial y}(vC) = v\frac{\partial C}{\partial y}+C\frac{\partial v}{\partial y} \tag{4.15b}$$

$$\frac{\partial}{\partial z}(wC) = w\frac{\partial C}{\partial z}+C\frac{\partial w}{\partial z} \tag{4.15c}$$

This may not seem like much help, because we have expanded three terms into six. However, if the flow is assumed to be incompressible, a derivation given in fluid mechanics texts (the continuity equation) is:

$$\rho\left(\frac{\partial u}{\partial x}+\frac{\partial v}{\partial y}+\frac{\partial w}{\partial z}\right)=0 \tag{4.16}$$

where ρ is the density of the fluid. Since Eqs. 4.15a, b, and c are added together in the mass balance equation, the incompressible assumption means that the terms on the far right-hand side of these equations will sum to zero, or:

$$\frac{\partial}{\partial x}(uC)+\frac{\partial}{\partial y}(vC)+\frac{\partial}{\partial z}(wC)=u\frac{\partial C}{\partial x}+v\frac{\partial C}{\partial y}+w\frac{\partial C}{\partial z} \tag{4.17}$$

The incompressible flow assumption is most always accurate for water in environmental applications, and is often a good assumption for air. Air flow is close to incompressible as long as the Mach number (flow velocity/speed of sound) is below 0.3. A Mach number of 0.3 corresponds to an air flow velocity of approximately 110 m/s.

Equation 4.14 then becomes

$$\begin{aligned}\frac{\partial C}{\partial t}+u\frac{\partial C}{\partial x}+v\frac{\partial C}{\partial y}+w\frac{\partial C}{\partial z}\\ =D\left(\frac{\partial^2 C}{\partial x^2}+\frac{\partial^2 C}{\partial y^2}+\frac{\partial^2 C}{\partial z^2}\right)+S\end{aligned} \tag{4.18}$$

The only assumptions made in developing Eq. 4.18 are (1) that diffusivity does not change with spatial coordinate and (2) incompressible flow. Equation 4.18 will be further simplified in order to develop analytical solutions for mass transport problems. In some cases, all that needs to be done is orient the flow direction so that it corresponds with one of the coordinate axes. There would then be only one convection term.

Adsorption and Desorption in Sediments and Soils

Sorption relates to a compound sticking to the surface of a particle. Adsorption relates to the process of compound attachment to a particle surface, and desorption relates to the process of detachment. Sorption processes will now be reviewed because there are many compounds that are sorptive and subject to spills. Then the solutions of the diffusion equation can be examined as they apply to highly sorptive compounds.

Environmental chemicals are generally classified by the Greek terms hydrophilic (likes water) and hydrophobic (hates water). Water is a polar molecule, in that it has two hydrogen atoms on one side, and an oxygen atom on the other. Solutes with a polarity or charge, therefore, will have water molecules surrounding them with the tendency to have the proper charge of atom adjacent to the solute. Most amides and alcohols are strongly polar, and also soluble in water. These are generally hydrophilic compounds. Other organic compounds with larger molecular weights, especially with aromatic rings, are generally nonpolar and are classified as hydrophobic compounds. It makes sense that these hydrophobic compounds would adsorb to the nonpolar organic material in the sediments or soils. There are handbooks [7] that can be used to estimate the chemical thermodynamics of a water-particle system.

How can sorption be handled in our transport equation? For particles that are not transported with the flow field, like sediments and groundwater flow, we are interested in the water concentrations. The sorbed portion of the compound is not in the solute phase, and should not be considered in the transport equation except when transfer of the compound between the water and particles occur. Adsorption would then be a sink of the compound and desorption would be a source.

Let us assign S_p to be the mass of chemical sorbed to particles per mass of solids contained in our control volume, and C to be the concentration of the compound in solution. Then, the source term in the diffusion equation is equal to the rate of change of mass due to adsorption and desorption per unit volume, or:

$$S = \frac{\rho_b}{\varepsilon}\frac{\partial S_p}{\partial t} \tag{4.19}$$

where ρ_b is the bulk density of the solid (mass of solid/volume of fluid and solid), ε is the porosity of the media (volume of fluid/volume of fluid and solid), and $\partial S_p/\partial t$ is the rate of sorption relative to the mass of solid (mass adsorbed/mass of solid/time). If the sorption rate is negative, desorption is occurring. The units of S in Eq. 4.19 are mass adsorbed/volume of fluid/time. This is similar to the units for the $\partial C/\partial t$ term, which are a change of mass/volume of fluid/time.

The source term in Eq. 4.19 requires a separate differential equation for S_p, which would incorporate the concentration of the compound in solution. There would thus be two equations that need to be solved simultaneously. However, most sorption rates are high, relative to the transport rates in sediments and soil. Thus, *local equilibrium in adsorption and desorption is often a good assumption.* It also simplifies the solution to a transport problem considerably. If that assumption is made, S_p changes in proportion to C alone, or:

$$S_p = S_p(C) \tag{4.20}$$

and

$$\frac{\partial S_p}{\partial t} = \frac{\partial S_p}{\partial C}\frac{\partial C}{\partial t} \tag{4.21}$$

Now, if Eq. 4.21 is substituted into 4.19, we get:

$$S = \frac{\rho_b}{\varepsilon} \frac{\partial S_p}{\partial C} \frac{\partial C}{\partial t} \tag{4.22}$$

The $\partial S_p / \partial C$ term can be found from the equilibrium relationship of Freundlich isotherms, expressed as:

$$S_p = K_d C^{\beta} \tag{4.23}$$

where K_d is an equilibrium-partitioning coefficient between the fluid and sorption to the solid and β is a coefficient fit to measured data. Then,

$$\frac{\partial S_p}{\partial C} = \beta K_d C^{\beta - 1} \tag{4.24}$$

At the lower concentrations normally found in the environment, $\beta = 1$ is a valid assumption. Then Eq. 4.24 becomes

$$\frac{\partial S_p}{\partial C} = K_d \ (\beta = 1) \tag{4.25}$$

Substituting Eq. 4.25 into 4.22 now results in a source term that no longer contains the variable S_p, and keeps the partial differential equation (PDE) of our mass balance linear:

$$S = \frac{\rho_b}{\varepsilon} K_d \frac{\partial C}{\partial t} \tag{4.26}$$

Now, if Eq. 4.26 is substituted into our mass transport Eqs. 4.15a–c for the source term, the result is a PDE where the only dependent variable is C:

$$\begin{aligned} &\frac{\partial C}{\partial t} + u\frac{\partial C}{\partial x} + v\frac{\partial C}{\partial y} + w\frac{\partial C}{\partial z} \\ &\quad = D\left(\frac{\partial^2 C}{\partial x^2} + \frac{\partial^2 C}{\partial y^2} + \frac{\partial^2 C}{\partial z^2}\right) - \frac{\rho_b}{\varepsilon} K_d \frac{\partial C}{\partial t} \end{aligned} \tag{4.27}$$

or

$$\begin{aligned} &\left(1 + \frac{\rho_b}{\varepsilon} K_d\right) \frac{\partial C}{\partial t} + u\frac{\partial C}{\partial x} + v\frac{\partial C}{\partial y} + w\frac{\partial C}{\partial z} \\ &\quad = D\left(\frac{\partial^2 C}{\partial x^2} + \frac{\partial^2 C}{\partial y^2} + \frac{\partial^2 C}{\partial z^2}\right) \end{aligned} \tag{4.28}$$

If we divide Eq. 4.28 by the term $(1 + K_d \rho_b / \varepsilon)$, we can see that all convective and diffusive transport is *retarded* by equilibrium adsorption and desorption. Thus, a *retardation factor* is defined:

$$R = \text{retardation factor} = 1 + K_d \rho_b / \varepsilon \tag{4.29}$$

and Eq. 4.28 becomes:

$$\frac{\partial C}{\partial t}+\frac{u}{R}\frac{\partial C}{\partial x}+\frac{v}{R}\frac{\partial C}{\partial y}+\frac{w}{R}\frac{\partial C}{\partial z} = \frac{D}{R}\left(\frac{\partial^2 C}{\partial x^2}+\frac{\partial^2 C}{\partial y^2}+\frac{\partial^2 C}{\partial z^2}\right) \tag{4.30}$$

Equation 4.30 indicates that as long as it can be assumed that the sorption rates are fast compared to our transport rates and the equilibrium partitioning is linearly related to concentration, the retardation factor can utilize and simply convert all of the transport terms through dividing by R. Thus, if there is a spill into the groundwater table that is highly hydrophobic, it would transport through the soil more slowly than one which is hydrophilic. Both the convective and the diffusive flux would be "retarded" for the hydrophobic compound. If both hydrophilic and hydrophobic compounds are contained in the spill, the hydrophilic compound would show up first at a downstream location. The similarity to the manner in which a chromatographic column separates compounds is not fortuitous, because the column is separating compounds through their sorption to the column's media.

Determination of K_d from octanol-water partitioning coefficient. There have been a number of empirical equations developed to determine the water-solid partitioning coefficient, K_d [7]. These are primarily for the many organic chemicals that exist in the environment, usually due to human impacts. Many of them use the octanol-water partitioning coefficient for the compound as an indicator of hydrophobicity. Octanol is a relatively insoluble organic compound. Since most organic compounds tend to adsorb to the organic portion of the particles, a hydrophobic organic compound placed in an octanol-water solution will tend toward the octanol. The ratio of concentration in the octanol over concentration in the water will indicate the degree of the hydrophobicity. It is a straightforward and relatively easy measurement to make, so most organic compounds of interest in the environment have an octanol-water partitioning coefficient that has been measured.

Karikhoff et al. [2] developed a simple empirical equation for equilibrium partitioning of organic compounds that will be used in this text (other equations are given in Lehman et al. [7]):

$$K_d = \beta f K_{ow} \tag{4.31}$$

where K_{ow} is the dimensionless octanol-water partitioning coefficient, f is the fraction of soil that is organic matter (usually from zero in sand to 0.01 in sandy soil to 0.10 in muck), and β is an empirical coefficient, estimated by Karikhoff to be 0.41 cm^3/g. It is generally the organic matter in the medium to which organic compounds adsorb, hence the use of organic fraction.

The other parameters required to compute a retardation coefficient are the bulk density, ρ_b, and the porosity of the media, ε. The bulk density of the water and soil is typically 1.6–2.1 g/cm^3. The porosity of the soil or sediments is typically 0.2–0.4. Thus ρ_b/ε is typically between 4 and 10 g/cm^3.

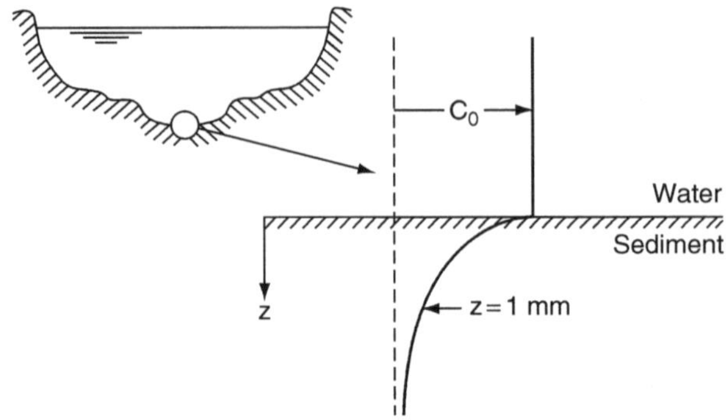

Fig. 4.5 Illustration of dissolved oxygen profile in lake sediments. (From [1])

Example Applications of the Diffusion Equation

The first application of the diffusive is transport of oxygen into lake sediments and the use of oxygen by the bacteria to result in a steady-state oxygen concentration profile.

Example 1: Steady O_2 concentration profile in lake sediments (steady-state solution with a first-order sink) Given a concentration, C_o, in the overlying water, and a first-order sink of oxygen in the sediments, develop an equation to describe the dissolved oxygen concentration profile in the sediments (Fig. 4.5).

Assume:

- Steady: $\frac{\partial}{\partial t} \Rightarrow 0$
- No flow: $u, v, w \Rightarrow 0$
- Small horizontal variation: $\frac{\partial^2 C}{\partial z^2} \gg \frac{\partial^2 C}{\partial x^2}, \frac{\partial^2 C}{\partial y^2}$
- No sorption: $R = 1$ (accurate for O_2 in sediments)
- First-order sink: $S = -kC$, where k is a rate constant

Then, the diffusive mass transport Eqs. 4.15a–c becomes:

$$0 = D\frac{\partial^2 C}{\partial z^2} - kC$$

or, since $C = C(z)$

$$0 = D\frac{d^2 C}{\partial z^2} - kC$$

A solution to this equation requires two boundary conditions because it is a second-order equation. These two are:

B.C.#1: @ $z = 0$, $C = C_{;0}$
B.C.#2: @ $z \rightarrow \infty$, $C \Rightarrow 0$

This solution may be achieved by: (1) separating variables and integrating or (2) solving the equation as a second-order, linear ordinary differential equation (ODE). The latter will be used since the solution technique is more general.

1. Assign λ to be the $\frac{d}{dz}$ operator. Then, the equation becomes

$$\left(\lambda^2 - \frac{k}{D}\right) C = 0$$

2. Solve for λ

$$\lambda = \pm \sqrt{k/D}$$

3. The solution, developed in texts on solving ordinary differential equations [6], is

$$C = \beta_1 e^{\lambda_1 z} + \beta_2 e^{\lambda_1 z} \quad \lambda_1 = +\sqrt{k/D}$$

$$\lambda_2 = -\sqrt{k/D}$$

4. β_1 and β_2 are determined from boundary conditions

   Apply B.C. #2:

$$C = 0 = \beta_1 \, e^{\sqrt{k/D}\infty} + \beta_2 \, e^{-\sqrt{k/D}\,\infty}$$

   This is only possible if $\beta_1 = 0$. Apply B.C. #1:

$$C_0 = 0 + \beta_2 e^{-0} = \beta_2$$

   Thus, the solution is:

$$C = C_o \, \mathrm{e}^{-\sqrt{k/D}z}$$

which is plotted in Fig. 4.6.

At steady state, the oxygen profile is a balance between diffusion from the sediment surface and bacterial use of oxygen in the sediments. If the sediments are mostly sand, the depth of the layer with oxygen can be 10 cm or more. If the sediments have a substantial organic content (like a mud), the aerobic layer (>0.1 g/m^3 oxygen concentration) can be less than 1 mm.

Example 2: Unsteady dissolution of a highly soluble pollutant **(Herbicides, Pesticides, Ammonia, Alcohols, etc.)** ***into groundwater (unsteady, one-dimensional solution with pulse boundary conditions)*** A tanker truck carrying a highly soluble compound in Mississippi tried to avoid an armadillo at night, ran

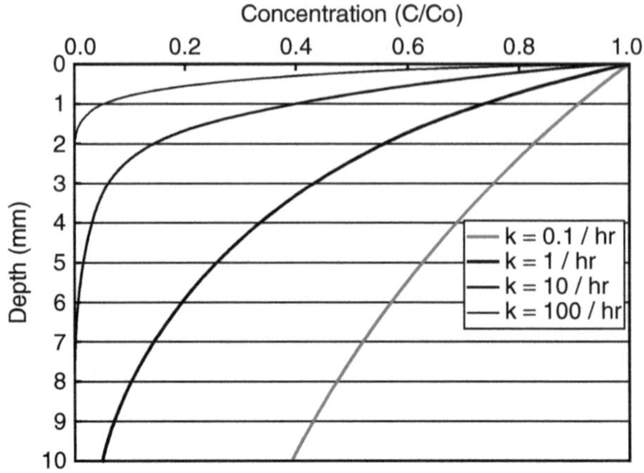

Fig. 4.6 Solution to Example 1

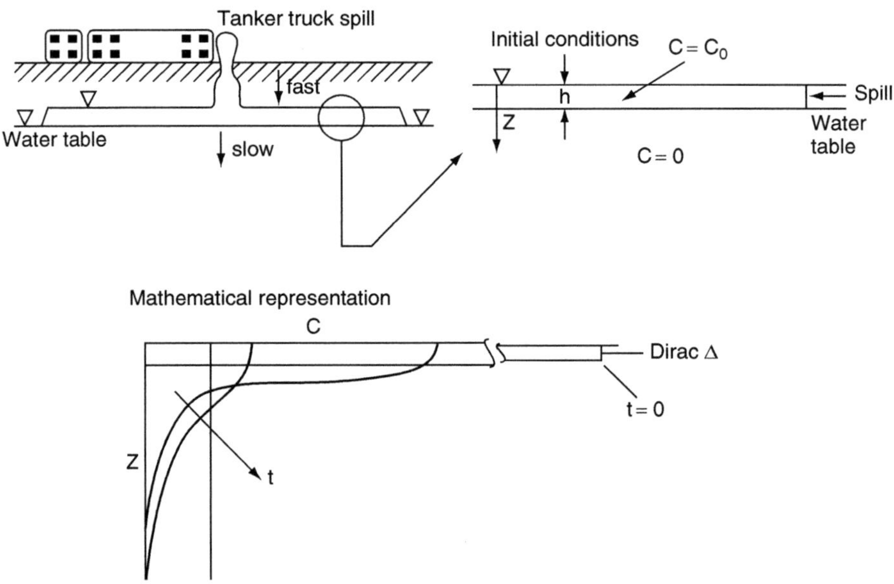

Fig. 4.7 Illustration of the tanker truck spill. (From [1])

off the interstate at a high speed, turned over in the drainage ditch, and spilled a soluble compound. The compound has infiltrated into the ground, and much of it has reached and temporarily spread out over the groundwater table, as illustrated in Fig. 4.7. As part of a spill response team, you need to estimate the groundwater contamination. Predict concentrations over time in the groundwater table.

The mass transport equation for this example is:

$$\frac{\partial C}{\partial t} + u\frac{\partial C}{\partial x} + v\frac{\partial C}{dy} + w\frac{\partial C}{\partial z} = D\left(\frac{\partial^2 C}{\partial x^2} + \frac{\partial^2 C}{\partial y^2} + \frac{\partial^2 C}{\partial z^2}\right) + S$$

Assume:

1. Minimal horizontal variations

$$0 \cong \frac{\partial C}{\partial x} = \frac{\partial^2 C}{\partial x^2} \cong \frac{\partial C}{\partial y} \cong \frac{\partial^2 C}{\partial y^2}$$

2. No flow in the vertical direction, $w = 0$
3. No reactions, including adsorption and desorption, such that $S = 0$.

Then with these three assumptions, the governing equation becomes:

$$\frac{\partial C}{\partial t} = D\frac{\partial^2 C}{\partial z^2}$$

The initial conditions will be simulated with these boundary conditions:

1. The mass of chemical is assumed to be spread instantaneously across a very thin layer at $t = 0$ (a Dirac delta in z and t). At $z = 0^+$, $t = 0$, the total mass $= M$ and the total surface area is A.
2. At $z \Rightarrow \infty$, $C \Rightarrow 0$.

The above equation, with boundary conditions (1) and (2), has the solution:

$$C = \frac{2M/A}{\sqrt{4\pi Dt}}e^{-z^2/4Dt}$$

What does the solution look like? The solution can be made dimensionless by assuming that the initial thickness of the spill layer is Δh. Then, a new variable $z = k\,\Delta h$ will be used in assigning:

$$\eta = \frac{\Delta h}{\sqrt{4Dt}}$$

with

$$C^* = \frac{CA\,\Delta h}{2M}$$

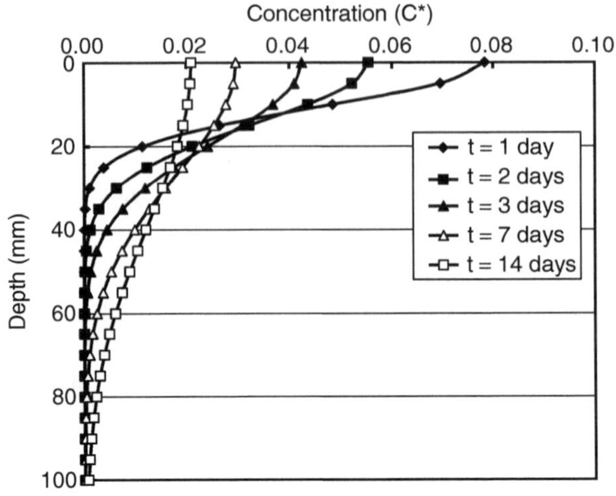

Fig. 4.8 Solution to the tanker truck spill illustrating groundwater concentration versus distance at various times. $\Delta h = 2$ mm, $D = 6 \times 10^{-4}$ mm^2/s

Substituting these equations into the solution gives:

$$C^* = \frac{\eta}{\sqrt{\pi}} e^{-(k\eta)^2}$$

which is plotted versus depth at various times in Fig. 4.8. The concentration at $z = 0$ decreases as the initial mass is diffused. At low values of time, the concentration at and close to $z = 0$ is strongly dependent upon the Δh chosen. At larger times and deeper depths, however, this dependency decreases, and the solution becomes independent of Δh.

It is interesting to note that the solution is very similar to a Gaussian probability distribution, with the following relationship for $P(z)$:

$$P(z) = \frac{1}{\sigma\sqrt{2\pi}} e^{-(z-z_m)^2/2\sigma^2}$$

where z_m is the depth of the maximum concentration (or the center of concentration mass).

Comparing the probability distribution and the solution to this problem, we can see that:

$$2\sigma^2 \Leftrightarrow 4Dt$$

or:

$$D = \sigma^2/2t$$

Note that if we measure σ, we can determine D.

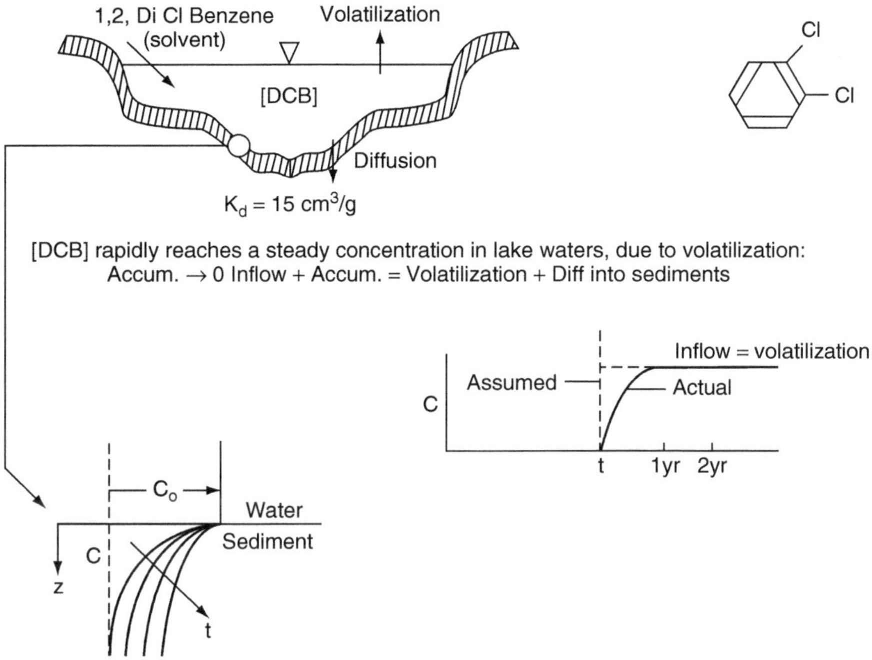

Fig. 4.9 Illustration of concentration front moving down into the sediments of a lake. (From [1])

Example 3*: *Dichlorobenzene concentration in lake sediments due to a plating facility discharge (solution to a concentration front) Sometimes the boundary conditions can be approximated as a step in concentration. This difference in boundary conditions changes the solution from one which is related to pulse boundaries (known mass release) to one resulting from a concentration front with a known concentration at one boundary.

For many years, a plating facility for a telecommunications company let their rinse waters flow into an adjacent lake. The compounds used in their rinse included dichlorobenzene, which is a semi-volatile compound that also has a fairly high tendency to adsorb to organic compounds in the sediments. Within a few years of the plating facility opening, the dichlorobenzene concentration reached a steady-state value in the lake waters as illustrated in Fig. 4.9. Estimate the buildup of dichlorobenzene in the sediments during the 50 years since the facility opened until it stopped discharging its untreated waste water.

Assumptions:

1. Biodegradation is small. $\therefore$ $S \Rightarrow 0$ except for sorption.
2. Variation in *x* and *y* are small

$$\frac{\partial^2 C}{\partial x^2}, \frac{\partial^2 C}{\partial y^2} \ll \frac{\partial^2 C}{\partial z^2}$$

3. No flow in sediments under lake: $u = v = w = 0$
4. $D \cong 6 \times 10^{-10}$ m^2/s
5. $$\frac{\rho_b}{\varepsilon} = 6.3 \;\Rightarrow\; R = 1 + \frac{\rho_b}{\varepsilon} K_d = 96$$

Then the diffusion equation for the sediments becomes:

$$\frac{\partial C}{\partial t} = \frac{D}{R} \frac{\partial^2 C}{\partial z^2}$$

with boundary conditions:

(a) $t > 0$, $z = 0$; $C = C_0$
(b) $t = 0$, $z \neq 0$; $C = 0$

There are three known techniques to solve this governing equation:

Laplace transforms, Fourier transforms, and change of variables, which incorporates both luck and skill. We will use change of variables:

Assign $\eta = \frac{z}{\sqrt{4Dt/R}}$

$$\frac{\partial C}{\partial t} = \frac{\partial C}{\partial \eta} \frac{\partial \eta}{\partial t} = \frac{-1}{4} \frac{z}{\sqrt{Dt/Rt}} \frac{\partial C}{\partial \eta} = \frac{-\eta}{2t} \frac{\partial C}{\partial \eta}$$

$$\frac{\partial C}{\partial z} = \frac{\partial C}{\partial \eta} \frac{\partial \eta}{\partial z} = \frac{1}{2\sqrt{Dt/R}} \frac{\partial C}{\partial \eta}$$

$$\frac{\partial^2 C}{\partial z^2} = \frac{\partial}{\partial \eta}\left(\frac{\partial C}{\partial z}\right) \frac{\partial \eta}{\partial z} = \frac{R}{4Dt} \frac{\partial^2 C}{\partial \eta^2}$$

Then the governing equation becomes:

$$\frac{-\eta}{2t} \frac{\partial C}{\partial \eta} + \frac{D}{R}\left(\frac{R}{4Dt} \frac{\partial^2 C}{\partial \eta^2}\right) = 0$$

or

$$\frac{d^2 C}{d\eta^2} + 2\eta \frac{dC}{d\eta} = 0 \tag{4.32}$$

This equation may be written as: $\frac{dC'}{d\eta} + 2\eta C' = 0$

where $C' = \frac{\mathrm{d}C}{d\eta}$ or

$$\frac{1}{C'} \, dC' = -2\eta \, d\eta$$

We can integrate this:

$$\ell n\, C' = -\eta^2 + \beta_o$$

or

$$C' = e^{\beta o}\, e^{-\eta^2} = \beta_1\, e^{-\eta^2}$$

Now integrate again:

$$C = \beta_1 \int_0^{\eta} e^{-\eta^2} d\eta + \beta_2 \Rightarrow \beta_1 \int_0^{\eta} e^{-\phi^2} d\phi + \beta_2$$

Now, note that the error function is given as:

$$erf\ (\eta) = \frac{2}{\sqrt{\pi}} \int_0^{\eta} e^{-\phi^2} d\phi$$

and the complementary error function is erfc(η) = 1−erf(η). Values of the error function and complimentary error function for various values of η may be found in an Internet search. The error function is designed such that erf(∞) = 1, erfc(∞) = 0, erf(0) = 0, and erfc(0) = 1. The solution may therefore be written as:

$$C = \beta_1\, erf\ (\eta) + \beta_2$$

Now we need to determine our boundary conditions in terms of η:

1. $t > 0,\ z = 0,\quad \eta = 0,\quad C = C_0$
2. $t = 0,\ z = 0,\quad \eta = \infty,\quad C = 0$

Checking other boundary conditions:
$t \to \infty,\quad \eta \Rightarrow 0,\quad C = C_0$
$z \to \infty,\quad \eta \Rightarrow \infty,\quad C = 0$

Now, at $\eta = 0$, $C = C_0$, thus:

$$C_0 = \beta_1 0 + \beta_2$$

or

$$\beta_2 = C_0$$

At $\eta = \infty$, $C = 0$

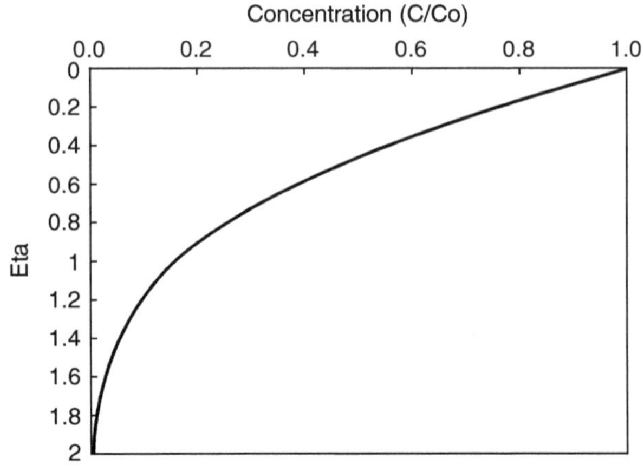

Fig. 4.10 Illustration of the effect of the change of variables used in Example 3. Eta = η

Table 4.2 Penetration of dichlorobenzene into the sediment over time

Time, z	1 year, C/C_0	4 years, C/C_0	10 years, C/C_0	50 years, C/C_0
1 mm	0.96	0.98	0.988	0.994
1 cm	0.62	0.803	0.87	0.94
10 cm	0	0.015	0.11	0.48
20 cm	0	0	0.01	0.16
30 cm	0	0	0	0.03
100 cm	0	0	0	0

$$0 = \beta_1 1 + C_0$$

or

$$\beta_1 = -C_0$$

Then, our solution is:

$$C = C_0\left(1 - erf\left(\frac{z}{\sqrt{4Dt/R}}\right)\right) = C_0\, erfc(\eta) \tag{4.33}$$

which is illustrated in Fig. 4.10.

We will now apply Eq. 4.33 to estimate the dichlorobenzene penetration versus time from spillage. The results are given in Table 4.2, which gives the interstitial dichlorobenzene concentrations.

The total concentration (TDCB(z)) includes compound adsorbed to the sediments.

$$TDCB(z) = C(z) + (1 - \varepsilon)\rho_s S$$

where ε = porosity

$1-\varepsilon$ = % by volume sediment $\cong 0.6$
ρ_s = density of sediment $\cong 2.5\ \text{g/cm}^3$
S = concentration of sorbed compound (g DCB/g sediment)

Since $S = K_d C_1$, the above equation becomes:

$$\begin{aligned} TDCB(z) &= C(z)(1 + (1-\varepsilon)\rho_s K_d) \\ &= C(z)(1 + 0.6(2.5\ \text{g/cm}^3)(15\ \text{cm}^3/\text{g}) \end{aligned}$$

or

$$TDCB = 23.5\, C(z)$$

Thus, the total dichlorobenzene per volume of sediment and water would be 23.5 times the concentrations given in Table 4.2.

Conclusion

The purpose of this section of the volume is to introduce the reader to the equations and mathematics used in developing approximate solutions (due to simplified boundary conditions) to convection-diffusion processes. One may say that, with computational capabilities, there is no longer any need to develop these approximate solutions. However, these approximate solutions are useful in the following manners:

1. A quick, back of the envelope solution is always much quicker and more reliable than a computational solution. Computational solutions require substantial time to develop and are often wrong until they are fully vetted.
2. A computational solution always requires vetting, which means that a computational solution is compared to an analytical solution, hopefully in a similar condition with simplified boundary conditions. This means that some analytical solution is always needed, and as close to the real simulation as possible, to make sure that the computational solution is doing what the user desires.
3. Developing analytical solutions are an excellent means of getting a feel for solutions to the convection-diffusion equations. It is a knowledge-building practice that is difficult to surpass.

Note that the examples given in this section do not include any with convection. That is because convection in the environment most always includes either turbulence (surface waters and the atmosphere) or dispersion (groundwater). These will be dealt with in other sections.

Future Directions

Most of the future directions with regard to transport in the environment (without turbulent transport) will involve transport across interfaces, such as the air–water and solid–fluid interfaces. While research has been conducted on describing the predominant transport mechanisms for these two cases (McCready et al. [8], [3–5]), there is more to be done. An especially vexing problem is transport in the vadose zone of soils (unsaturated zone). The multiplicity of three interfaces, air, water, and soil, and the heterogeneities in the soil make this a complex problem to handle in a deterministic manner. However, meaningful relationships for an effective diffusion coefficient still need to be developed.

Bibliography

Primary Literature

1. Gulliver JS (2007) An introduction to chemical transport in the environment. Cambridge University Press, Cambridge
2. Karikhoff SW, Brown DS, Scott TA (1979) Sorption of hydrophobic pollutants on natural sediments. Water Res 13:241
3. Brunley BH, Jirka GH (1987) Near-surface turbulence in a grid-stirred tank. J Fluid Mech 183:235–263
4. Campbell JA, Hanratty TJ (1982) Mass transfer between a turbulent fluid and a solid boundary: linear theory. AICHE J 28:988
5. Janzen J, Jirka H, Jirka G, Schulz H, Gulliver JS (2010) Estimation of mass transfer velocity based on measured turbulence parameters. Am Inst Chem Eng J 56(8):2005–2017
6. Kreyszig E (1982) Advanced engineering mathematics, 4th edn. Wiley, New York
7. Lehman WJ, Reehl WF, Rosenblatt DH (1990) Handbook of chemical property estimation. American Chemical Society, Washington, DC
8. McCready MA, Vassiliadou E, Hanratty T (1986) Computer simulation of turbulent mass transfer at a mobile interface. AICHEJ 32(7):1108
9. Fick AE (1855) Ueber Diffusion, Annelen der Physik 170(1):59–86

Books and Reviews

Crank J (1975) The mathematics of diffusion, 2nd edn. Oxford University Press, Oxford
Cussler EL (1997) Diffusion: mass transfer in fluid systems, 2nd edn. Cambridge University Press, Cambridge

Chapter 5
Chemicals in the Environment, Turbulent Transport

John S. Gulliver

Glossary

Diffusion	The spreading of fluid constituents through the motion inherent to atoms and molecules.
Diffusion coefficient	A coefficient that describes the tendency of molecules to spread a constituent mass
Dirac delta	An impulse of a given quantity (mass) that occurs over an infinitely short time or space.
Kinematic viscosity	The fluid viscosity divided by the fluid density, resulting in units that are similar to a diffusion coefficient, or length squared per time.
Laminar flow	Flow that has no turbulent eddies, where the fluid flows in laminas and diffusion creates the mixing of the fluid.
Prandtl's mixing length	The mean length that the turbulence in the flow will transport mass, momentum, or energy.
Reynolds number	The ratio of inertial to viscous forces, resulting in a meaningful velocity times a meaningful distance divided by kinematic viscosity.
Turbulent diffusion	The mixing of fluids through turbulent eddies created by convection.

This chapter was originally published as part of the Encyclopedia of Sustainability Science and Technology edited by Robert A. Meyers. DOI:10.1007/978-1-4419-0851-3

J.S. Gulliver
Department of Civil Engineering, University of Minnesota,
500 Pillsbury Drive S.E., Minneapolis, MN 55454, USA
e-mail: gulli003@umn.edu

J.S. Gulliver (ed.), *Transport and Fate of Chemicals in the Environment: Selected Entries from the Encyclopedia of Sustainability Science and Technology*,
DOI 10.1007/978-1-4614-5731-2_5, © Springer Science+Business Media New York 2012

Turbulent diffusion coefficient — A coefficient that comes from the multiplication of two turbulent velocities of the flow, divided by density of the fluid. The coefficient's location in the mass transport equation is similar to diffusion coefficients, and the units are similar; so it is called a "turbulent diffusion coefficient."

Definition of Turbulent Transport in the Environment

It is fairly safe to state that, except for flow through porous media, the environment experiences turbulent flow. To emphasize this point, the constriction of a water flow or airflow that would be required will be considered to have the other option, laminar flow.

An experimentally based rule of thumb is that laminar flow typically occurs when the pipe Reynolds number, Vd/ν, is less than roughly 2,000, or when an open-channel Reynolds number, Vh/ν, is less than roughly 500, where V is the cross-sectional mean velocity, d is the pipe diameter, ν is the kinematic viscosity of the fluid, and h is the channel depth. The diameter or depth that would not be exceeded to have laminar flow by these experimental criteria is given in Table 5.1.

Table 5.1 shows that with the boundary conditions present in most environmental flows, that is, the earth's surface, ocean top and bottom, river or lake bottom, etc., turbulent flow would be the predominant condition. One exception that is important for interfacial mass transfer would be very close to an interface, such as air–solid, solid–liquid, or air–water interfaces, where the distance from the interface is too small for turbulence to occur due to the high viscous dissipation. Because turbulence is an important source of mass transfer, the lack of turbulence very near the interface is also significant for mass transfer, where diffusion once again becomes the predominant transport mechanism.

Table 5.1 Maximum diameter or depth to have laminar flow, with the transition Reynolds number for a pipe at 2,000

Water ($\nu = 10^{-6}$ m²/s)		Air ($\nu \sim 2 \times 10^{-5}$ m²/s)	
V (m/s)	D (m)	h (m)	d (m)
10	2×10^{-4}	5×10^{-5}	0.004
3	7×10^{-4}	1.5×10^{-4}	0.014
1	0.002	0.0005	0.04
0.3	0.007	0.0015	0.14
0.1	0.02	0.005	0.4
0.03	0.07	0.015	1.4
0.01	0.2	0.05	4.0

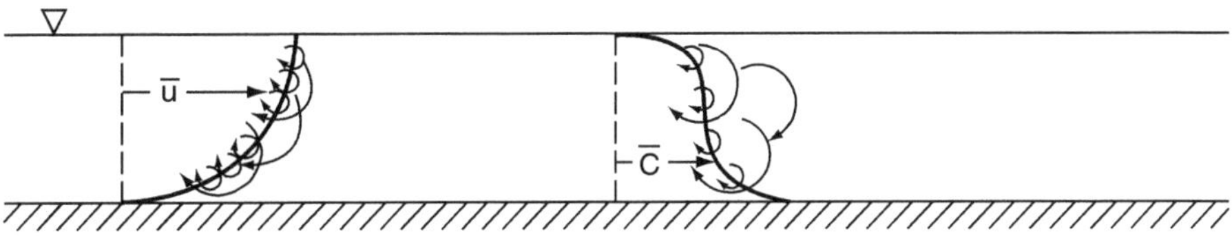

Fig. 5.1 Turbulent eddies superimposed on a temporal-mean velocity and temporal-mean concentration profiles (From Gulliver [1])

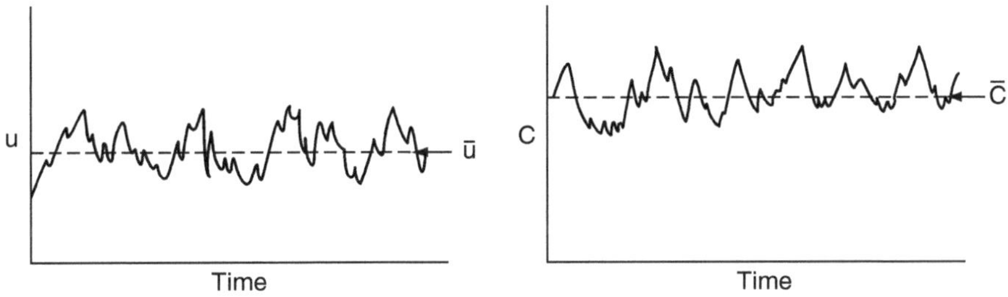

Fig. 5.2 Time traces of typical measurements of velocity and concentration in a turbulent flow

Introduction

What is turbulent flow? The simple illustration of a free-surface flow given in Fig. 5.1 is used to describe the essential points of the turbulence phenomena. Turbulent open-channel flow can be described with a temporal mean velocity profile which reaches a steady value with turbulent eddies superimposed upon it. These turbulent eddies are continually moving about in three dimensions, only restricted by the boundaries of the flow, such that they are eliminated from the temporal mean velocity profile, $\bar{u}$ in Fig. 2.1. It is this temporal mean velocity profile that is normally sketched in turbulent flows.

There will also be a temporal mean concentration. If there is a source or sink in the flow, or transport across the boundaries as in Fig. 5.1, then the temporal mean concentration profile will eventually reach a value such as that given in Fig. 5.1. This flux of compound seems to be from the bottom toward the top of the flow. Superimposed upon this temporal mean concentration profile will be short-term variations in concentration caused by turbulent transport. The concentration profile is "flatter" in the middle of the flow because the large turbulent eddies that transport mass quickly are not as constrained by the flow boundaries in this region. Now, if a concentration-velocity probe is placed into the flow at one location, the two traces of velocity and concentration versus time would look something like that shown in Fig. 5.2.

Turbulent diffusion is thus not really diffusion, but the mixing of chemicals through turbulent eddies created by convection. Turbulent diffusion is thus a form of convection. Although it has the appearance of diffusion in the end, that is, random mixing similar to diffusion, the causes of diffusion and turbulent diffusion are very different. Since the end products are similar, diffusion coefficients and turbulent diffusion coefficients are often simply added together.

It is convenient to divide the velocity and concentration traces into temporal mean values and fluctuating components:

$$u = \bar{u} + u' \tag{5.1}$$

and

$$C = \bar{C} + C' \tag{5.2}$$

where $\bar{u}$ is the temporal mean velocity at a point location, u' is the fluctuating component of velocity (variable over time),$\bar{C}$ is the temporal mean concentration at a point location, and C' is the fluctuating concentration component of concentration which is also variable over time. Formal definitions of $\bar{u}$ and$\bar{C}$ are as follows:

$$\bar{u} = \frac{1}{\Delta t} \int_0^{\Delta t} u \, dt \tag{5.3}$$

and

$$\bar{C} = \frac{1}{\Delta t} \int_0^{\Delta t} C \, dt \tag{5.4}$$

where Δt is long compared to the time period of the oscillating components.

Mass Transport Equation with Turbulent Diffusion Coefficients

In this section the most common equations for dealing with mass transport in a turbulent flow will be derived. Beginning with the mass transport equation developed in the entry "▾Transport in the Environment,"

$$\begin{aligned} &\frac{\partial C}{\partial t} + \frac{\partial (uC)}{\partial x} + \frac{\partial (vC)}{\partial y} + \frac{\partial (wC)}{\partial z} \\ &= \frac{\partial}{\partial x}\left(D\frac{\partial C}{\partial x}\right) + \frac{\partial}{\partial y}\left(D\frac{\partial C}{\partial y}\right) + \frac{\partial}{\partial z}\left(D\frac{\partial C}{\partial z}\right) + S \end{aligned} \tag{5.5}$$

the temporal mean of the entire equation will be taken and eventually one will end up with an equation that incorporates turbulent diffusion coefficients.

In a turbulent flow field, Eq. 5.5 is difficult to apply because C, u, v, and w are all highly variable functions of time and space. Osborne Reynolds [2] reduced the complexities of applying Eq. 5.5 to a turbulent flow by taking the temporal mean of each term (e.g., the entire equation). Then, the mean value of a fluctuating component will be equal to zero, or

$$\overline{\frac{\partial C}{\partial t}} = \overline{\frac{\partial(\bar{C}+C')}{\partial t}} = \overline{\frac{\partial \bar{C}}{\partial t}} + \overline{\frac{\partial C'}{\partial t}} = \frac{\partial \bar{C}}{\partial t} + 0 \tag{5.6}$$

Equation 5.6, the change of a temporal mean over time, may seem like a misnomer, but it will be left in to identify changes in$\bar{C}$ over a longer time period than Δt. Continuing,

$$\overline{\frac{\partial C}{\partial x}} = \overline{\frac{\partial(\bar{C}+C')}{\partial x}} = \overline{\frac{\partial \bar{C}}{\partial x}} + \overline{\frac{\partial C'}{\partial x}} = \frac{\partial \bar{C}}{\partial x} \tag{5.7}$$

$$\overline{\frac{\partial C}{\partial y}} = \overline{\frac{\partial(\bar{C}+C')}{\partial y}} = \overline{\frac{\partial \bar{C}}{\partial y}} + \overline{\frac{\partial C'}{\partial y}} = \frac{\partial \bar{C}}{\partial y} \tag{5.8}$$

$$\overline{\frac{\partial C}{\partial z}} = \overline{\frac{\partial(\bar{C}+C')}{\partial z}} = \overline{\frac{\partial \bar{C}}{\partial z}} + \overline{\frac{\partial C'}{\partial z}} = \frac{\partial \bar{C}}{\partial z} \tag{5.9}$$

However, the temporal mean value of two fluctuating components, multiplied by each other, will not necessarily be zero:

$$\overline{u'C'} \neq \bar{u}'\bar{C}' \tag{5.10}$$

This is similar to a least-square regression, where the mean error is zero, but the sum of square error is not. The x-component of our convective transport terms will be dealt with first:

$$\overline{uC} = \overline{(\bar{u}+u')(\bar{C}+c')} = \overline{\bar{u}\bar{C}} + \overline{\bar{u}C'} + \overline{u'\bar{C}} + \overline{u'C'} \tag{5.11}$$

Three of the four terms in Eq. 5.11 may be reduced to something known:

$$\overline{\bar{u}\bar{C}} = \bar{u}\bar{C} \tag{5.12}$$

$$\overline{\bar{u}C'} = 0 \tag{5.13}$$

$$\overline{u'\bar{C}} = 0 \tag{5.14}$$

but, the fourth term will take some additional consideration, because it is not equal to zero:

$$\overline{u'C'} \neq 0 \tag{5.15}$$

By inference, the following can be written for all three convective transport terms:

$$\overline{uC} = \bar{u}\bar{C} + \overline{u'C'} \tag{5.16}$$

$$\overline{vC} = \bar{v}\bar{C} + \overline{v'C'} \tag{5.17}$$

and

$$\overline{wC} = \bar{w}\bar{C} + \overline{w'C'} \tag{5.18}$$

Finally, applying continuity $(\bar{u} + \bar{v} + \bar{w} = 0)$ to Eq. 5.5 and taking the temporal mean results of Eqs. 5.6, 5.7, 5.8, 5.9, 5.16, 5.17, and 5.18

$$\begin{aligned} &\frac{\partial \bar{C}}{\partial t} + \bar{u}\frac{\partial \bar{C}}{\partial x} + \bar{v}\frac{\partial \bar{C}}{\partial y} + \bar{w}\frac{\partial \bar{C}}{\partial z} \\ &= -\frac{\partial}{\partial x}\overline{u'C'} - \frac{\partial}{\partial y}\overline{v'C'} - \frac{\partial}{\partial z}\overline{w'C'} + \frac{\partial}{\partial x}\left(D\frac{\partial \bar{C}}{\partial x}\right) \\ &+ \frac{\partial}{\partial y}\left(D\frac{\partial \bar{C}}{\partial y}\right) + \frac{\partial}{\partial z}\left(D\frac{\partial \bar{C}}{\partial z}\right) + \bar{S} \end{aligned} \tag{5.19}$$

where the turbulent convective transport term can be moved to the right-hand side, because the concentration distribution that results from these terms looks similar to diffusion.

With this temporal mean process, we have reduced the terms for which we will have difficulty defining boundary conditions in turbulent flow fields from seven in Eq. 5.5 to three in Eq. 5.19. We will now deal with these three terms.

The diffusion equation is a useful and convenient equation to describe mixing in environmental flows, where the boundaries are often not easily defined. It also lends itself to analytical solutions and is fairly straightforward in numerical solutions. Although an alternative technique for solutions to mixing problems is the mixed cell method described in the entry "▾Chemicals in the Environment, Dispersive Transport," there are complications when applied to multiple dimensions and to flows that vary with space and time. Finally, we are comfortable with the diffusion equation, so we would prefer to use that to describe turbulent mixing if possible.

Therefore, let us consider the following thought process: if the end result of turbulence, when visualized from sufficient distance, looks like diffusion with seemingly random fluctuations, then we should be able to identify the terms causing these fluctuations in Eq. 5.19. Once identified, they can be related to a "turbulent diffusion coefficient" that describes the diffusion caused by turbulent eddies. Looking over the terms in Eq. 5.19 from left to right, we see an unsteady term, three mean convective terms, the three "unknown" terms, the diffusive terms and the source/sink rate terms. The "unknown" terms are the only possibility to describe turbulent diffusion.

In the late nineteenth century, Boussinesq [3] probably went through something similar to the thought process described above. The end result was the *Boussinesq eddy diffusion coefficient*:

$$-\overline{u'C'} = \varepsilon_x \frac{\partial \bar{C}}{\partial x} \tag{5.20a}$$

$$-\overline{v'C'} = \varepsilon_y \frac{\partial \bar{C}}{\partial y} \tag{5.20b}$$

$$-\overline{w'C'} = \varepsilon_z \frac{\partial \bar{C}}{\partial z} \tag{5.20c}$$

where ε_x, ε_y, and ε_z are the turbulent (or eddy) diffusion coefficients, with units of m^2/s similar to the (molecular) diffusion coefficients.

Then Eq. 5.19 with Eqs. 5.20a, 5.20b, and 5.20c becomes

$$\begin{aligned} \frac{\partial \bar{C}}{\partial t} + \bar{u}\frac{\partial \bar{C}}{\partial x} + \bar{v}\frac{\partial \bar{C}}{\partial y} + \bar{w}\frac{\partial \bar{C}}{\partial z} &= \frac{\partial}{\partial x}\left[(D+\varepsilon_x)\frac{\partial \bar{C}}{\partial x}\right] \\ &+ \frac{\partial}{\partial y}\left[(D+\varepsilon_y)\frac{\partial \bar{C}}{\partial y}\right] + \frac{\partial}{\partial z}\left[(D+\varepsilon_z)\frac{\partial \bar{C}}{\partial z}\right] + \bar{S} \end{aligned} \tag{5.21}$$

Turbulent diffusion is created by the flow field, which can vary with distance. Hence, turbulent diffusion coefficient cannot be assumed constant with distance. Removing that assumption leaves turbulent diffusion coefficient inside of the brackets.

Character of Turbulent Diffusion Coefficients

A turbulent eddy can be visualized as a large number of differently sized rotating spheres or ellipsoids. Each sphere has sub-spheres, and so on until the smallest eddy size is reached. The smallest eddies are dissipated by viscosity, which explains why turbulence does not occur in narrow passages: there is simply no room for eddies that will not be dissipated by viscosity.

The cause of the rotation is shear forces created by solid boundaries or variations in velocity lateral to the primary flow direction. A buoyant plume of smoke or steam, for example, will have a temporal mean velocity profile develop laterally to the plume, as the rising plume mixes with the ambient air. Turbulent eddies are formed by this velocity gradient, and can be seen at the edge of the smoke or steam plume. The magnitude of turbulent diffusion coefficients is primarily dependent

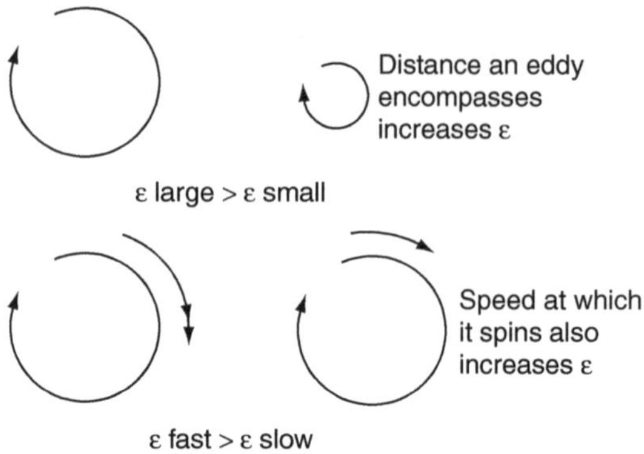

Fig. 5.3 Character of turbulent diffusion coefficients (From Gulliver [1])

upon the scale of turbulent eddies and the speed of the eddy rotation. As illustrated in Fig. 5.3, a large eddy will have greater eddy diffusion coefficient than a small eddy because it will transport a compound (or solute) farther in one rotation. Likewise, a faster spinning eddy will have a larger eddy diffusion coefficient than one which is the same size but spinning more slowly because the solute simply gets there faster. These two facts provide meaning to the following observations:

1. The largest scale of turbulence is roughly equal to the smallest overall scale of the flow field. This may be seen in comparing the size of eddies at the edge of the smoke or steam plume to the width of the plume.
2. The rotational eddy velocity is roughly proportional to the velocity gradient times the eddy scale.
3. Eddy size decreases near boundaries to the flow field. Since the eddy size is zero at a solid boundary, and often close to zero at a fluid density interface (like an air–water interface), the turbulent eddy size has to decrease as one approaches the boundaries. In addition, since the flow cannot go through a boundary, the largest eddy size cannot be greater than the distance from the center of the eddy to the boundary.
4. Turbulent diffusion occurs because turbulent eddies are transporting mass, momentum, and energy over the eddy scale at the rotational velocity. This transport rate is generally orders of magnitude greater than the transport rate due to molecular motion. Thus, when a flow is turbulent, diffusion is normally ignored because $\varepsilon \gg D$. The exception is very near the flow boundaries, where the eddy size (and turbulent diffusion coefficient) decreases to zero.

Thus, what influences the velocity and scale of eddies? For the most part, it is the velocity gradients and scale of the flow. Velocity gradients are the change in velocity over distance. If we have a high velocity, we typically have a large velocity gradient somewhere in the flow field. At solid walls, for example, the velocity must go to zero. Thus, *the large velocity difference results in large velocity gradients, which results in faster spinning eddies and a larger turbulent diffusion coefficient.* This process is illustrated in Fig. 5.4.

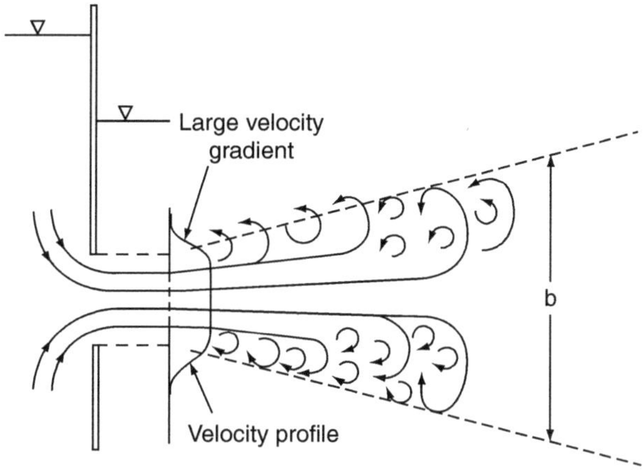

Fig. 5.4 Eddy formation at the edge of a jet issuing into a tank illustrates the importance of velocity gradients in eddy diffusion coefficient (From Gulliver [1])

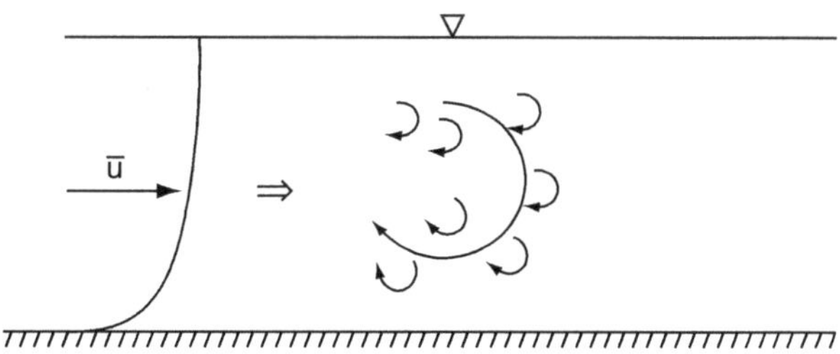

Fig. 5.5 Large and small eddies in an open-channel flow. The large eddies perform most of the top-to-bottom transport (From Gulliver [1])

The scale of the flow field is also important because *the larger eddies perform most of the transport*. The small eddies are always there in a turbulent flow, and their existence is important for local mixing. It is the large eddies, however, that are the most responsible for transport, as illustrated in Fig. 5.5.

The four observations, listed above, were enough for Ludwig Prandtl [4] to hypothesize a simple model for describing turbulent transport that works surprisingly well, considering the complexity of turbulent flow.

Prandtl's Mixing Length Hypothesis for Turbulent Flow

Prandtl's mixing length hypothesis was developed for momentum transport, instead of mass transport. The end result was a turbulent viscosity, instead of a turbulent diffusivity. However, since both turbulent viscosity and turbulent diffusion coefficient are properties of the flow field, they are related. Turbulent viscosity describes the transport of momentum by turbulence, and turbulent diffusivity describes the transport of mass by the same turbulence. Thus,

$$\varepsilon_x = \mu_{tx}/\rho, \ \varepsilon_y = \mu_{ty}/\rho, \text{ and } \ \varepsilon_z = \mu_{tz}/\rho$$

where μ_{tx}, μ_{ty}, and μ_{tz} are the turbulent viscosity in the x, y, and z directions. Now, for the x-component of momentum (ρu), the Boussinesq approximation is

$$-\rho\,\overline{u'u'} = \mu_{tx}\frac{\partial \bar{u}}{\partial x} \tag{5.22}$$

$$-\rho\,\overline{v'u'} = \mu_{ty}\frac{\partial \bar{u}}{\partial y} \tag{5.23}$$

$$-\rho\,\overline{w'u'} = \mu_{tz}\frac{\partial \bar{u}}{\partial z} \tag{5.24}$$

Let us consider the fully developed velocity profile in the middle of a wide-open channel, with x-, y-, and z-components in the longitudinal, lateral, and vertical directions, respectively. It is fully developed because $\partial \bar{u}/\partial x$ is close to zero. The fact that it is a wide channel means that $\partial \bar{u}/\partial y$ also is very small in the middle. From Eqs. 5.22 and 5.23, we can see that the turbulent transport of momentum in the x- and y-directions will be small because the gradients are small. Equation 5.24 indicates that there will be a net turbulent transport of momentum in the z-direction.

$$-\overline{w'u'} = \varepsilon_z\frac{\partial \bar{u}}{\partial z} \neq 0 \tag{5.25}$$

Now, half of the w' values will be positive, and the other half will be negative. We will use this criterion to divide them into two parts:

$$\overline{w'u'} = \overline{w'u'}^{+} + \overline{w'u'}^{-} \tag{5.26}$$

where $w'u'^{+}$ has a value when w' is positive and is equal to zero when w' is negative. $w'u'^{-}$ has a value when w' is negative and is equal to zero when w' is positive. Consider the cases when w' is positive. Then Eq. 5.26 becomes

$$\overline{w'u'} = \overline{w'u'}^{+} + 0 \tag{5.27}$$

Let us assume that an eddy of length L is pulling a blob of fluid upward, as illustrated in Fig. 5.6. On average, the blob will have an x-component of velocity equal to $\bar{u}(z - L/2)$, where z is the location where u' is to be estimated. Thus, the eddy pulls up, on average, the u value that is at $z - L/2$. This will become the deviation from the temporal mean velocity at location z:

$$u' = u - \bar{u} \approx \bar{u}(z - L/2) - \bar{u}(z) \cong \frac{1}{2}(\bar{u}(z - L) - \bar{u}(z)) \tag{5.28}$$

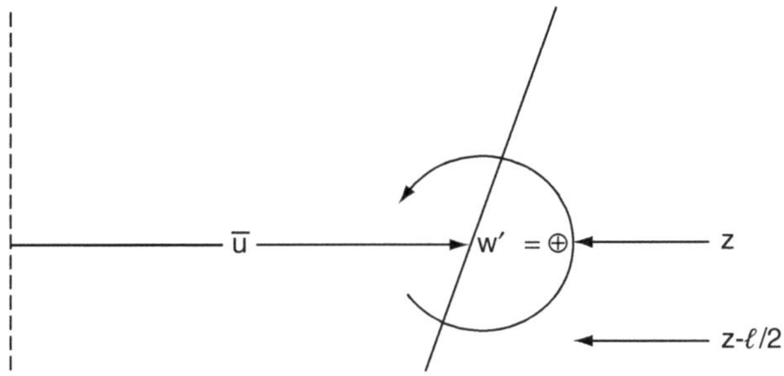

Fig. 5.6 Illustration of the relationship between velocity profile, turbulent eddies, and mixing length

Equation 5.28 is a relation for a difference in velocity, which can be written as a velocity gradient times a distance:

$$u' = -\frac{\partial \bar{u}}{\partial x}\left(\frac{L}{2}\right) \tag{5.29}$$

Velocity difference	=	velocity gradient	×	distance

Then,

$$\overline{w'u'}^{+} \approx \overline{\frac{w'}{2}\left[\bar{u}(z-L) - \bar{u}(z)\right]} \approx -\frac{\overline{w'}}{2} L \frac{\partial \bar{u}}{\partial z} \tag{5.30}$$

the development is similar for $\overline{w'u'}^{-}$:

$$\overline{w'u'}^{-} \approx \overline{w'u'}^{+} \approx -\frac{\overline{w'}}{2} L \frac{\partial \bar{u}}{\partial z} \tag{5.31}$$

Now combining Eqs. 5.26, 5.30, and 5.31 gives

$$\overline{w'u'} = -\overline{w' L \frac{\partial \bar{u}}{\partial z}} \tag{5.32}$$

Because turbulent eddies tend to be close to spherical in shape:

$$|w'| \approx |u'| \tag{5.33}$$

and from Eq. 5.29:

$$w' \sim L \frac{\partial \bar{u'}}{\partial z} \tag{5.34}$$

Table 5.2 Dynamic roughness lengths, z_0, for typical atmospheric surfaces (Turner [7])

Surface Type	z_0 (m)
Urban	1.-3
Forest	1.3
Deciduous forest in winter	0.5
Desert shrubland	0.3
Wetland	0.3
Cropland (summer)	0.2
Cropland (winter)	0.01
Grassland (summer)	0.1
Grassland (winter)	0.001
Water	~0.0001

If we substitute Eq. 5.34 into Eq. 5.32, and then substitute the result into Eq. 5.24, we get

$$-\overline{w'u'} = \varepsilon_z \frac{\partial \bar{u}}{\partial z} = L^2 \left(\frac{\partial \bar{u}}{\partial z} \right)^2 \tag{5.35}$$

or

$$\varepsilon_z = L^2 \left| \frac{\partial \bar{u}}{\partial z} \right| \tag{5.36}$$

Equation 5.36 is Prandtl's *mixing length hypothesis*, and it works well, considering that the basis for the equation is so empirical. However, Eq. 5.36 does present a challenge for us that mixing length, L, still needs to be specified. Measurements have shown us the following:

1. Near a wall, $L = \kappa z$, where κ is von Kármán's constant [5] and is very close to 0.4, and z is the distance from the closest wall.

   Prandtl also made another assumption in this region, that $w'u'$ could be approximated by a constant equal to the mean wall shear stress, or

$$-\overline{w'u'} = \tau / \rho = u_*^2 \tag{5.37}$$

   Then, eliminating $\overline{w'u'}$ from Eqs. 5.35 and 5.37 results in the well-known *logarithmic velocity profile*:

$$\frac{\bar{u}}{u_*} = \frac{1}{k} \ell n \left(\frac{z}{z_0} \right) \tag{5.38}$$

where u_* is the shear velocity at the wall, τ is the wall shear stress, and z_0 *is* an integration constant, often *called the dynamic roughness*. Table 5.2 provides some typical dynamic roughness lengths for atmospheric boundary layers. Applying Eq. 5.36 to 5.37 results in an equation for ε_z in this region:

$$\varepsilon_z = \kappa u_* z \tag{5.39}$$

2. Very near a wall (approaching the laminar sublayer where the turbulence is so small that it is eliminated by the viscosity of the fluid), that is, for $zu_*/\nu < 35$, $L \sim y^2$ [6].

   Making the same assumption that $u'w'$ is approximately equal to wall shear stress, this relation for L results in the following relation for velocity profile very near the wall:

$$\frac{\bar{u}}{u_*} = \beta \frac{\nu}{u_{**} z} \tag{5.40}$$

   Equation 5.40 is not used in mass transport calculations near a wall or interface because the unsteady character of mass transport in this region is very important, and Eq. 5.39 is for a temporal mean velocity profile.
3. Away from a wall, where the closest wall does not influence the velocity profile, L is a function of another variable of the flow field (Prandtl [8]). For example, consider the jet mixer given in Fig. 5.4. In this case, the mixing length, L, is a function of the width of the jet or plume. As the jet/plume grows larger, the value of L is larger.

   Here, it is easier to simply give the experimental relation for eddy diffusivity:

$$\varepsilon_z = \beta\, u_{\max} b \tag{5.41}$$

where b is the width of the mixing zone, β is a constant, and $u_{\max}$ is the maximum velocity in the jet at the given location, x.

Figure 5.7 gives some relationships for eddy diffusion coefficient profiles under different conditions that will be handy in applications of turbulent diffusive transport.

Example Applications

Example 5.1: Profile of eddy diffusion coefficient Estimate the eddy diffusivity profile for a wind velocity of 18 m/s measured at 10 m over a large lake (Fig. 5.8), and calculate the elevation above the water surface where $\varepsilon_z = D$ for water vapor. There is only one assumption needed:

1. The wind fetch is sufficient so that $\overline{U}_{10}$ is influenced by shear at the water surface (10 m is inside the boundary layer of the lake surface at this point).

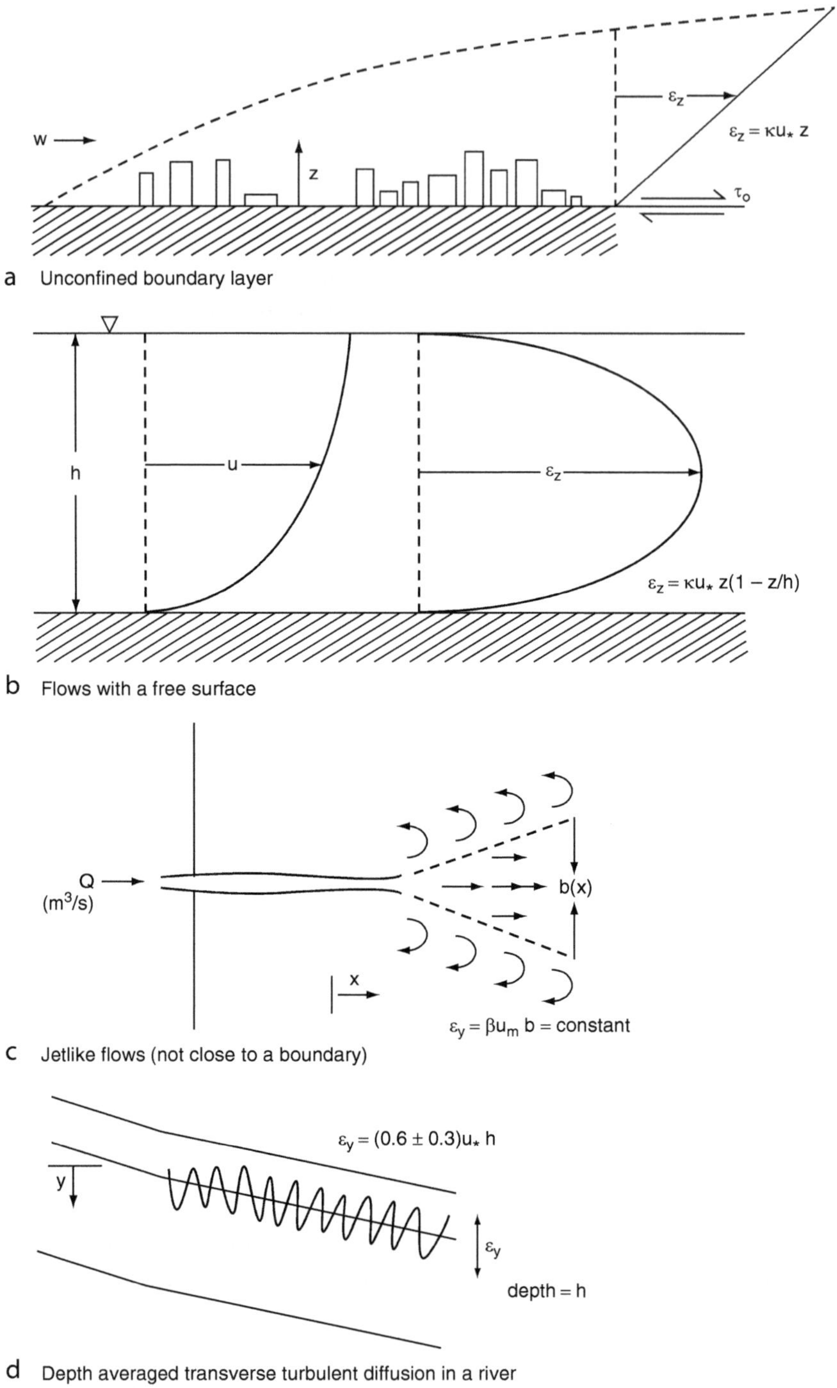

Fig. 5.7 Profiles of eddy diffusion coefficient for various types of applications (From Gulliver [1])

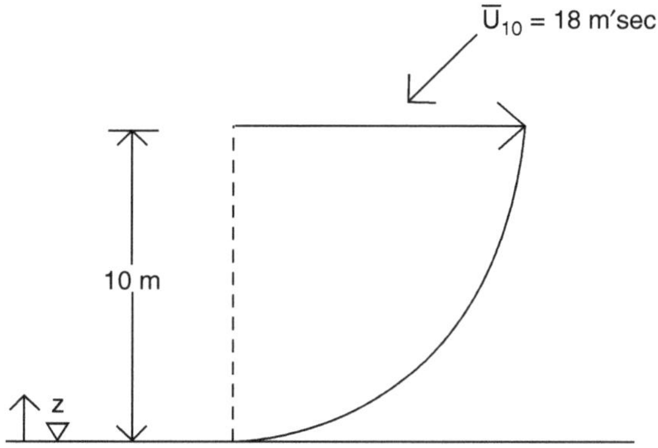

Fig. 5.8 Velocity profile over a large lake

Then, mixing length theory may be used with momentum transport to derive:

$$\frac{\partial \bar{u}}{\partial z} = \frac{u_*}{\kappa z} \tag{5.42}$$

and,

$$\varepsilon_z = \kappa\, u_* z \tag{5.43}$$

Now, Wu [9] has provided the following equation from a fit of field data:

$$u_* = 0.01\, \overline{U}_{10}\left(8 + 0.65\, \overline{U}_{10}\right)^{1/2} \tag{5.44}$$

which indicates that as the waves get larger at high wind speeds the boundary roughness effect upon u_* increases by the factor $\left(8 + 0.65\, \overline{U}_{10}\right)^{1/2}$, where $\overline{U}_{10}$ is given in m/s.

Then,

$$\varepsilon_z(\mathrm{m}^2/\mathrm{s}) = 0.01\ \kappa\, \bar{u}_{10} z (8 + 0.65\, \bar{u}_{10})^{1/2} = 0.32\, z$$

when z is given in meters. Now, the diffusion coefficient of water vapor in air is calculated to be $D = 2.6 \times 10^{-5}\ \mathrm{m}^2/\mathrm{s}$. Then, the elevation at which the diffusivity of water vapor would equal eddy diffusivity in this case would be,

$$0.32z = 2.6 \times 10^{-5}$$

or,

$$z = 8 \times 10^{-5}\ \mathrm{m} = 0.08\ \mathrm{mm} = 80\ \ \mu\mathrm{m}$$

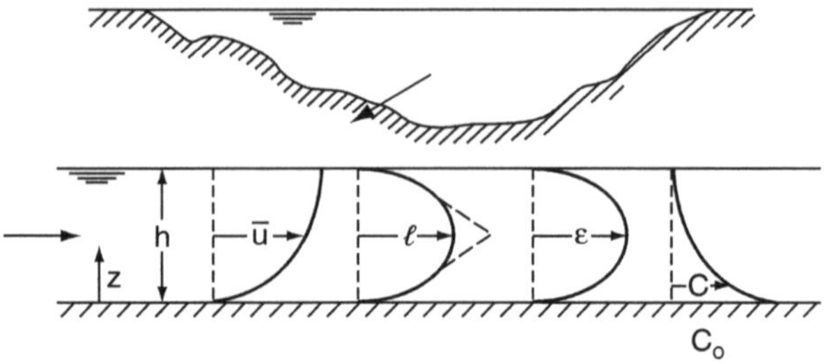

Fig. 5.9 Lateral and longitudinal cross sections of a typical river

Thus at $z = 80$ μm elevation above the water surface, eddy diffusivity will be equal to the diffusivity of water. A similarly small elevation would result for almost any environmentally relevant compound. We can thus see that both ε and D need to be considered simultaneously in Eq. 5.20 only very close to surfaces in turbulent flow, where ε approaches the diffusion coefficient. Otherwise, diffusivities can be ignored in solving turbulent flow transport problems, since $\varepsilon + D$ is essentially equal to ε.

Example 5.2: Concentration profile of suspended sediment in a river (assuming ε_z is constant) We will apply Eq. 5.20 to solve for the concentration profile of suspended sediment in a river, with some simplifying assumptions. Suspended sediment is generally considered similar to a solute, in that it is a scalar quantity in Eq. 5.20, except that it has a settling velocity. We will also change our notation, in that the bars over the temporal mean values will be dropped. This is a common protocol in turbulent transport, and will be followed here for conformity. Thus, if an eddy diffusion coefficient,ε, is in the transport equation,

$$u \quad \text{means} \quad \bar{u}$$

$$v \quad \text{means} \quad \bar{v}$$

$$w \quad \text{means} \quad \bar{w} \text{ and}$$

$$C \quad \text{means} \quad \bar{C}$$

throughout the remainder of this entry. Fig. 5.9 gives a longitudinal and lateral cross section of our river. We will make the following assumptions:

1. The flow is steady over the long term, so that $\partial C/\partial t = 0$.
2. The flow is fully developed, such that any gradient with respect to x is equal to zero ($\partial C/\partial x = 0$).
3. The river can be divided into a series of longitudinal planes with no significant interaction, such that $v = 0$ and $\varepsilon_y = 0$ (this is the assumption of the stream-tube computational models).
4. The vertical eddy diffusivity, ε_z, is a constant value.

Assumptions 3 and 4 are more difficult to justify.

The solute will have a vertical velocity, $w = -v_s$, where v_s is the settling velocity of the suspended sediment.

Then, Eq. 5.20 becomes

$$-v_s\frac{\partial C}{\partial z} = \frac{\partial}{\partial z}\left[(D + \varepsilon_z)\frac{\partial C}{\partial z}\right] \tag{5.45}$$

where we have not yet applied assumption 4. We can move the settling velocity into the partial term:

$$\frac{\partial(-v_s C)}{\partial z} = \frac{\partial}{\partial z}\left[(D + \varepsilon_z)\frac{\partial C}{\partial z}\right] \tag{5.46}$$

and since both sides of Eq. 5.46 are a gradient with respect to z, the terms inside of the gradients must also be equal:

$$-v_s C = (D + \varepsilon_z)\frac{dC}{dz} \tag{5.47}$$

Equation 5.47 is converted to an ordinary differential equation because all variables are only a function of z. Now, we will deal with assumption 4. Fig. 5.7 gives the equation developed by Rouse [10] for ε_z:

$$\varepsilon_z = \kappa u_* z(1 - z/h) \tag{5.48}$$

where u_* is the shear velocity at the bottom of the channel, or

$$u_* = \sqrt{\tau/\rho} \tag{5.49}$$

where τ is the shear stress at the wall. For a fully developed open-channel flow in a wide channel, the following relation is easily derived:

$$u_* = \sqrt{ghS} \tag{5.50}$$

This derivation can be found in a text on fluid mechanics or open-channel flow. Assumption 4 states that $\varepsilon_z = \bar{\varepsilon}_z$ for all values of z, where ε_z is the depth average, or

$$\bar{\varepsilon}_z = \frac{1}{h}\int_0^h \varepsilon_z dz = \frac{\kappa u_*}{h}\int_o^h z(1 - z/h)dz = 0.067\, u_* h \tag{5.51}$$

where h is the depth of the stream. The term $\overline{\varepsilon}_z$ is almost always much greater than D in a turbulent flow. Thus,

$$D + \overline{\varepsilon_z} \cong \overline{\varepsilon_z}$$

Now, substituting these equations into Eq. 5.47 results in

$$\overline{\varepsilon_z}\frac{dC}{dz} + v_s C = 0$$

We will solve this by separating variables,

$$\frac{dC}{C} = \frac{-v_s}{\overline{\varepsilon_z}}dz$$

integrating, and taking both sides of the solution to the power of e:

$$C = \beta_1\, e^{\frac{-v_s}{\overline{\varepsilon_z}}z}$$

Now, we need a boundary condition to determine β_1. This is difficult with suspended sediment profiles. We can develop a fairly good estimate of the distribution of suspended sediment once we have a known concentration at some location in the flow field. In the sediment transport field bed load and suspended load are often discussed. The relation between the two, and some experience and measurements of both simultaneously, can be used to predict an equivalent suspended sediment concentration at the bed. Then, the relevant boundary condition is

1. At

$$z = 0,\ C = C_0.$$

where C_0 is the concentration that has been determined from the bed load–suspended load relationship. Applying this boundary condition gives β_1 = C_0, and our solution is

$$C = C_0\, e^{-v_s z/\overline{\varepsilon}_z}$$

The result is illustrated in Fig. 5.10. This problem can also be solved *without* assumption 4 [10].

Example 5.3: Concentration of organic compounds released into the air by an industrial plant (application of the product rule to error function solutions) There is some concern about the emissions from the adhesives produced in an industrial plant. Specifically, the town of Scream Hollow is 1 km away from the plant, where citizens have begun to complain about odors from the plant and of headaches.

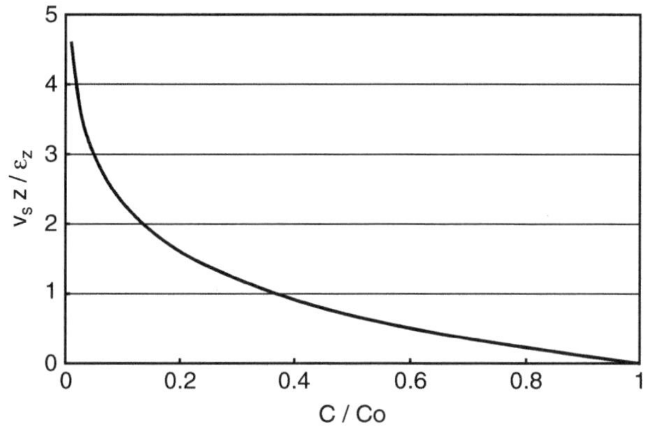

Fig. 5.10 Suspended sediment concentration profile for Example 2

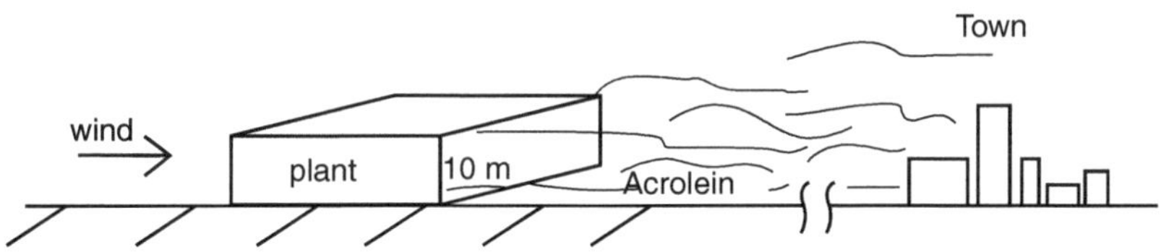

Fig. 5.11 Illustration of toxic chemical release into the atmosphere, with the wind blowing toward a town

One culprit, aside from a haunting, may be the release of Acrolein, C_3H_4O, a priority pollutant that is an intermediary of many organic reactions. The average release from the 200 m × 200 m × 10 m plant sketched in Fig. 5.11 is assumed to be 20 g/h. If the wind is blowing directly toward Scream Hollow, at 3 m/s measured at 3 m height, with a dynamic roughness of 0.2 m for the farmland, what concentrations will the Scream Hollow inhabitants experience? Is this above the EPA threshold limit of 0.1 ppm(v)?

We will need to make some assumptions to formulate this problem. They are:

1. The Acrolein release is distributed over the most downwind plane of the building. With the important concentration being 1 km away, this is not a bad assumption. Then, the Acrolein will be released over the plane that is 200 m × 10 m. If 20 g/h = 0.0056 g/s are released into a wind moving at 3 m/s, the initial concentration is

$$C_0 = \frac{0.0056 \text{ g/s}}{3 \text{ m/s}(200 \text{ m})(10 \text{ m})} = 9.3 \times 10^{-7} \text{ g/m}^3$$

2. We will use a cross-sectional mean velocity of $U = \bar{u}$ at 3 m height, or $U = 3$ m/s.
3. We will use $\bar{\varepsilon}_z = \bar{\varepsilon}_y = \varepsilon_z$ at 3 m height.
4. We will not consider any of the source or sink terms for Acrolein.

Fig. 5.12 Illustration of the coordinate system for Example 5

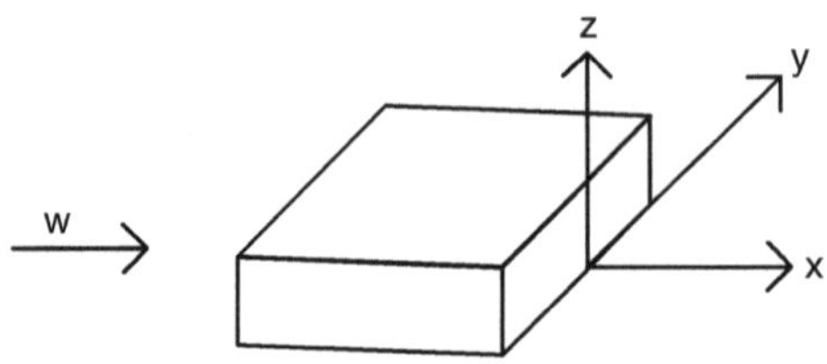

We will also set up the coordinates so that $(x,y,z) = (0,0,0)$ occurs on the ground at mid-plant width, and will orient the wind in the x-direction.

With these assumptions, the governing equation becomes

$$U\frac{\partial C}{\partial x} = \bar{\varepsilon}_y \frac{\partial^2 C}{\partial y^2} + \bar{\varepsilon}_z \frac{\partial^2 C}{\partial z^2} \tag{5.52}$$

The boundary conditions are

1. At $(x,y,z) = (0, -100 \text{ m} \Rightarrow 100 \text{ m}, 0 \Rightarrow 10 \text{ m})$, $C = C_0$.
2. As $x \Rightarrow \infty$, $y \Rightarrow \infty$, or $z \Rightarrow \infty$, $C \Rightarrow 0$.
3. Zero mass flux at $z = 0$.

These boundary conditions, illustrated in Fig. 5.12, will give us a concentration front, but in two dimensions. In addition, we have a zero flux condition that will require an image solution. We will use the solution of Example 5 in the entry "▾Transport in the Environment" to develop a solution for this problem. The solution, before applying boundary conditions, was

$$C = \beta_o + \beta_1 erf\left(\frac{z}{\sqrt{4Dt}}\right)$$

Now, we need an image to the concentration front about the $z = 0$ plane. In the y-direction we have a step-up at $y = -\Delta y$ and a step down at $y = \Delta y$. We will also use the product rule (Example 3, Transport in the Environment) to indicate that the solution to our governing equation for the y-direction should be multiplied times the solution in the z-direction. Then the solution can be given as

$$\frac{C}{C_o} = \beta_o + \left[\beta_1\, erf\left(\frac{(z+\Delta z)}{\sqrt{4\bar{\varepsilon}_z x/U}}\right) + \beta_2\, erf\left(\frac{(z-\Delta z)}{\sqrt{4\bar{\varepsilon}_z x/U}}\right)\right] \times \left[\beta_3\, erf\left(\frac{(y+\Delta y)}{\sqrt{4\bar{\varepsilon}_y x/U}}\right) + \beta_4\, erf\left(\frac{(y-\Delta y)}{\sqrt{4\bar{\varepsilon}_y x/U}}\right)\right]$$

where $\Delta y = 100$ m and $\Delta z = 10$ m.

Now, to see if our boundary conditions can be satisfied with the form of the solution:

$$@x \Rightarrow \infty, C \Rightarrow 0. \quad \text{Thus} \quad \beta_0 = 0.$$

$$@z \Rightarrow \infty, C \Rightarrow 0. \quad \text{Thus} \quad \beta_1 = -\beta_2.$$

$$@y \Rightarrow \infty, C \Rightarrow 0. \quad \text{Thus} \quad \beta_3 = -\beta_4.$$

$$@x \Rightarrow 0, \text{ and } (y, z) = (0, 0), C/C_0 = 1.$$

With the last boundary condition, Eq. E5.5.3 becomes

$$1 = (\beta_1 - \beta_2)(\beta_3 - \beta_4)$$

or,

$$1 = 2\beta_1 \times 2\beta_3$$

or,

$$1 = 2\beta_2 \times 2\beta_4$$

Finally, @ $x \Rightarrow 0$ and $(y,z) = (0,\Delta z)$, $C/C_0 = ½$. Thus $\beta_1 = ½$.

Applying this last boundary condition results in $\beta_3 = ½$, $\beta_2 = -½$, and $\beta_4 = -½$. Thus, the solution to Eq. 5.52 is

$$\frac{C}{C_o} = \frac{1}{2}\left\{ erf\left(\frac{(z+\Delta z)}{\sqrt{4\bar{\varepsilon}_z x/U}}\right) - erf\left(\frac{(z-\Delta z)}{\sqrt{4\bar{\varepsilon}_z x/U}}\right)\right\} \times \left\{ erf\left(\frac{(y+\Delta y)}{\sqrt{4\bar{\varepsilon}_y x/U}}\right) - erf\left(\frac{(y-\Delta y)}{\sqrt{4\bar{\varepsilon}_y x/U}}\right)\right\} \tag{5.53}$$

Now, if we use $\Delta z = 10$ m, $\Delta y = 100$ m, $U = 3$ m/s, the only remaining parameter to find is $\bar{\varepsilon}$. Using Eq. 5.43 given in Example 1:

$$\bar{\varepsilon}_z = \bar{\varepsilon}_y = \kappa u_* z \tag{5.43}$$

Note that the logarithmic boundary equation can be written as

$$\frac{\bar{u}}{u_*} = \ell n\left(\frac{z}{z_o}\right) \tag{5.54}$$

where z_0 is the dynamic roughness, assumed to be 0.2 m for the crop land between the plant and Scream Hollow. Then,

$$u_* = \frac{\bar{u}}{\ell n(z/z_o)} = \frac{3\,\mathrm{m/s}}{\ell n\left(\frac{3\,\mathrm{m}}{0.2\,\mathrm{m}}\right)} = 1.1\ \mathrm{m/s}$$

and

$$\bar{\varepsilon}_z = 0.4(1.1m/s)(3m) = 1.3\ m^2/s$$

If we now plug all of the parameters for the industrial plant into Eq. 5.53, we get $C = 0.25\ \mu g/m^3 = 2.5 \times 10^{-7}\ g/m^3$. In terms of ppm(v), we will use $\rho_{air} = 1.2\ g/m^3$, and the molecular weights of air and Acrolein of 29 and 56 g/mole, respectively. Then,

$$\begin{aligned} C &= \frac{2.5 \times 10^{-7}\,\mathrm{g/m^3}}{\rho_{\mathrm{air}}} \frac{MW_{\mathrm{air}}}{MW_{C_3H_40}} \\ &= \frac{2.5 \times 10^{-7}\,\mathrm{g/m^3}}{1.2\,\mathrm{g/m^3}} \frac{20\,\mathrm{g/mole}}{56\,\mathrm{g/mole}} \\ &= 1.08 \times 10^{-7} \frac{\mathrm{moles}\, C_3H_40}{\mathrm{mole\ air}} \end{aligned}$$

This is right at the threshold for continuous exposure, and the pollution from the plant should be investigated in more detail.

Conclusions

1. Although turbulent diffusion is a convection transport, and not a diffusive transport, the result looks similar to diffusion, and can be described by a turbulent diffusion coefficient.
2. Most environmental flows are turbulent. The exceptions are flow through porous media and flows that are very close to an interface.
3. Reynolds averaging and the Boussinesq assumption result in a turbulent transport equation that contains many features of the diffusive mass transport equation, and can be solved by similar techniques.
4. Prandtl's mixing length is a relatively accurate simplification for many turbulent flows.

Future Directions

The future for turbulent transport in the environment is in the direction of computational mass transport. This requires a simultaneous fluid dynamics–mass transport solution. On typical environmental scales, the computational power of our computers still must be advanced to solve these large problems while resolving the scale of the smallest turbulent eddies. Direct numerical simulation cannot deal with the scale of these problems, and large eddy simulation cannot keep both the scale and grid refinement required.

Bibliography

Primary Literature

1. Gulliver JS (2007) Introduction to chemical transport in the environment. Cambridge University Press, Cambridge, UK, 288 pp
2. Reynolds O (1895) On the dynamical theory of incompressible viscous fluids and the determination of the criterion. Philos Trans R Soc 186:123–164
3. Boussinesq J (1877) Essai sur la théorie des eaux courantes. Mem Pres Acad Sci Paris 23:46
4. Prandtl L (1925) Bericht iiber Untersuchungen zur ausgebildeten Turbulenz. Z Angew Math Mech 5:136–139
5. von Kármán T (1930) Mechanische Ahnlichkeit und Turbulenz. Nachr Ges Wiss Gottingen 5:58–76. Also Proceedings of third international congress on applied mechanics, vol I, Stockholm, pp 85–93, 1930
6. Reichardt H (1951) Vollstandige Darstellung der turbulenten. Geschwindigkeitsverteilung. Ann Angew Math Mech 31:7
7. Turner DB (1994) Workbook of atmospheric dispersion estimates, 2nd edn. Lewis, Boca Raton
8. Prandtl L (1942) Bemerkungen zur Theorie der freien Turbulenz. Z Angew Math Mech 22:241–243
9. Wu J (1980) Wind-stress coefficients over sea surface. J Geophys Res 74:444
10. Rouse H (1937) Modern conceptions of the mechanics of fluid turbulence. Trans Am Soc Civil Eng 102(1965):463–543

Books and Reviews

Nezu I, Nakagawa H (1993) Turbulence in open channel flow. Balkema, Rotterdam
White FM (1974) Viscous fluid flow. McGraw-Hill, New York

Chapter 6
Chemicals in the Environment, Dispersive Transport

John S. Gulliver

Glossary

Convection	The movement of a constituent with movement of the fluid.
Diffusion	The spreading of fluid constituents through the motion inherent to atoms and molecules.
Diffusion coefficient	A coefficient that describes the tendency of molecules to spread a constituent mass.
Dirac delta	An impulse of a given quantity (mass) that occurs over an infinitely short time or space.
Dispersion	The process of mixing caused by a variation in velocity and transverse diffusion or turbulent diffusion.
Dispersion coefficient	A coefficient that can describe the mixing caused by a transverse velocity profile and transverse diffusion or turbulent diffusion. A dispersion coefficient means that some sort of spatial mean velocity is being used to describe the flow. Then, the mixing lateral or longitudinal to the spatial mean velocity due to a combination of a velocity profile and diffusion or turbulent diffusion is described by the dispersion coefficient. The coefficient's location in the mass transport

This chapter was originally published as part of the Encyclopedia of Sustainability Science and Technology edited by Robert A. Meyers. DOI:10.1007/978-1-4419-0851-3

J.S. Gulliver (✉)
Department of Civil Engineering, University of Minnesota, 500 Pillsbury Drive S.E., Minneapolis, MN 55454, USA
e-mail: gulli003@umn.edu

J.S. Gulliver (ed.), *Transport and Fate of Chemicals in the Environment: Selected Entries from the Encyclopedia of Sustainability Science and Technology*, DOI 10.1007/978-1-4614-5731-2_6, © Springer Science+Business Media New York 2012

equation is similar to diffusion coefficients, and the units are similar.

Laminar flow — Flow that has no turbulent eddies, where the fluid flows in laminas and diffusion creates the mixing of the fluid.

Turbulent diffusion — The mixing of fluids through turbulent eddies created by convection.

Turbulent diffusion coefficient — A coefficient that comes from the multiplication of two turbulent velocities of the flow, divided by density of the fluid. The coefficient's location in the mass transport equation is similar to diffusion coefficients, and the units are similar, so it is called a "turbulent diffusion coefficient."

Definition of Dispersive Transport in the Environment

Dispersion is the enhanced mixing of material through spatial variations in velocity. When it is of interest (when we are not keeping track of the three-dimensional mixing), dispersion is typically one or two orders of magnitude greater than turbulent diffusion. The process of dispersion is associated with a spatial mean velocity, the assumption of plug flow, and a velocity profile. The means used in association with diffusion, turbulent diffusion, and dispersion are identified in Table 6.1.

The means by which diffusion and possibly turbulent diffusion are combined with a spatial mean velocity to result in dispersion is illustrated in Fig. 6.1. A velocity profile over space with mixing due to diffusion (and possibly turbulent

Table 6.1 Temporal or spatial means and scales used in association with various mixing processes

Process	Variable representing *p*rocess	Mean	Scale of mean
Diffusion	Diffusion coefficient	Temporal	Molecular
Turbulent diffusion	Turbulent diffusion *co*efficient	Temporal	Minutes
Dispersion	Dispersion coefficient	Spatial	Scale of flow

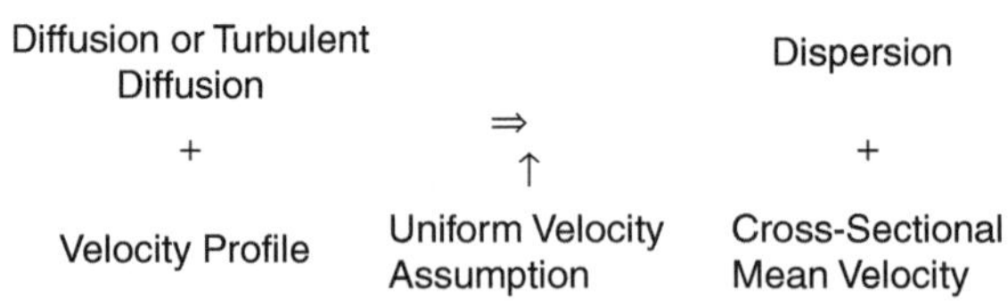

Fig. 6.1 Representation of the process by which diffusion or turbulent diffusion is related to dispersion

diffusion) is combined into a cross-sectional mean velocity and a dispersion coefficient. Without the cross-sectional mean velocity, there *is no* dispersion coefficient.

Introduction

Dispersion was first developed as a means of dealing with reactors, where there was little interest in the processes creating mixing inside of the reactor, but great need to appropriately describe the output from the reactor. The physics of the mixing process is lost in the conversion to a spatial mean velocity profile, but the end result can still be modeled by dispersion.

A similar spatial mean velocity (bulk mean velocity) is used for the plug flow reactor model. Thus, plug flow with dispersion is a natural match, where the mixing that truly occurs in any reactor or environmental flow is modeled as dispersion. This is the model that will be applied to utilize dispersion as a mixing model.

Dispersion in Laminar Flow

Any flow with a nonuniform velocity profile will, when spatial mean velocity and concentration are taken, result in dispersion of the chemical. For laminar flow, the well-described velocity profile means that we can describe dispersion analytically for some flows. Beginning with the diffusion equation in cylindrical coordinates (laminar flow typically occurs in small tubes):

$$\frac{\partial C}{\partial t} + U\frac{\partial C}{\partial x} = D\left[\frac{\partial^2 C}{\partial r^2} + \frac{1}{r}\frac{\partial C}{\partial r} + \frac{\partial^2 C}{\partial x^2}\right] + S \tag{6.1}$$

where the x-coordinate is aligned with the flow velocity, and $v = w = 0$. We will outline the development of a dispersion coefficient for a fully developed laminar pipe flow. This means that we are far enough from the entrance that the velocity profile is essentially in equilibrium with the loss of pressure along the pipe. This flow has a velocity profile (developed in most fluid mechanics texts):

$$U = U_{\max}(1 - r^2/R^2) \tag{6.2}$$

as illustrated in Fig. 6.2.

To convert Eq. 6.1 to cross-sectional mean values, we will assign:

$$\hat{C} = \frac{1}{\pi R^2}\int_0^R C 2\pi r \, dr \tag{6.3}$$

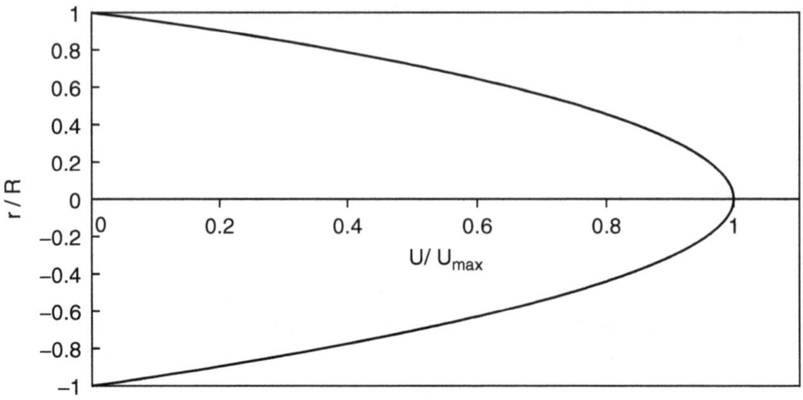

Fig. 6.2 The velocity profile in a fully developed tubular flow is a paraboloid

$$U = \frac{1}{\pi R^2}\int_0^R U\,2\pi r\,dr = \frac{U_{\max}}{2} \tag{6.4}$$

Then, in Eq. 6.1:

$$C = \hat{C} + C'(r) \tag{6.5}$$

and

$$u = U + u'(r) \tag{6.6}$$

Now, if we equate Eq. 6.2 and 6.6, we get

$$u'(r) = u_m\left(\frac{1}{2} - \frac{r^2}{R^2}\right) \tag{6.7}$$

Substituting Eq. 6.5, 6.6, and 6.7 into Eq. 6.1:

$$\begin{aligned}&\frac{\partial \hat{C}}{\partial t} + \frac{\partial C'}{\partial t} + (U + u')\frac{\partial(\hat{C} + C')}{\partial x}\\ &= D\left[\frac{\partial^2(\hat{C} + C')}{\partial r^2} + \frac{1}{r}\frac{\partial(\hat{C} + C')}{\partial r} + \frac{\partial^2(\hat{C} + C')}{\partial x^2}\right] + S\end{aligned} \tag{6.8}$$

An order of magnitude analysis will tell us that $\partial^2/\partial r^2 >> \partial^2/\partial x^2$. In addition, $\partial\hat{C}/\partial r = 0$, by definition. Now, if we put everything that is known on the left-hand side of Eq. 6.8, the result will be:

$$\frac{\partial \hat{C}}{\partial t} + U\frac{\partial \hat{C}}{\partial x} = -\left[\frac{\partial C'}{\partial t} + U\frac{\partial C'}{\partial x} + U'\frac{\partial \hat{C}}{\partial x} + U'\frac{\partial C'}{\partial x}\right]$$
$$+D\frac{\partial^2 C'}{\partial r^2} + \frac{D}{r}\frac{\partial C'}{\partial r} + S = D_L\frac{\partial^2 \hat{C}}{\partial x^2} + S \quad (6.9)$$

The second equality in Eq. 6.9 is a definition of longitudinal dispersion coefficient, D_L. G. I. Taylor [1] assumed that some of the terms in Eq. 6.9 would cancel and that longitudinal convective transport would achieve a balance with transverse diffusive transport. He then solved the second equality in Eq. 6.9 for a fully developed tubular flow, resulting in the relation,

$$D_L = \frac{R^2 U^2}{48 D} = \frac{d^2 U^2}{192 D} \quad (6.10)$$

A similar relation can be developed for laminar flow down an inclined plate:

$$D_L = \frac{32 h^2 U^2}{945 D} \quad (6.11)$$

The longitudinal dispersion coefficient is proportional to the square of the flow scale (d or h), proportional to the square of the velocity scale (U), and *inversely* proportional to the diffusion coefficient. The greater the diffusion, the less severe the spread of the chemical by the velocity profile because of local mixing, and the smaller the longitudinal dispersion coefficient. This result may seem illogical, but can be explained by the following: Longitudinal dispersion describes mixing only in terms of the cross-sectional mean concentration, and transverse mixing actually slows down longitudinal dispersion. The governing Eq. 6.9 does not concern itself with local mixing issues.

Knowing the relations given in Eq. 6.10 and 6.11, we no longer need the more cumbersome middle portion of Eq. 6.8, and we can work to solve the equation,

$$\frac{\partial \hat{C}}{\partial t} + U\frac{\partial \hat{C}}{\partial x} = D_L\frac{\partial^2 \hat{C}}{\partial x^2} + S \quad (6.12)$$

In expressing the equations for longitudinal dispersion, we will drop the "hat" above the cross-sectional mean values. Thus, if D_L is involved, we are discussing cross-sectional mean concentrations.

Dispersion in Turbulent Flow

The dispersion that occurs in turbulent flow can also be calculated, as long as the velocity profile is given. This was done by Taylor [2] for a tubular flow and by Elder

[3] for a two-dimensional, open-channel flow. Both investigators assumed that a logarithmic velocity profile would apply in the entire flow field. The logarithmic velocity profile is a fairly good description where shear stress can be assumed constant. It is a good assumption for a fully developed turbulent flow field, because the locations where the logarithmic profile applies are those with the greatest change in velocity.

For a fully developed tubular flow, assuming a logarithmic velocity profile, Taylor derived the equation,

$$D_{\mathrm{L}} = 5.05\ d\ u_* \tag{6.13}$$

where u_* is the shear velocity at the wall. Elder [3] derived the following equation for a two-dimensional, open-channel flow:

$$D_{\mathrm{L}} = 5.93\ h\ u_* \tag{6.14}$$

It is interesting to compare Eq. 6.13 and 6.14 with those for a fully developed laminar flow, Eq. 6.10 and 6.11. In the chapter on Turbulent Transport, eddy diffusion coefficient in a turbulent boundary layer was found to be linearly dependent upon distance from the wall and on the wall shear velocity. If we replace the diffusion coefficient in Eq. 6.11 with an eddy diffusivity that is proportional to $h\ u_*$, we get,

$$D_{\mathrm{L}} \sim \frac{h^2\, U^2}{h u_*} \tag{6.15}$$

Noting that for a given boundary roughness we can generally say that $U \sim u_*$, Eq. 6.15 becomes,

$$D_{\mathrm{L}} \sim h u_* \tag{6.16}$$

which is what we have in Eq. 6.14.

The relations developed for longitudinal dispersion coefficient are given in Table 6.2. The experimental results in rivers tend to have a large range because of the variety of lateral velocity profiles that exist in natural rivers and streams.

Solutions to Transport with Convection

Diffusive transport with convection occurring simultaneously can be solved more easily if we orient our coordinate system properly. First, we must orient one axis in the direction of the flow. In this case, we will choose the x-coordinate so that u is nonzero and v and w are zero. Second, we must assume a uniform velocity profile, $u = U =$ constant with y and z. Then Eq. 6.30 of the Transport in the Environment chapter becomes

Table 6.2 Relationships for longitudinal dispersion coefficient in pipes and channels developed from theory and experiments

Flow conditions	D_L	Notation
Laminar flow in a pipe [1]	$\frac{R^2U^2}{48D}$	R = radius of tube U = cross-sectional mean velocity D = diffusivity
Laminar flow down an inclined plate	$\frac{32h^2U^2}{945D}$	h = depth V_p = velocity of upper plate Δ_z = spacing of plates
Laminar flow-linear velocity profile (Couette Flow)	$\frac{V_p^2\Delta_z^2}{120D}$	u_* = shear velocity
Turbulent flow in a pipe, assuming logarithmic velocity profile [2]	$10.1Ru_*$	
turbulent flow down an inclined plate, assuming logarithmic velocity profile [3]	$5.93hu_*$	
Open flume (experimental)	8–400 hu_*	
Canals (experimental)	8–20 hu_*	
Rivers (experimental)	8–7,500 hu_*	

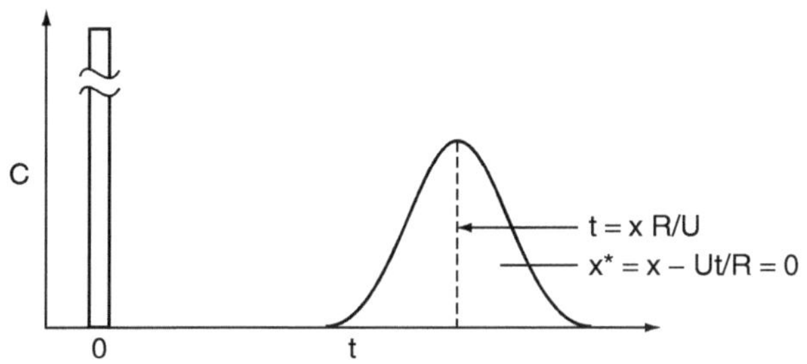

Fig. 6.3 Pulse response in fixed and moving coordinate systems

$$\frac{\partial C}{\partial t} + \frac{U}{R}\frac{\partial C}{\partial x} = \frac{D}{R}\left[\frac{\partial^2 C}{\partial x^2} + \frac{\partial^2 C}{\partial y^2} + \frac{\partial^2 C}{\partial z^2}\right] + \frac{S}{R} \tag{6.17}$$

where S is a source or sink term other than adsorption or desorption. In Eq. 6.17, we are assuming that chemical reaction does not take place on the surface of a solid, i.e., while the chemical is sorbed to solids. Now, we will convert our Eulerian (fixed) coordinate system to one that moves (Lagrangian) with velocity U/R and assign an independent variable

$$x^* = x - Ut/R \tag{6.18}$$

such that $x^* = 0$ at $x = U\,t/R$. The response of the system to a conservative pulse, given in Fig. 6.3, indicates that in the Lagrangian coordinate system specified, there is no convection term, only diffusion. Then Eq. 6.17 becomes

$$\frac{\partial C}{\partial t} = \frac{D}{R}\left[\frac{\partial^2 C}{\partial x^{*2}} + \frac{\partial^2 C}{\partial y^2} + \frac{\partial^2 C}{\partial z^2}\right] + \frac{S}{R} \tag{6.19}$$

Determination of Dispersion Coefficient from Tracer Clouds

The one-dimensional mass transport equation for plug flow with dispersion, and a retardation coefficient of 1, is

$$\frac{\partial C}{\partial t} + U\frac{\partial C}{\partial x} = D_{\mathrm{L}}\frac{\partial^2 C}{\partial x^2} + S \tag{6.20}$$

We will convert our fixed coordinate system to a coordinate system moving at velocity U through the change of variables, $x^* = x - Ut$. Then Eq. 6.20 is given as

$$\frac{\partial C}{\partial t} = D_{\mathrm{L}}\frac{\partial^2 C}{\partial x^{*2}} + S \tag{6.21}$$

Now, if we are determining the dispersion coefficient through the use of a pulse tracer cloud, the boundary conditions are those of a Dirac delta:

1. At $t = 0$, a pulse of mass M is released at $x^* = 0$.
2. As $t \Rightarrow \infty$, $C \Rightarrow 0$.

The solution to Eq. 6.21 with these boundary conditions [4] is:

$$C = \frac{M/A}{\sqrt{4\pi D_{\mathrm{L}} t}}\, e^{\frac{-(L-Ut)^2}{4D_{\mathrm{L}} t}} \tag{6.22}$$

or, in dimensionless variables,

$$\frac{C}{C_0} = \frac{C}{M/(AL)} = \frac{1}{\sqrt{4\pi Cou/Pe}}\exp\left[-\frac{(1-Cou)^2}{4Cou/Pe}\right] \tag{6.23}$$

Where $Cou = U\,t/L = t/t_r$ is a Courant number and $Pe = U\,L/D_{\mathrm{L}}$ is a Peclet number. Comparing Eq. 6.23 to a Gaussian probability distribution,

$$P(\phi) = \frac{1}{\sigma\sqrt{2\pi}}\exp\left[-\frac{\phi^2}{2\sigma^2}\right]$$

we can see that the plug flow with dispersion, when $Pe > 10$, can be fit to a Gaussian distribution in terms of C/C_0 and $(1-\theta)$, with the relationships,

$$\sigma^2 = \frac{\sigma_t^2}{t_r^2} = 2\,Cou/Pe \tag{6.24}$$

where

$$\sigma_t^2 = \frac{\int_0^\infty t^2\, C/C_0\, dt}{\int_0^\infty C/C_0\, dt} - t_r{}^2 \tag{6.25}$$

and

$$t_r = \frac{\int_0^\infty t^2\, C/C_0\, dt}{\int_0^\infty C/C_0\, dt} \tag{6.26}$$

In addition, if we could measure the tracer cloud over distance at one time, we would use the relation,

$$\sigma^2 = \frac{\sigma_x^2}{U^2 t_r{}^2} \tag{6.27}$$

where

$$\sigma_x^2 = \frac{\int_0^\infty x^2\, C/C_0\, dx}{\int_0^\infty C/C_0\, dx} - \bar{x}^2 \quad and \quad \bar{x} = \frac{\int_0^\infty x\, C/C_0\, dx}{\int_0^\infty C/C_0\, dx} \tag{6.28}$$

The response of a plug flow with dispersion model to a pulse input, Eq. 6.23, is given in Fig. 6.4 for various values of the Peclet number, $Pe = U\,L/D_L$.

Dispersion in Groundwater Flow

Dispersion in a flow through porous media occurs due to heterogeneity in the media, i.e., the conductivity of the soil varies with space. This is shown on three levels in Fig. 6.5. On the particle scale, a thread of tracer will be split a number of times as it moves through the media. Each split of the tracer thread will move through the media at a speed corresponding to the resistance that it encounters. If

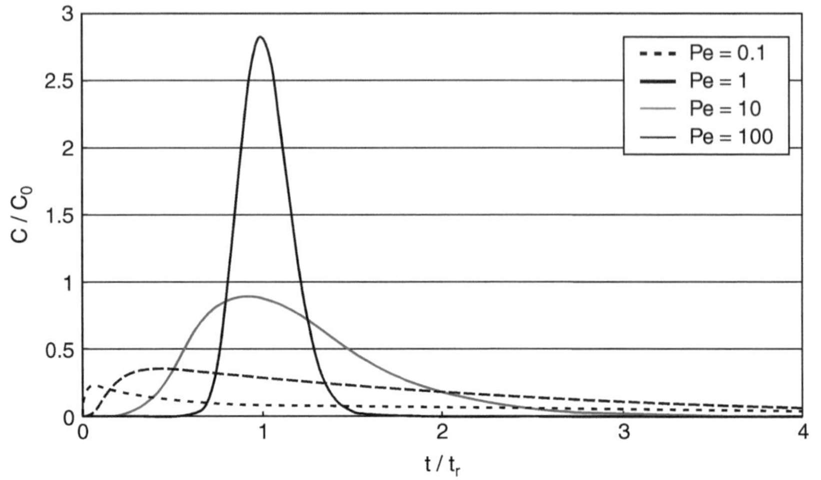

Fig. 6.4 Response of the plug flow with dispersion model to a pulse input

you take a number of tracer threads coming out of the media at different times, and collected them in an outlet pipe, what you would see at the end of the pipe would be a dispersed pulse. This dispersion would be much greater than the diffusion that would occur. A lateral dispersion would also occur because the media would move some of the tracer threads laterally. Thus, in groundwater flow, there generally is longitudinal and lateral dispersion, created by the character of the media.

On a larger scale, a similar process can occur. Fingering of the tracer is created by layered beds with a low conductivity and lenses with a high conductivity. The tracer that ends up in a lens travels at a relatively high speed. Those tracer molecules will reach the measuring point sooner than the tracer molecules stuck in the low conductivity beds, creating longitudinal dispersion. As the lenses are not all parallel to each other, they will also create a lateral dispersion of the tracer.

Finally, on a still larger scale, about the scale of a small town, there will be isopleths (lines of constant concentration) of our tracer that look something like those given in the last illustration of Fig. 6.4. If enough particle and lens effects have occurred with an apparent randomness to our tracer cloud, then the cloud disperses in a manner similar to that illustrated.

Incorporating the retardation coefficient for the chemical, discussed in the chapter on Transport in the Environment, the mass transport equation is then written as:

$$\begin{aligned}\frac{\partial C}{\partial t} + U/R\frac{\partial C}{\partial x} = \frac{(D+D_x)}{R}\frac{\partial^2 C}{\partial x^2} + \frac{(D+D_y)}{R}\frac{\partial^2 C}{\partial y^2} \\ + \frac{(D+D_z)}{R}\frac{\partial^2 C}{\partial z^2}\end{aligned} \tag{6.29}$$

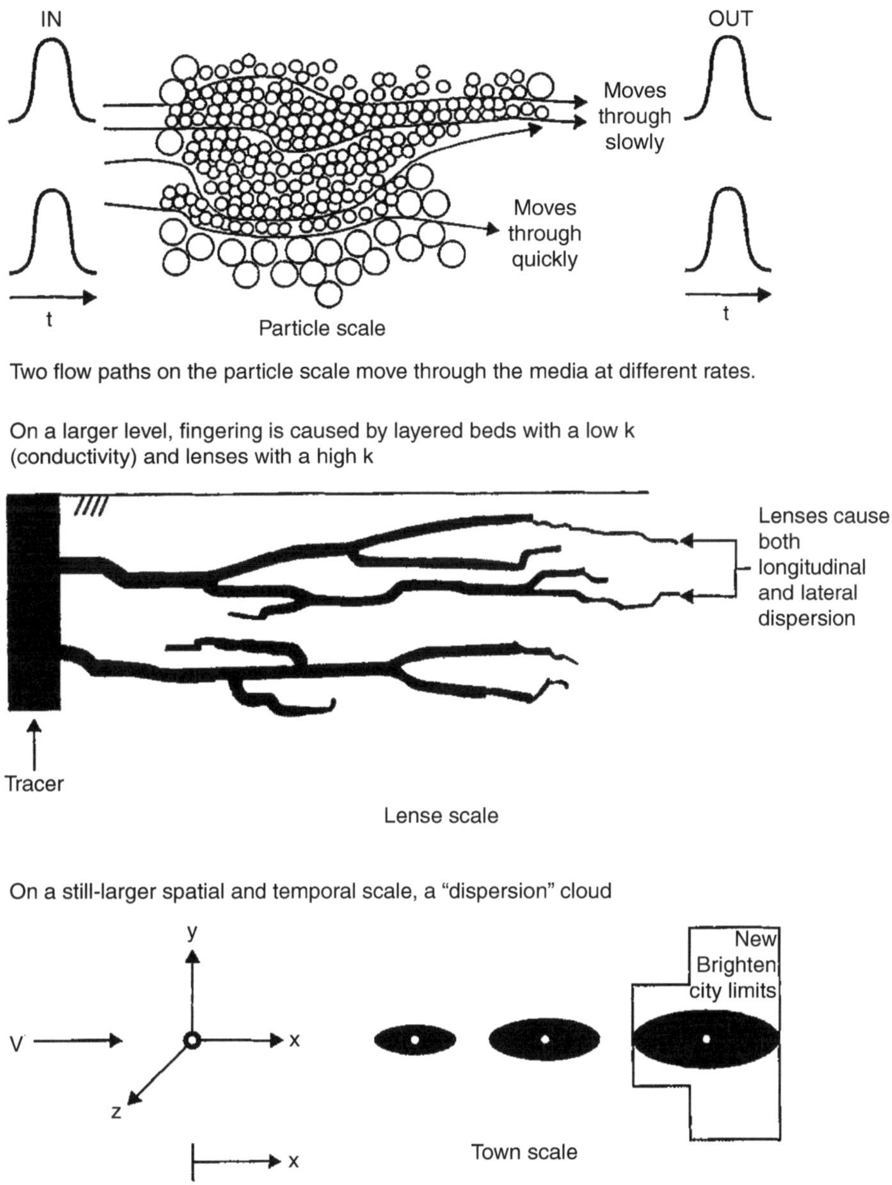

Fig. 6.5 Illustration of dispersion in groundwater flow at various scales (From [5])

where R is a retardation coefficient, and D_x, D_y, and D_z are the dispersion coefficients in the x-, y-, and z-directions, respectively.

The dispersion coefficients are dependent upon the character of the media and the flow velocity. It is difficult to predict these coefficients within an order of magnitude, so they are normally measured or fit to measured data in the field. For example, if a concentration variance in the x, y, and z directions can be measured in response to a pulse release, then an approximate solution would be:

$$C = \frac{MR^{1.5}}{8(\pi^3 t^3 D_x D_y D_z)^{0.5}} \exp\left[\frac{-Rx^{*2}}{4D_x t}\frac{-Ry^2}{4D_y t}\frac{-Rz^2}{4D_z t}\right] \tag{6.30}$$

where x^* is $x - Ut/R$. Comparing Eq. 6.30 to that for a three variable Gaussian distribution, we can see that:

$$D_x = 9\ R\sigma_x{}^2/\ (2t) \tag{6.31a}$$

$$D_y = 9R\ \sigma_y{}^2/(2t) \tag{6.31b}$$

$$D_z = 9\ R\sigma_z{}^2/(2t) \tag{6.31c}$$

where σ_x^2, σ_y^2 and σ_z^2 are the variance of concentration in the x, y, and z directions, respectively. They are given by the equations,

$$\sigma_x^2 = \frac{\int x^2 C\, dx}{\int C\, dx} - \left(\frac{\int x C\, dx}{\int C\, dx}\right)^2 \tag{6.32a}$$

$$\sigma_y^2 = \frac{\int y^2 C\, dx}{\int C\, dx} \tag{6.32b}$$

$$\sigma_x^2 = \frac{\int z^2 C\, dx}{\int C\, dx} \tag{6.32c}$$

The last term in Eq. 6.32a is the distance to the center of mass. In the diffusion equation, it is equal to Ut/R.

Dispersion coefficients in groundwater flow. In a uniform media of particles, the longitudinal dispersion coefficient, D_L, and the transverse dispersion coefficient, D_t, are both functions of the grain diameter and velocity. (In our previous example, D_L was D_x, and D_t would indicate D_y and D_z). The relevance of longitudinal and transverse dispersion relative to diffusion may therefore be *very roughly* characterized by a Peclet number, Pe:

$$Pe = Ud/D \tag{6.33}$$

where d is the grain diameter and U is a bulk velocity, Q/A, where the cross section includes porous media. If the consideration is not of uniform media but is for a heterogeneous region of high and low groundwater flow permeability, then the appropriate length scale would be the size of these permeability regions, normal to the flow. The characterization is [6];

$$D_L/D \sim Pe \tag{6.34}$$

Koch and Brady [7] have characterized transverse dispersion coefficient as a fraction of longitudinal dispersion coefficient:

$$D_t/D_L \sim 0.1 \tag{6.35}$$

Thus, longitudinal dispersion coefficient is roughly 10 times the value of transverse dispersion coefficient in a uniform media.

In the field, however, all media are heterogeneous, resulting in far greater dispersion than in uniform porous media. Because of the heterogeneities, the velocity profile can be highly variable over long distances, creating a much greater dispersion. Gelhar et al. [8] provided a plot of field data that can be manipulated to result in an equation that applies between a scale of 1 and 100,000 m:

$$Pe_{D_L} = \frac{UL}{D_L} = 2.5\,X\,10^{1\pm1.4} \tag{6.36}$$

where L is the horizontal scale of the measurement. The field data are given in Fig. 6.6, with the curves of Eq. 6.36.

Dispersion in Rivers

Rivers are an excellent environmental flow for describing the flow as a mean velocity with dispersion. The flow is confined in the transverse and vertical directions, such that a cross-sectional mean velocity and concentration can be easily defined. In addition, there is less variation in rivers than there is, for example, in estuaries or reactors, both of which are also described by the plug flow with dispersion model. For that reason, the numerous tracer tests that have been made in rivers are useful to characterize longitudinal dispersion coefficient for use in untested river reaches. A sampling of the dispersion coefficients at various river reaches which were determined from tracer tests is given in Table 6.2. Also given are the relevant mean parameters for each reach.

The question that we need to ask ourselves is whether the longitudinal dispersion can be predicted accurately for these rivers. Equation 6.16, which predicts that $D_L/(u_* h)$ = constant, is shown in Table 6.2 to have a large range of constants, probably because of the variations in cross section and morphology seen in natural streams. Fischer [9] observed that this constant seemed to depend upon mean surface width, W, and substituted W for h in the numerator of Eq. 6.16 to develop

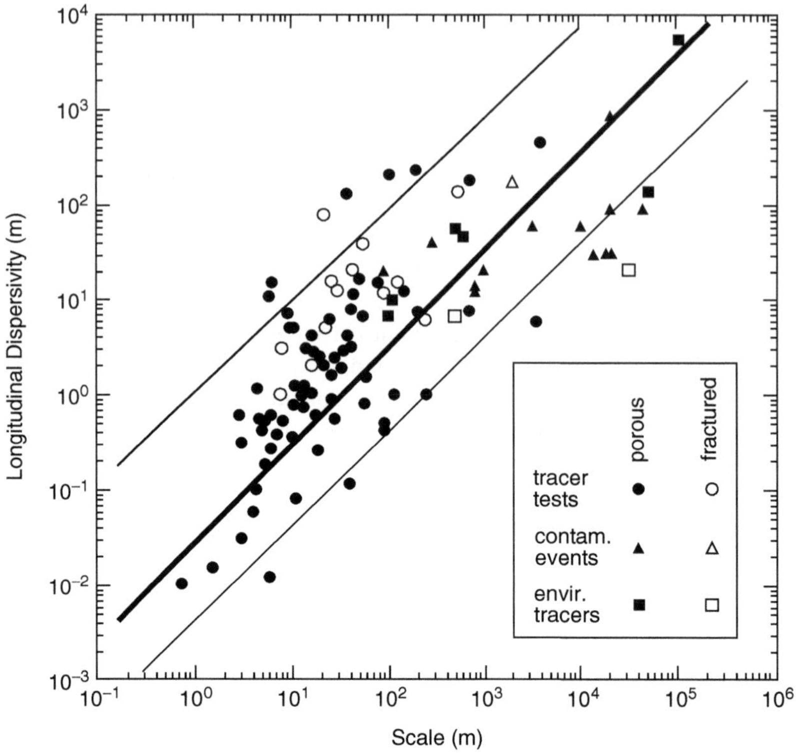

Fig. 6.6 Field data on dispersion coefficients, taken from Gelhar et al. [8] with Eq. 6.36 added. Longitudinal dispersivity is D_L/U

the following empirical equation to characterize longitudinal dispersion coefficient in rivers:

$$D_{LP} = 0.011 \frac{U^2 W^2}{u_* h} \tag{6.37}$$

where D_{Lp} is the predicted value of D_L, h the mean depth, U the cross-sectional mean velocity, and u_* the mean shear velocity of the river reach. When compared to the data given in Table 6.3 the root mean square error relative to the measurement, given by the equation,

$$\begin{matrix}\text{Relative} \\ \text{rms Errors}\end{matrix} = \sqrt{\frac{1}{n}\sum_{n}\left(1 - \frac{D_{LP}}{D_L}\right)^2} \tag{6.38}$$

where n, the number of measurements, is 1.71. This relative rms error means that roughly 67% of the predictions are within a factor of 1.71 (within 71%) of the observed.

Table 6.3 Measurements of longitudinal dispersion coefficient in laboratory flumes and rivers (after Fisher et al., 1979)

References	Channel	Depth. h(m)	Width. W(m)	Mean velocity, U (m/sec)	Shear velocity, u^* (m/sec)	Observed despersion coefficient, D_L (m^2/sec)	$D_L/(hu^*)$	D_{LR} predicted by Eq. (6.37) (m^2/sec)
Thomas (1958)	Chicago Ship Canal	8.07	48.8	0.27	0.0191	3.0	20	12.4
State of California (1962)	Sacramento River	4.00		0.53	0.051	15	74	
Owens et al. (1964)	River Derwent	0.25		0.38	0.14	4.6	131	
Glover (1964)	South Platte River	0.46		0.66	0.069	16.2	510	
Schuster (1965)	Yuma Mesa A Canal	3.45		0.68	0.345	0.76	8.6	
Fischer (1967)	Trapezoidal laboratory channel with roughened sides	0.035	0.40	0.25	0.0202	0.123	174	0.156
		0.047	0.43	0.45	0.0359	0.253	150	0.244
		0.035	0.40	0.45	0.0351	0.415	338	0.298
		0.035	0.34	0.44	0.0348	0.250	205	0.202
		0.021	0.33	0.45	0.0328	0.400	392	0.248
		0.021	0.19	0.46	0.0388	0.220	270	0.103
Fischer (1968b)	Green-Duwamish River, Washington	1.10	20		0.049	6.5–8.5	120–160	
Yotsukura et al. (1970)	Missouri River	2.70	200	1.55	0.074	1500	7500	3440
Godfrey and Frederick (1970)	Copper Creek, Virginia (below gauge)	0.49	16	0.27	0.080	20	500	5.24
		0.85		0.60	0.100	21	250	15.1
		0.49	16	0.26	0.080	9.5	245	4.86
	Clinch River, Tennessee	0.85	47	0.32	0.067	14	235	22
		2.10	60	0.94	0.104	54	245	73
		2.10	53	0.83	0.107	47	210	28
	Copper Creek, Virginia (above gauge)	0.40	19	0.16	0.116	9.9	220	2.19
	Powell River, Tennessee	0.85	34	0.15	0.055	9.5	200	6.12
	Cinch River, Virginia	0.58	36	0.21	0.049	8.1	280	22.1

(continued)

Table 6.3 (continued)

References	Channel	Depth. h(m)	Width. W(m)	Mean velocity, U (m/sec)	Shear velocity, u^* (m/sec)	Observed despersion coefficient, D_L (m^2/sec)	$D_L/(hu^*)$	D_{LR} predicted by Eq. (6.37) (m^2/sec)
	Coachella Canal, California	1.56	24	0.71	0.043	9.6	140	47.6
McQuivey and Keefer (1974)	Bayou Anacoco	0.94	26	0.34	0.067	33	520	13
		0.91	37	0.40	0.067	39	690	38
	Nooksack River	0.76	64	0.67	0.27	35	170	98
	Wind Bighorn Rivers	1.10	59	0.88	0.12	42	330	232
		2.16	69	1.55	0.17	160	440	340
	John Day River	0.58	25	1.01	0.14	14	170	88
		2.47	34	0.82	0.18	65	150	20
	Comite River	0.43	16	0.37	0.05	14	650	16
	Sabine River	2.04	104	0.58	0.05	315	3090	330
		4.75	127	0.64	0.08	670	1760	190
	Yadkin River	2.35	70	0.43	0.10	110	470	44
		3.84	72	0.76	0.13	260	520	68
Gulliver (1977)	MERS Experimental Streams	0.32	32	0.95	0.52	0.11	17	0.61

Tracer Determination of Longitudinal Dispersion Coefficient in Rivers

Tracers are generally used to determine longitudinal dispersion coefficient in rivers. Some distance is required, however, before the lateral turbulent diffusion is balanced by longitudinal convection, similar to Taylor's [1] analysis of dispersion in a laminar flow. This transport balancing distance, X^* is given by the equation:

$$X^* = \frac{0.2\ UW^2}{\varepsilon_y} \tag{6.39}$$

The region $x < x^*$ can be visualized as a mixing region, which can skew the results of a tracer study. Downstream of this region, where turbulent diffusion is balanced by longitudinal convection, the variance of a tracer pulse grows linearly with distance. It is best to begin the measurements a distance X^* below the tracer release. Once dispersion coefficient at one or more river discharges have been measured, Eq. 6.37 can be used to adjust the dispersion coefficient to all discharges:

$$\frac{D_{LQ_1}}{D_{LQ_2}} = \left(\frac{U^2W^2}{u_*h}\right)_{Q_1} \left(\frac{u_*h}{U^2W^2}\right)_{Q_2} \tag{6.40}$$

The technique used to perform the analysis of tracer studies will be the subject of Example 4.

Example Applications

Example 1: Determination of retardation coefficient As part of a forensic investigation of a continuous Malathion spill, you need to determine the retardation coefficient of the soil at the site for Malathion. You have decided to do so in a column experiment with the soil, illustrated in Fig. 6.7. Also given in the figure are the results of a pulse test with the non-sorptive tracer, chloride, and the results of a pulse test with Malathion. What is the retardation coefficient, R?
Chloride:
$t_r = t_{r1} = 20$ min.
$\sigma_{t1} = 3$ min.
Malathion:
$t_r = t_{r2} = 400$ min.
$\sigma_{t2} = 66$ min.
This test can result in a retardation coefficient from a comparison of both residence times and the variance.
Residence times: The residence time of the chloride can be given as $t_{r1} = L/U$, and the residence time of the Malathion can be given as $t_{r2} = R\ L/U$. Thus,

Fig. 6.7 Illustration of the column test for retardation coefficient and results of the tracer tests

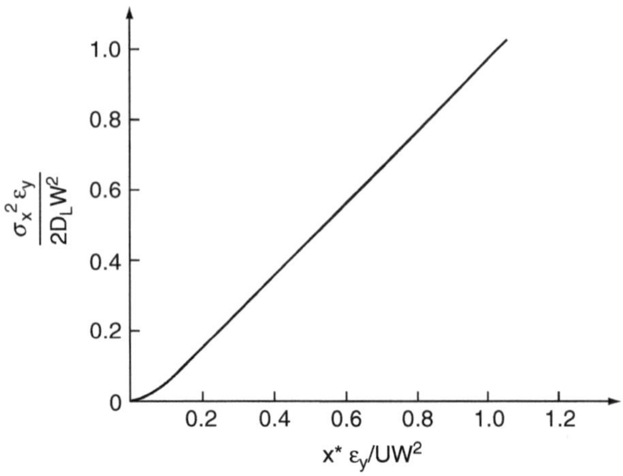

$$R = \frac{tr_2}{tr_1} = \frac{400 \text{ min}}{20 \text{ min}} = 20$$

Variance of tracer curves: Eq. 6.27 provides the relationships:

$$\sigma^2 = \sigma_t{}^2/t_r{}^2 = 2D_L Cou/(UL).$$

Therefore, assuming that $Cou = 1$,

$$\sigma_{t2}^2 = \frac{2D_L}{LR}\frac{R}{U} tr_2^2 = \frac{2D_L}{UL} tr_2^2 = \sigma_{t1}^2 \frac{tr_2^2}{tr_1^2} = R^2 \sigma_{t1}^2$$

or,

$$R = \sigma_{t2}/\sigma_{t1} = \ 66 \text{ min}/3 \text{ min} = 22$$

Our two means of determining the retardation coefficient in the column gave $R = 20$ and $R = 22$. We can also check whether the organic carbon content of the soil fits what is generally known from the literature. First, there is the relation for R:

$$R = 1 + \frac{\rho_B}{\varepsilon} K_d \tag{6.41}$$

Second, we have the equation from Karikhoff et al. [10]:

$$k_d = 0.41 \left(\frac{\text{cm}^3}{\text{g}}\right) f K_{ow} \tag{6.42}$$

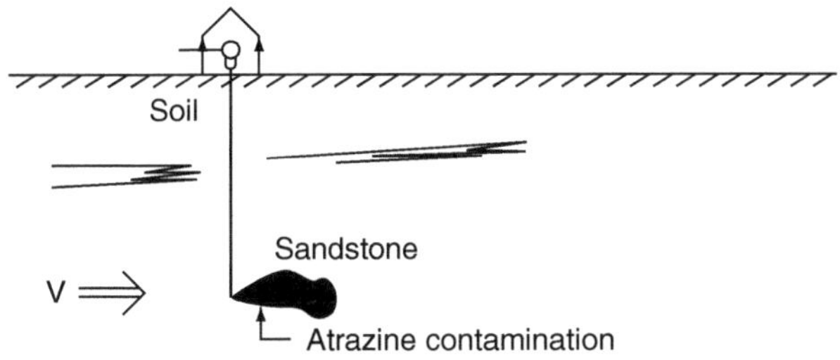

Fig. 6.8 Illustration of well and Atrazine spill (From [5])

Combining Eq. 6.29 and 6.30 results in an equation for the organic fraction:

$$f = \frac{(R-1)\varepsilon}{0.41\,\rho_B\,(\mathrm{g/cm^3})\,K_{ow}} = 0.030 - 0.033 \tag{6.43}$$

For Malathion, $K_{ow} = 230$. From soil tests, $\varepsilon = 0.3$, and $\rho_\beta = 2.0$ g/cm^3. Then Eq. 6.43 gives $f = 0.030 - 0.033$. This number is about right for the organic-rich soil that was tested. Thus, we know that our column tests are of the right order.

Example 2: Atrazine spill into an irrigation well (three-dimensional dispersion with convection)Three kilograms of 1,000 g/m^3 Atrazine pesticide is accidentally dumped down an old farm irrigation well, placed to pump water out of porous sandstone, as illustrated in Fig. 6.8. Estimate the movement of the Atrazine plume over time and the concentrations in the plume.

Given:

U = 10–4 m/s	d (sandstone) ≅ 1 mm
f = 10–4 (not soil)	D = 10–1^0 m^2/s
ε = 0.3	ρ_b/ε = 6 g/cm^3

Our solution is [4]:

$$C = \frac{M}{8\varepsilon(\pi t/R)^{3/2}\sqrt{D_x D_y D_z}} \exp\left[-\frac{R\left(x-\frac{U}{R}t\right)^2}{4D_x t} - \frac{Ry^2}{4D_y t} - \frac{Rz^2}{4D_y t}\right] \tag{6.44}$$

Now, estimate the maximum concentration location of this maximum and spread of the Atrazine cloud as a function of time.

Dispersion coefficients: As a first guess, let us use the empirical relations provided in Eq. 6.34 and 6.35. We will assume no heterogeneity in rock porosity and no lenses.

$$\frac{D_x}{D} = \frac{D_L}{D} \cong Pe = \frac{Ud}{D} \quad \text{or}$$
$$D_x = (10^{-4}\,\text{m/s})(10^{-3}\text{m}) = 10^{-7}\,\text{m}^2/\text{s}$$

$$\frac{D_y}{D} = \frac{D_z}{D} \cong 0.1\,Pe \quad \text{or}$$
$$D_y, D_z = 0.1\,(10^{-7}\,\text{m}^2/\text{s}) = 10^{-8}\,\text{m}^2/\text{s}$$

Retardation Coefficient:
Lehman et al. [11] gives:
log (K_{Ow}) = 2.75 for Atrazine
$\therefore K_{0w} = 10^{2.75} = 562$
Using Karikhoff et al. [10] relationship:
$K_d = \beta f K_{Ow}$ where $\beta = 0.41$ cm^3/g and $f = 10^{-4}$
Then

$K_d = 0.41$ cm^3/g $(10^{-4})(562) = 0.023$ cm^3/g

and

$$R = 1 + \frac{\rho_b}{\varepsilon} K_d = 1 + 6\,\frac{\text{g}}{\text{cm}^3}(0.023\ \text{cm}^3/\text{g}) = 1.14$$

Even with a fairly sorptive organic compound, the retardation coefficient in rock is not much different from 1.0. Eq. 6.44 gives a maximum at $y = 0$, $z = 0$, $x = U\,t/R$. Thus,

$$C_{\max} = \frac{M}{8\varepsilon(\pi t/R)^{3/2}\sqrt{D_x D_y D_z}}$$

and

$$X_{\max} = \frac{Ut}{R} = 0.88 \times 10^{-4}\,(m/s)\,t\,(s)$$

We will indicate spread by 4σ, which corresponds to 95% of the total mass of the cloud, for a Gaussian distribution like this equation provides.

For 1-D diffusion: $\sigma^2 = 2\,Dt/R$

For 3-D diffusion: $\sigma^2 = \frac{2}{9}\,Dt/R \quad \sigma = \frac{\sqrt{2Dt/R}}{3}$

and

$$4\sigma_x = \frac{4\sqrt{2D_x t/R}}{3}$$

$$4\sigma_z = 4\sigma_y = \frac{4\sqrt{2D_y\,t/R}}{3}$$

Table 6.4 Estimated concentration over time and space with transport through a uniformly porous sandstone

Time	C_{max} (g/m^3)	X_{max} (m)	$4\sigma_x$ (m)	$4\sigma_y$ (m)
1 hr	4.0×10^8	0.32	0.04	0.014
1 day	3.4×10^6	7.6	0.17	0.054
1 month	2.1×10^4	229	0.90	0.29
1 year	500	2,800	3.1	0.98
2 years	180	5,500	4.4	1.39
10 years	16	27,000	9.9	3.13

Table 6.5 Estimated concentration over time and space with transport through a sandstone media with 1 m heterogeneities

Time	C_{max}(g/m^3)	X_{max}(m)	$4\sigma_x$(m)	$4\sigma_y$(m)
1 hr	1.25	0.32	1.3	0.4
1 day	.011	7.6	5.5	1.7
1 month	6.6×10^{-5}	229	29	9.2
1 year	1.57×10^{-6}	2,800	99	31
2 years	5.6×10^{-7}	5,500	140	44
10 years	5.0×10^{-8}	27,000	320	99

Note that in Table 6.4, the concentrations at 1 h, 1 day, and 1 month are above the initial Atrazine concentration (1,000 g/m^3). This is one problem with Dirac delta boundary conditions because they initially have no volume, only mass. At greater elapsed time, however, the inaccuracies of the Dirac delta solution have a minimal impact on the resulting concentration.

However, virtually all media have heterogeneities of high and low porosity regions, as well as lenses that form around cracks. If we assume that the spacing of these regions is a mean of 1 m, instead of the 1 mm grain size, then our estimated dispersion coefficients are increased by a factor of 10^3. Applying the above equations to these parameters results in a reduction in C_{max} by a factor of 3.2×10^4 and an increase in both 4σ values by a factor of 32. The resulting estimates of the pertinent parameters are given in Table 6.5. These values are likely to be more realistic for transport through a groundwater aquifer.

The unknown dispersion coefficient is not uncommon in groundwater transport problems. It is typically one of the parameters fitted to measurements in groundwater transport.

Example 3: Drinking water pollution by trichloroethylene (steady state groundwater transport with lateral dispersion) A military ammunition plant in Arden Hills, Minnesota, used trichloroethylene (TCE) as a metal cleaning solvent for many years. Trichloroethylene is currently believed to be a carcinogen. Unaware of the hazardous nature of TCE, plant personnel placed the waste grease and TCE in a trench to burn (the grease) or soak into the ground (the TCE) and disappear from sight for many years. What was not known, however, is that they were placing the TCE into an aquifer that surfaces near the armory. Four kilometers downstream, the City of New Brighton used this aquifer as a source of municipal water supply. What is the expected TCE concentration in the New Brighton water supply and what should be done in the adjacent cities? Is the TCE plume sufficiently captured by the New Brighton wells? The following conditions were approximated from available data:

Supply of TCE = 100 kg/day
$U = 1.6 \times 10^{-5}$ m/s
Aquifer thickness, $H = 30$ m
$R \cong 1.0$ in the aquifer for TCE.
Aquifer porosity, $\varepsilon = 0.3$
New Brighton extraction, $Q = 0.25$ m^3/s
Drinking water recommended limit for TCE = 5 μg/L (5×10^{-3} g/m^3)

As a first assumption, we will assume that the New Brighton well was located in the center of the plume, compute the capture zone, and then the concentrations within this capture zone. The capture zone is given by:

$$Y = \frac{Q}{\varepsilon HU} = 1740\, m \tag{6.45}$$

At a velocity of 1.6×10^{-5} m/s, the 4 km distance would be covered in 8 years, which is short compared to the ~40 years of dumping TCE. We will therefore assume that the system is at steady state. We have these boundary conditions:

At $x = 0$, $\dot{M} = 100\,\text{kg/day} = 1.2\,\text{g/s}$
At $y = \infty$, $C = 0$
At $x = \infty$, $C = 0$.

The solution with these boundary conditions is:

$$C = \frac{\dot{M}}{H\varepsilon\,(4\pi x U D_t)^{1/2}} \exp\left(\frac{-Uy^2}{4\,D_t\,x}\right)$$

with a capture zone mean of

$$\bar{C} = \frac{2}{Y}\int_0^{Y/2} C dy$$

where Y is the width of the capture zone, and a leakage from the capture zone, M_L of

$$M_L = 2\varepsilon H \int_{Y/2}^{\infty} C dy$$

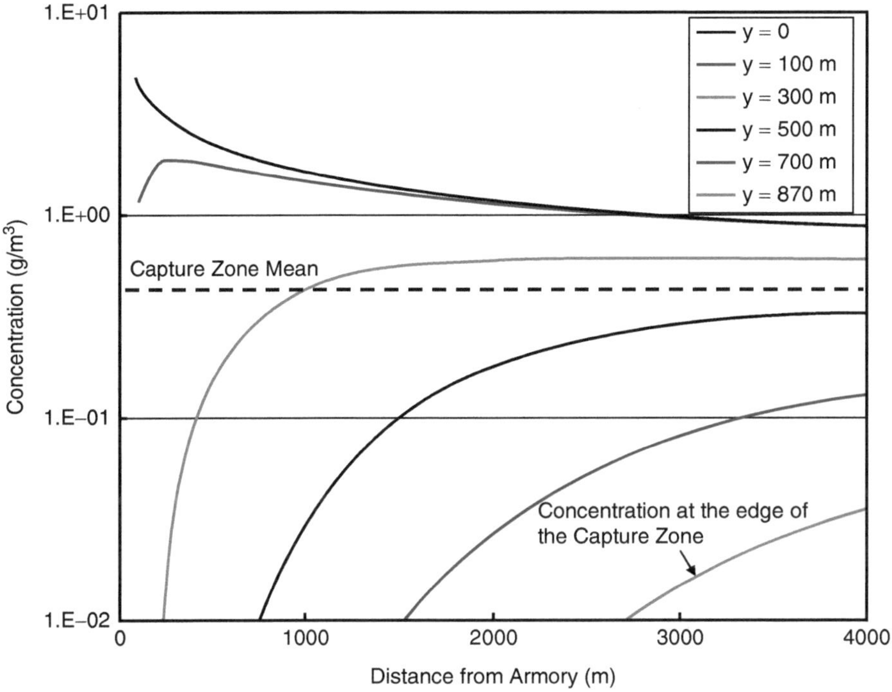

Fig. 6.9 Predicted concentration of trichloroethylene in groundwater plume versus distance for the ammunitions plant release – City of New Brighton case study

Using the rough approximations of Eq. 6.35 and 6.36, we get

$$D_L = 0.0026 \text{m}^2/\text{s}$$

and

$$D_t = 2.6 \,\text{X}\, 10^{-4} \text{m}^2/\text{s}.$$

Note that Eq. 6.36 gives $D_L = 0.0026$ m^2/s with a 67% confidence interval of between 0.064 and 1×10^{-4} m^2/s, or 1.4 orders of magnitude. This variation would need to be considered in any preliminary analysis of this problem.

The solution for this application is given in Fig. 6.9 at various lateral distances from the peak concentration. The capture zone mean is 0.42 g/m^3, or almost 100 times the recommended limit, which would raise concern in New Brighton. The leakage from the capture zone at $x = 4{,}000$ m for this scenario is computed to be 47 g, or sufficient mass to result in a concentration of 9×10^{-4} g/m^3 for a similar capture zone. It is possible, then, that almost all of the plume was captured by the City of New Brighton, so the problem may not cover a wider area than the immediate downstream cities. This, at least, is one positive result of the low transverse dispersion of groundwater plumes.

Example 4: Determination of D_L in a river A tanker car carrying a solvent derailed on a bridge and fell into the Nemadji river, Wisconsin. The forensic investigation team has a computational model that will simulate the spill, if some coefficients are determined, including D_{L} and t_r. The most cost-effective means of determining these parameters would be to perform a conservative tracer pulse test and adjust the parameters from discharge on the day of the tracer test (30 m^3/s) to discharge on the day of the spill (70 m^3/s) with some predictive equations that have been developed.

The location of the pulse input is identified as $x = 0$. The measured tracer concentrations and other relevant data are given in Fig. 6.10. From this data, determine the D_L and t_r parameters on the day of the test for the reach from $x =$ 8,000 m to $x =$ 35,000 m.

Because the variance of the tracer curve grows linearly with distance after $x = x^*$, we can make the following statements:

$$\Delta t_r = t_{r_2} - t_{r_1}$$

and

$$\Delta\sigma_t^2 = \sigma_t^2|_2 - \sigma_t^2|_1$$

and finally,

$$\frac{\Delta\sigma_t^2}{\Delta t_r^2} = \frac{2D_L}{U\Delta x} = \frac{2D_L\,\Delta t_r}{\Delta x^2}$$

At $x =$ 8,000 m,

$$t_{r_1} = \frac{\int Ct\,dt}{\int C\,dt} \cong \frac{\sum Ct\,\Delta t}{\sum C\,\Delta t}$$

Using all of the Δt values set equal to 0.1 h, Δt can be cancelled out of the equation, and

$$t_{r_1} = \frac{1.30\,\mathrm{h\,g/m^3}}{0.67\,\mathrm{g/m^3}} = 1.94\,\mathrm{h}$$

Also,

$$\sigma_t^2|_1 = \frac{\int Ct^2dt}{\int C\,dt} - t_{r1}^2 \cong \frac{\sum Ct^2\,\Delta t}{\sum C\,\Delta t} - t_{r1}^2$$

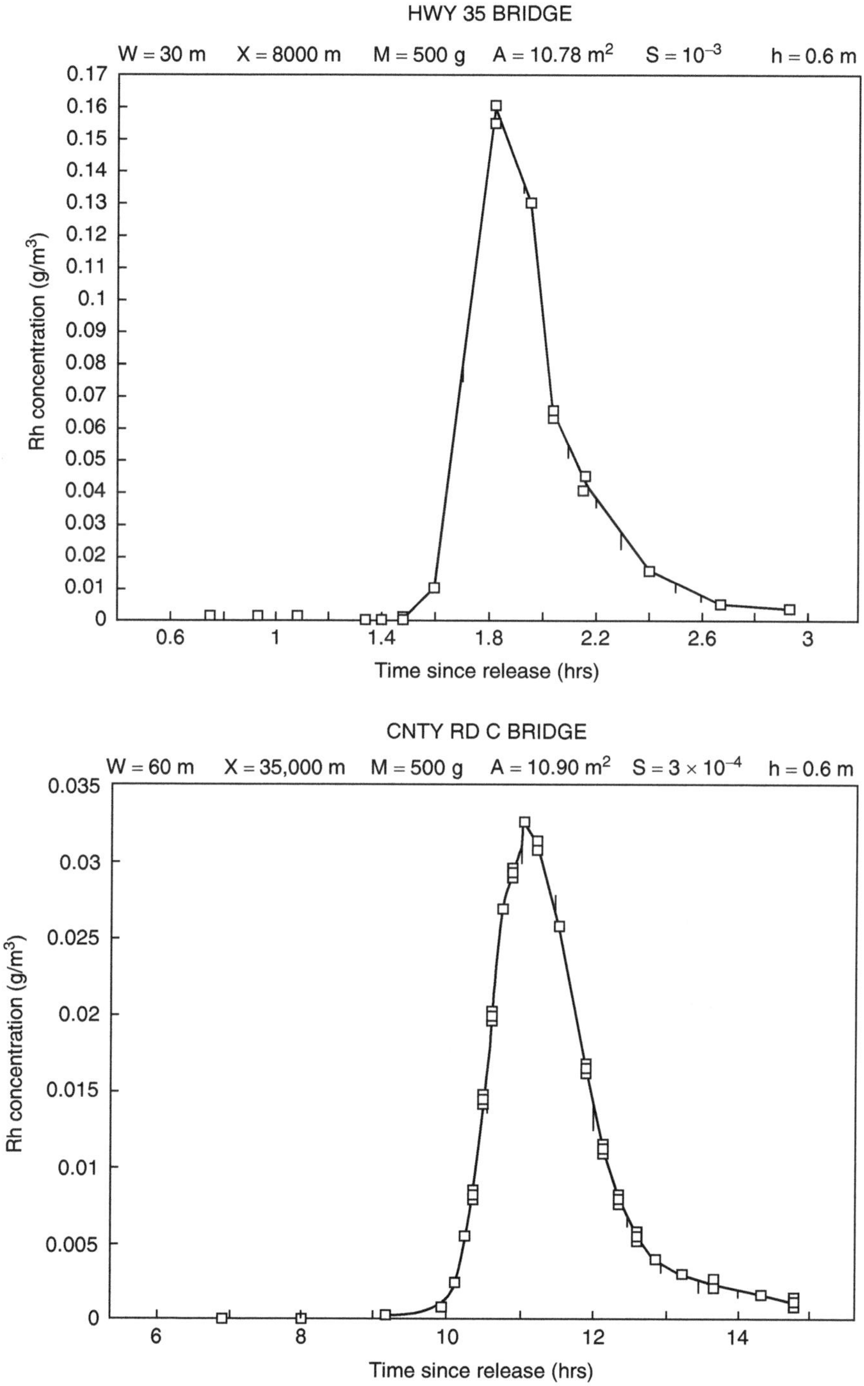

Fig. 6.10 Tracer measurements taken at the Highway 35 and County Road C bridges

or, again cancelling the equal Δt's:

$$\sigma_t^2|_1 = \frac{2.59\,\mathrm{h}^2\mathrm{g/m}^3}{0.666\,\mathrm{g/m}^3} - (1.94\,\mathrm{h})^2 = 0.125\ \mathrm{h}^2$$

We have assumed that the location 8,000 m downstream of the pulse injection will be out of the "mixing" region, as specified by Eq. 6.39. We are now ready to check this assumption with the parameters of the stream flow at x = 8,000 m. First, $U = Q/A = 8\ \mathrm{m}^3/\mathrm{s}/10.8\ \mathrm{m}^2 = 0.74$ m/s, and from the chapter on Turbulent Transport:

$$\varepsilon_y \cong (0.6 \pm 0.3)\, u_* h \tag{6.46}$$

where

$$u_* = \sqrt{ghS} = \sqrt{(9.8\,\mathrm{m/s^2})(0.6\,\mathrm{m})\,(10^{-3})} = 0.077\,\mathrm{m/s}$$

Then,

$$\begin{aligned}\varepsilon_y &= (0.6 \pm 0.3)\ (0.077\,\mathrm{m/s})(0.6\,\mathrm{m})\\ &= 0.028\ \pm\ 0.014\ \mathrm{m}^2/\mathrm{s}\end{aligned}$$

and Eq. 6.41 gives,

$$X_* = \frac{0.2\,(0.74\,\mathrm{m/s})\,(30\,\mathrm{m})^2}{0.028\,\mathrm{m}^2/\mathrm{s}} = 4760\,\mathrm{m}$$

With the approximations in Eq. 6.39, x = 8,000 m will be assumed sufficient to begin our determination of D_L, especially since that is one of the few access points (a bridge) into this reach of the river.

We will now perform a similar calculation on the tracer cloud at x = 35,000 m.

$$t_{r_2} = 11.54\ \mathrm{hr}$$

$$\sigma_t^2|_2 = \frac{14\ \mathrm{hr}^2\ \mathrm{g/m}^3}{0.105\,\mathrm{g/m}^3} - (11.54\ \mathrm{hr})^2 = 0.162\ \mathrm{hr}^2$$

$$\Delta x = 35,000\ \mathrm{m} - 8,000\ \mathrm{m} = 27,000\ \mathrm{m}$$

Now,

$$D_\mathrm{L} = \frac{\Delta\sigma_t^2\,\Delta x^2}{2\Delta t_r^3} = \frac{(0.162 - 0.125)\,\mathrm{hr}^2\,(27,000\,\mathrm{m})^2}{2\,(11.54 - 1.94)^3\,\mathrm{hr}^3}$$

or,

$$D_{\mathrm{L}} = 15,200\mathrm{m}^2/\mathrm{hr} \ = 4.2\mathrm{m}^2/\mathrm{s}$$

We can use Eq. 6.37 to adjust our dispersion coefficient from the 8 m^3/s with 0.6 m mean depth on the day of the tracer test to the 3 m^3/s with 0.4 m mean depth that existed as the river discharge on the day of the spill:

$$D_{\mathrm{L}} = 0.011\,(Q/Wh)^2\ W^2/(\sqrt{ghS}\,h) \tag{6.47}$$

and then, assuming that the slope does not change and that the banks are fairly steep, such that $dA = W\,dh$, and assigning the subscripts t and s to indicate tracer and spill:

$$\begin{aligned} D_{\mathrm{L}s} &= D_{\mathrm{L}t}\left(\frac{Q_s}{Q_t}\right)^2\left(\frac{h_t}{h_s}\right)^{7/2} \\ &= 4.2\,\frac{\mathrm{m}^2}{\mathrm{s}}\left(\frac{3\,\mathrm{m}^3/\mathrm{s}}{8\,\mathrm{m}^3/\mathrm{s}}\right)\left(\frac{0.6\,\mathrm{m}}{0.4\,\mathrm{m}}\right) = 2.4\ \mathrm{m}^2/\mathrm{s} \end{aligned}$$

In addition, t_r can be adjusted as well:

$$\Delta t_{rs} = \Delta t_{r_t}\frac{Q_t}{Q_s}\frac{h_s}{h_t} = (11.54 - 1.94)\,\mathrm{h}\,\frac{8\mathrm{m}^3/\mathrm{s}}{3\,\mathrm{m}^3/\mathrm{s}}\,\frac{0.4\,\mathrm{m}}{0.6\,\mathrm{m}}$$

so

$$\Delta t_{r_s} = 17\,\mathrm{h}$$

The use of an empirical relation, such as Eq. 6.37, to adjust parameters for discharge is more accurate than simply using the equation itself, because the coefficient and other variables that can have substantial uncertainty are eliminated from the equation.

Conclusions

Dispersive transport is handled in a manner similar to turbulent diffusion and diffusion, with almost the same equations and solutions. The primary difference is that a dispersion coefficient is one or two orders of magnitude greater than turbulent diffusion coefficients and about 10 orders of magnitude greater than diffusion coefficients. Assuming that the dispersion coefficient will result in a Gaussian distribution of a chemical is only a rough approximation, however, which can be inaccurate at the trailing edge of the chemical cloud.

Future Directions

As computational power of our computers continues to improve, less attention will be paid to dispersive transport in the environment. The merging of velocity profiles and mixing to make a dispersion coefficient will be of less value when a more exact solution is available on desktop computers. The simplicity of the analysis, however, still will make the dispersion equations valuable in developing an understanding of mixing and transport problems in environmental settings.

Bibliography

Primary Literature

1. Taylor GI (1953) Dispersion of soluble matter in solvent flowing slowly through a tube. Proc R Soc Lond Ser A 219:186
2. Taylor GI (1954) The dispersion of matter in turbulent flow through a pipe. Proc R Soc Lond Ser A 223:446
3. Elder JW (1959) The dispersion of marked fluid in turbulent shear flow. J Fluid Mech 5:544
4. Crank J (1975) The mathematics of diffusion, 2nd edn. Oxford University Press, Oxford
5. Gulliver JS (2007) An introduction to chemical transport in the environment. Cambridge University Press, Cambridge
6. Freeze RA, Cherry JA (1979) Groundwater. Prentice-Hall, Englewood Cliffs
7. Koch DL, Brady JF (1985) Dispersion in fixed beds. J Fluid Mech 154:399
8. Gelhar L, Welty C, Rehfeldt KR (1992) A critical-review of data on field-scale dispersion in aquifers. Water Resour Res 28(7):1955
9. Fisher HB (1973) Longitudinal dispersion and turbulent mixing in open-channel flow. Ann Rev Fluid Mech 5:59
10. Karikhoff SW, Brown DS, Scott TA (1979) Sorption of hydrophobic pollutants on natural sediments. Water Res 13:241
11. Lehman WJ, Reehl WF, Rosenblatt DH (1990) Handbook of chemical property estimation. American Chemical Society, Washington, DC

Books and Reviews

Fisher HB, List JE, Koh RCY, Imberger J, Brooks NH (1979) Mixing in inland and coastal waters. Academic, San Diego
Kreyszig E (1982) Advanced engineering mathematics, 4th edn. Wiley, New York
Levenspiel O (1962) Chemical reaction engineering. Wiley, New York

Chapter 7
Transport with Jets and Plumes of Chemicals in the Environment

Wenming Zhang, Nallamuthu Rajaratnam, and David Z. Zhu

Glossary

Bubbly jet	The jet produced by injecting gas-liquid mixture into a liquid.
Buoyant jet	The plume with momentum or jet with buoyancy.
Circular jet	The jet produced through a nozzle with a circular cross section.
Diffuser	The device which has multiple nozzles to quickly mix the discharged substances (e.g., effluent or air) with the surrounding ambient fluid.
Jet	The flow generated by the release of momentum usually through a nozzle or slot.
Jet in coflow	The jet discharged in the direction of a flowing ambient fluid.
Jet in crossflow	The jet discharged at an oblique angle to a flowing ambient fluid.
Plane jet	Also called "slot jet" or "two-dimensional jet", the jet produced through a slender slot.
Plume	The flow generated by the release of buoyancy.
Surface jet	The jet discharged at (or near) the surface of an ambient fluid.
Slurry jet	The jet produced by injecting solid-liquid mixture into a liquid.
Wall jet	The jet discharged tangentially or at a certain angle to a solid boundary (wall).

This chapter was originally published as part of the Encyclopedia of Sustainability Science and Technology edited by Robert A. Meyers. DOI:10.1007/978-1-4419-0851-3

W. Zhang • N. Rajaratnam • D.Z. Zhu (✉)
Department of Civil and Environmental Engineering, University of Alberta,
T6G 2W2 Edmonton, AB, Canada
e-mail: wenming@ualberta.ca; nrajaratnam@ualberta.ca; david.zhu@ualberta.ca

J.S. Gulliver (ed.), *Transport and Fate of Chemicals in the Environment: Selected Entries from the Encyclopedia of Sustainability Science and Technology*,
DOI 10.1007/978-1-4614-5731-2_7, © Springer Science+Business Media New York 2012

Definition of the Subject and Its Importance

Jets and plumes are common in our environment. Some examples of jets are: wastewater discharged from an outfall, emission from an aircraft or vehicle, and the eruption of volcano. Some examples of plumes are: the smoke from a chimney stack or cigarette, the thermal plumes from a fire, municipal wastewater or hot water discharged in deep water, and oil spill from sea bed. One of the most important features of jets or plumes is its ability of entraining ambient fluid to achieve self-dilution. This greatly triggers our interests to study jets and plumes. This book chapter is a review of the studies on turbulent jets and plumes, with a focus on the transport of conservative pollutants.

Introduction

The earliest experimental study of turbulent jets appears to be the work of Trupel on circular jets in 1915 [1]. Förthmann performed an experimental study of plane turbulent jets in 1934 and his work also considered plane turbulent wall jets [23]. The results of these investigations showed the similarity of the velocity profiles at different distances from the sources of the jets. These studies were followed by the extensive investigations of Hinze and Zijnen [28] on circular jets and Albertson et al. [4] for plane and circular jets. Turbulence characteristics of plane jets were studied by Heskestad [27] for plane jets and Wygnanski and Fielder [105] for circular jets. Theoretical solutions for plane jets were developed by Tollmien in 1926 for plane jets and by Goertler in 1942 for circular jets [73]. Numerical studies of turbulent jets followed, starting with the work of Rodi and Spalding [82]. Abramovich [1], Rajaratnam [73], and Fischer et al. [21] provided comprehensive treatment of jets.

Turner [97] provides an introduction to study turbulent plumes in his book on *Buoyancy Effects in Fluids*. Rouse et al. [84] performed an experimental study of plane and circular plumes wherein they found that the velocity and density defect profiles were similar if proper scales were chosen for velocity, width, and density defect. Morton et al. [58] published an integral study of plumes, wherein the concept of entrainment coefficient was introduced. Since then numerous studies have been conducted on turbulent plumes and forced turbulent plumes (buoyant jets). These studies have been summarized in Chen and Rodi [15] and Lee and Chu [45]. Turbulent jets and plumes have been studied extensively not only because these flows are very interesting but also that they are of considerable practical importance in the fields of hydraulic, mechanical, aeronautical, environmental, and chemical engineering and many other fields.

This book chapter will first review the most classic and well-established theories on simple jets or plumes in stagnant water, and then consider effects of boundaries including: the bed (wall) and the surface of ambient fluid; coflowing and cross-flowing ambient fluid; and the interaction of neighboring jets in the case of multiple jets. Next, two kind of multiphase jets and plumes – bubbly jets and plumes, and slurry jets and plumes – will be briefly introduced. Multiphase jets and plumes are much more complicated compared to the single-phase ones, but have gained more interest in recent years because of their wide applications. Finally, some directions for future researches will be highlighted.

Turbulent Jets and Plumes in Stagnant Environment

In this section, the focus is on the transport of conservative pollutants in a steady-state turbulent jet or plume issuing from a simple (plane or circular) nozzle into stagnant ambient fluid of large extent. Such jets or plumes are called simple jets or simple plumes. Theories in this area have been well established. Close to the nozzle exit, there is a wedge-like or cone-like region termed "potential core" where the width of initial velocity distribution decays to a point. The length of the potential core is very short, about $10.4b_0$ for a plane jet where b_0 is the half slot width, or $6.2d_0$ for a circular jet where d_0 is the nozzle diameter [45]. Therefore, for practical purposes, our attention will be limited on the flow beyond the potential core, i.e., in the "fully developed flow" region (see Fig. 7.1). The reader who is interested in flow development region may refer to Rajaratnam [73].

Simple Jets

The integral method is the most common method for analyzing simple jets or plumes. The following is a brief introduction on this method. For a plane jet as

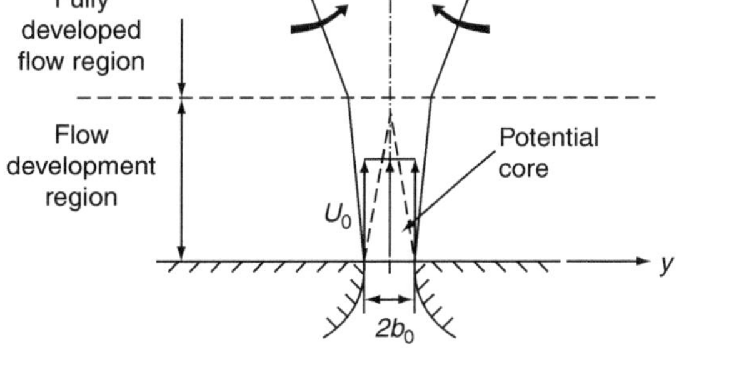

Fig. 7.1 Schematic of a simple plane jet (for circular jet, replace $2b_0$ by d_0 and y by r)

shown in Fig. 7.1, the Reynolds-averaged Navier-Stokes equation in x direction, continuity equation, and pollutant conservation equation, respectively, can be simplified as [73]:

$$u\frac{\partial u}{\partial x}+v\frac{\partial u}{\partial y}=\frac{1}{\rho}\frac{\partial \tau}{\partial y} \tag{7.1}$$

$$\frac{\partial u}{\partial x}+\frac{\partial v}{\partial y}=0 \tag{7.2}$$

$$\frac{\partial uC}{\partial x}+\frac{\partial vC}{\partial y}=\varepsilon\frac{\partial^2 C}{\partial y^2} \tag{7.3}$$

where u and v are the time-averaged velocities in x and y directions, respectively; C is the pollutant concentration; ρ is the density of the fluid; τ is the turbulent shear stress and ε is the mean value of turbulent diffusion coefficient. After multiplying Eq. 7.1 by ρ and then integrating from $y = 0$ to $y = \infty$, Eq. 7.1 becomes:

$$\frac{d}{dx}\int_o^\infty \rho u^2 dy=0 \tag{7.4}$$

Equation 7.4 states that the jet momentum flux at different x-sections is conserved. Using Eq. 7.2 and integrating the first term of Eq. 7.2 from $y = 0$ to $y = \infty$, we have:

$$\frac{dq_x}{dx}=\frac{d}{dx}\int_0^\infty u\,dy=-v_{y=\infty} \tag{7.5}$$

where q_x is the jet volume flux per unit slot length. The entrainment hypothesis assumes that the entrainment velocity $v_e = v_{y=\infty} = -\alpha_e u_m$, where α_e is called the jet entrainment coefficient, u_m is jet centerline (maximum) velocity, and the negative sign indicates the ambient fluid is entrained into the jet. Equation 7.5 says that the jet volume flux increases with traveling distance x due to the entrainment of ambient fluid, which explains the ability of jet in diluting pollutants. Similarly, integrating Eq. 7.3 from $y = 0$ to $y = \infty$, we have:

$$\frac{d}{dx}\int_0^\infty uC\,dy=0 \tag{7.6}$$

Equation 7.6 states that the mass fluxes of pollutants at different x-sections are conserved if the chemical or biological reactions of pollutants are not considered.

Numerous laboratory experiments and numerical simulations have confirmed that beyond the potential core, the jet velocity or concentration exhibits self-similarity. The most widely used expression for such similarity is the Gaussian distribution, which represents laboratory data satisfactorily:

$$\frac{u}{u_m} = \exp\left[-k_1\left(\frac{y}{b}\right)^2\right] \tag{7.7}$$

$$\frac{C}{C_m} = \exp\left[-k_1\left(\frac{y}{k_2 b}\right)^2\right] \tag{7.8}$$

where C_m is the jet time-averaged centerline (maximum) concentration, b is the jet velocity half-width where the velocity is 50% (if $k_1 = 0.693$, refer to Fig. 7.1) or 37% (if $k_1 = 1$) of u_m, $k_2 b$ defines the jet concentration half-width where the concentration is 50% or 37% of C_m, and k_2 is the ratio of concentration half-width to velocity half-width. Using Eqs. 7.4–7.8, the analytical solutions for plane jets can be derived, as shown in Table 7.1. The coefficients of the solutions are mainly determined from experimental results and integrations.

Using the same procedures as above, the equations of momentum flux, volume flux, and pollutant mass flux for a circular jet can be derived. These equations suggest that the jet momentum flux at any x-section is conserved and the jet volume flux across any x-section increases with traveling distance due to the entrainment of ambient fluid. Although the entrainment causes the pollutant to get diluted within the jet core, the mass flux of any conservative pollutant at any x-section is conserved. The solutions of these equations for circular jets are summarized in Table 7.1.

Simple Plumes

A plume is produced from a steady discharge of a fluid whose motion is controlled by its buoyancy, with negligible effect of initial momentum. First, a plane plume of density ρ_0 issued into a stagnant unstratified ambient fluid of density ρ_a is considered. It is assumed that $\frac{\Delta\rho_0}{\rho_a} \ll 1$ (true for most practical cases), where the initial density defect $\Delta\rho_0 = \rho_a - \rho_0$. After some manipulation, it can be shown that the Reynolds-averaged Navier-Stokes equation in x direction becomes:

$$u\frac{\partial u}{\partial x} + v\frac{\partial u}{\partial y} = \frac{1}{\rho_a}\frac{\partial \tau}{\partial y} + g\frac{\Delta\rho}{\rho_a} \tag{7.9}$$

where $\Delta\rho = \rho_a - \rho$ and ρ is the plume density. For a plane plume, the continuity equation and pollutant conservation equation can be simplified in the same form as for a plane jet (Eqs. 7.2 and 7.3).

Table 7.1 Summary of mean properties of simple plane jet and circular jet

Parameter	Plane jet	Circular jet
Maximum (centerline) concentration C_m	$\frac{C_m}{C_0} = \frac{\alpha_1}{\sqrt{\frac{x}{b_0}}}$, where α_1 = 3.37 in [21], 3.45 in [74], 3.21 in [45]	$\frac{C_m}{C_0} = \frac{\alpha_1}{\frac{x}{d_0}}$, where α_1 = 4.96 in [21], 5.34 in [74], 5.26 in [45]
Cross-sectional average concentration C_{avg}	$\frac{C_m}{C_{avg}} = \alpha_2$, where α_2 = 1.2 in [21], 1.25 in [45]	$\frac{C_m}{C_{avg}} = \alpha_2$, where α_2 = 1.4 in [21], 1.76 in [74], 1.68 in [45]
Maximum (centerline) velocity u_m	$\frac{u_m}{U_0} = \frac{\alpha_3}{\sqrt{\frac{x}{b_0}}}$, where α_3 = 3.50 in [73], 3.41 in [21], 3.65 in [45]	$\frac{u_m}{U_0} = \frac{\alpha_3}{\frac{x}{d_0}}$, where α_3 = 6.3 in [73], 6.2 in [21] and [45], 6.13 in [74]
Velocity half-width b[a]	$b = \alpha_4 x$, where α_4 = 0.10 in [73], 0.116 in [21], 0.097 in [74], 0.12 in [45]	$b = \alpha_4 x$, where α_4 = 0.10 in [73], 0.107 in [21], 0.096 in [74], 0.114 in [45]
Concentration half-width b_C[a]	$\frac{b_C}{b} = \alpha_5$, where α_5 = 1.35 in [21] and [45], 1.17 in [74]	$\frac{b_C}{b} = \alpha_5$, where α_5 = 1.19 in [21], 1.17 in [74], 1.2 in [45]
Entrainment coefficient α_e	α_e = 0.053 in [73] and [45]	α_e = 0.026 in [73], 0.028 in [74], 0.057 in [45]

Source: [21, 45, 73, 74].

[a]In [21] and [45], b (or b_C) is defined as where the velocity (or concentration) is 37% of u_m (or C_m); while in others, defined as 50% of u_m (or C_m).

Using the integral method, Eq. 7.9 can be reduced to:

$$\frac{d}{dx}\int_0^\infty \rho u^2 dy = \int_0^\infty g\Delta\rho dy \tag{7.10}$$

Equation 7.10 says that axial momentum flux increases in x direction, and the increase rate is equal to the buoyancy per unit length (in x direction) of the plume. For a plane plume, the continuity equation and pollutant conservation equation can be reduced the same as Eqs. 7.5 and 7.6. Equation 7.5 states that the volume flux of a plume increases due to the entrainment of ambient fluid, and Eq. 7.6 states that although the concentration of pollutant decreases, its total mass flux is conserved. Note that C in Eq. 7.6 can be also interpreted as $g\Delta\rho$, and then Eq. 7.6 becomes the integral form of buoyancy conservation equation.

Experimental results show that: similarly as for jets, the plume velocity or concentration also exhibit self-similarity and the Gaussian distribution can well describe it. Using the Gaussian profiles (Eqs. 7.7 and 7.8), the analytical solutions for a plane plume can be derived as shown in Table 7.2. The coefficients differ slightly in different references as they are determined using the results of different experiments. Here, to constitute the solutions, an useful dimensionless parameter is indroduced – the densimetric Froude number at the slot exit $F_0 = \frac{U_0}{\sqrt{g\left(\frac{\Delta\rho}{\rho_a}\right)b_0}}$ (for circular plume, $F_0 = \frac{U_0}{\sqrt{g\left(\frac{\Delta\rho}{\rho_a}\right)d_0}}$).

Table 7.2 Summary of mean properties of plane plume and circular plume

Parameter	Plane plume	Circular plume
Maximum (centerline) concentration C_m or density defect $\Delta\rho_m$	$\frac{C_m}{C_0}=\frac{\Delta\rho_m}{\Delta\rho_0}=\frac{\alpha_1 F_0^{\frac{2}{3}}}{x/b_0}$, where $F_0=\frac{U_0}{\sqrt{g\frac{\Delta\rho}{\rho_a}b_0}}$, $\alpha_1 = 3.78$ in [21], 3.84 in [74], 4.25 in [45]	$\frac{C_m}{C_0}=\frac{\Delta\rho_m}{\Delta\rho_0}=\frac{\alpha_1 F_0^{\frac{2}{3}}}{\left(\frac{x}{d_0}\right)^{\frac{5}{3}}}$, where $F_0=\frac{U_0}{\sqrt{g\frac{\Delta\rho}{\rho_a}d_0}}$, $\alpha_1 = 7.75$ in [21], 7.83 or 9.37 in (from different methods in Ref. [74]), 8.90 in [45]
Cross-sectional average concentration C_{avg}	$\frac{C_m}{C_{avg}}=\alpha_2$, where $\alpha_2 = 1.32$ in [74], 1.25 in [45]	$\frac{C_m}{C_{avg}}=\alpha_2$, where $\alpha_2 = 1.40$ in [21], 1.70 in [45]
Maximum (centerline) velocity u_m	$\frac{u_m}{u_0}=\frac{\alpha_3}{F_0^{\frac{2}{3}}}$, where $\alpha_3 = 2.09$ in [21], 2.52 in [74], 2.85 in [45]	$\frac{u_m}{U_0}=\frac{\alpha_3}{F_0^{\frac{2}{3}}\left(\frac{x}{d_0}\right)^{\frac{1}{3}}}$, where $\alpha_3 = 4.34$ in [21], 4.00 or 4.33 in (from different methods in Ref. [74]), 4.35 in [45]
Velocity half-width b[a]	$b = \alpha_4 x$, where $\alpha_4 = 0.116$ in [21] and [45], 0.128 in [74]	$b = \alpha_4 x$, where $\alpha_4 = 0.100$ in [21], 0.085 in [74], 0.105 in [45]
Concentration half-width b_C[a]	$b_C/b. = \alpha_5$, where α_5 =1.35 in [21] and [45], 1.17 in [74]	$b_C/b = \alpha_5$, where α_5 =1.20 in [21], 1.16 in [74], 1.19 in [45]
Entrainment coefficient α_e	$\alpha_e = 0.136$ in [74], 0.103 in [45]	$\alpha_e = 0.047$ in [74], 0.088 in [45]

Source: [21, 45, 74].
[a]In [21] and [45], b (or b_C) is defined as where the velocity (or concentration) is 37% of u_m (or C_m); while in [74], defined as 50% of u_m (or C_m).

For a circular plume, the Reynolds-averaged Navier-Stokes equations, continuity equation and pollutant conservation equation, can be simplified, and the integral method can be used to obtain their integral forms of the equations. These equations indicate that the momentum flux in the plume increases with axial (x) direction, and the increase rate is equal to the buoyant force per unit axial length; the volume flux in the plume increases due to the entrainment of ambient fluid; and the pollutant mass flux (or the flux of density defect) remains invariant in the axial direction. The analytical solutions for a circular plume are also presented in Table 7.2. The constants in these equations change slightly in different references where different experimental results were used.

Buoyant Jets

For a buoyant jet, the initial momentum flux cannot be neglected. Near the nozzle (slot) exit, it is expected that the buoyant flow will be like a jet; after some distance from the exit where the increase of momentum flux is much larger than the initial momentum flux, the buoyant flow will behave like a plume. For a plane buoyant jet, based on dimensional analysis, a characteristic length scale to judge whether the buoyant flow behaves like a jet or plume may be defined as:

$$L_M = \frac{M_0}{B_0^{\frac{2}{3}}} \tag{7.11}$$

where the initial specific momentum flux $M_0 = qU_0$; the initial specific buoyancy flux $B_0 = qg\frac{\Delta\rho_0}{\rho_a}$. If the jet centerline trajectory $l \ll l_M$, the buoyant jet can be treated as a pure jet and the simple jet equations can be used; if $l \gg l_M$, it can be treated as a pure plume and the simple plume equations are valid. Buoyant jets will be plume-like beyond $\frac{l}{l_M} \geq 4 \sim 5$ [41, 65].

Characteristic length scale of jet/plume can only roughly help us calculate the evolution of a buoyant jet. The Reynolds equations, continuity equation and pollutant mass conservation equations can be simplified and the integral method can be used to obtain the analytical solutions. For a plane buoyant jet, the integral forms of the momentum, continuity, pollutant mass (or buoyancy) conservation equations are the same as Eqs. 7.10, 7.5 and 7.6, respectively. Here, for a buoyant jet, the integral form of energy equation needs to be introduced (as a result of multiplying Eq. 7.9 by u and integrating from $y = 0$ to $y = \infty$):

$$\frac{d}{dx}\int_0^\infty \frac{\rho u^2}{2} u dy = -\int_0^\infty \tau \frac{\partial u}{\partial y} dy + \int_0^\infty g\Delta\rho u \, dy \tag{7.12}$$

Equation 7.12 states that the flux of kinetic energy in the jet plume is decreased by turbulence production (first term in the right hand side of Eq. 7.12) and increased by the work done by buoyancy (second term in the right hand side).

The Gaussian type self-similarity equations are still valid for plane buoyant jets. Using Eqs. 7.7 and 7.8, as well as the experimental results on the jet spreading rate $\frac{db}{dx}$ and on the ratio of the concentration half-width to the velocity half-width $\frac{b_C}{b}$, and after some mathematical manipulations, the solutions for a plane buoyant jet can be obtained as shown in Table 7.3. It is interesting to note that in Table 7.3, the jet centerline concentration or velocity equation are composed of two parts, corresponding to two limits (pure jet-like or pure plume-like conditions).

The characteristic length scale for a buoyant circular jet or plume is:

$$L_M = \frac{M_0^{\frac{3}{4}}}{B_0^{\frac{1}{2}}} \tag{7.13}$$

Similarly, for a buoyant circular jet, the integral equations of the momentum, continuity, pollutant mass (or buoyancy) conservation, and kinetic energy can be obtained. Using the Gaussian distribution for jet velocity or concentration and some experimental data, the analytical solutions for buoyant circular jets are shown in Table 7.3.

Table 7.3 Summary of mean properties of turbulent buoyant jet

Parameter	Plane buoyant jet	Circular buoyant jet
Characteristic length for jet/plume L_M	$L_M = \frac{M_0}{B_0^{\frac{2}{3}}}$	$L_M = \frac{M_0^{\frac{3}{4}}}{B_0^{\frac{1}{2}}}$
Maximum (centerline) concentration C_m or density defect $\Delta\rho_m$	$\frac{C_m}{C_0} = \frac{\Delta\rho_m}{\Delta\rho_0} = \frac{\alpha_{11}}{\left[\frac{\alpha_{12}}{F_0^2} + \frac{\alpha_{13}}{\left(\frac{x}{b_0}\right)^{\frac{3}{2}}}\right]^{\frac{1}{3}}} \frac{1}{\left(\frac{x}{b_0}\right)}$, where $F_0 = \frac{U_0}{\sqrt{g\frac{\Delta\rho}{\rho_a}b_0}}$, $\alpha_{11} = 12.75$, $\alpha_{12} = 21.19$, $\alpha_{13} = 50.0$ in [74]	$\frac{C_m}{C_0} = \frac{\Delta\rho_m}{\Delta\rho_0} = \frac{\alpha_{11}}{\left[\frac{\alpha_{12}}{F_0^2}\left(\frac{x}{d_0}\right)^5 + \alpha_{13}\left(\frac{x}{d_0}\right)^3\right]^{\frac{1}{3}}}$, where $\alpha_{11} = 100$, $\alpha_{12} = 1{,}920$, $\alpha_{13} = 6{,}720$ in [74]
Cross-sectional average concentration C_{avg}	$\frac{C_m}{C_{avg}} = \sqrt{1 + \frac{1}{k_2^2}} = 1.24$, where $k_2 = \frac{b_C}{b} = 1.35$ in [45]	$\frac{C_m}{C_{avg}} = 1.80$ in [74] $\frac{C_m}{C_{avg}} = 1 + \frac{1}{k_2^2} = 1.69$, where $k_2 = \frac{b_C}{b} = 1.2$ in [45]
Maximum (centerline) velocity u_m	$\frac{u_m}{U_0} = \left[\frac{\alpha_{31}}{F_0^2} + \frac{\alpha_{32}}{\left(\frac{x}{b_0}\right)^{\frac{3}{2}}}\right]^{\frac{1}{3}}$, where $\alpha_{31} = 21.2$, $\alpha_{32} = 50.7$ in [74]	$\frac{u_m}{U_0} = \left[\frac{\alpha_{31}}{F_0^2\left(\frac{x}{d_0}\right)} + \frac{\alpha_{32}}{\left(\frac{x}{d_0}\right)^3}\right]^{\frac{1}{3}}$, where $\alpha_{31} = 64.75$, $\alpha_{32} = 223.25$ in [74]
Velocity half-width b	$b = \alpha_4 x$, where $\alpha_4 = 0.097$ in [74]	$b = \alpha_4 x$, where $\alpha_4 = 0.097$ in [74]
Concentration half-width b_C	$\frac{b_C}{b} = \alpha_5$, where α_5 =1.18 in [74]	$\frac{b_C}{b} = \alpha_5$, where α_5 =1.16 in [74]
Entrainment Coefficient α_e	$\alpha_e = \alpha_{ej} + \frac{\alpha_{61}(\alpha_{ep} - \alpha_{ej})}{\alpha_{61} + \frac{\alpha_{62}F_0^2}{\left(\frac{x}{b_0}\right)^{\frac{3}{2}}}}$, where α_{ej} and α_{ep} are the entrainment coefficients for the plane jet and plume, respectively, $\alpha_{61} = 17.1$, $\alpha_{62} = 41.4$ in [74]	$\alpha_e = \alpha_{ej} + \frac{\alpha_{61}}{\alpha_{62} + \frac{\alpha_{63}F_0^3}{\left(\frac{x}{d_0}\right)^3}}$, where $\alpha_{61} = 0.44$, $\alpha_{62} = 20.74$, α_{63} = 2.38 in [74]

Source: [45, 74].

Effect of Boundaries on Jets and Plumes

In this section, the effect of different types of boundaries on jets and plumes will be considered, including the solid bed (wall), the free surface of ambient fluid, the coflowing or crossflowing ambient fluid, and the neighboring jets in the case of multiple jets.

Wall Jets

Wall jets are the jets discharged tangentially or at certain angles to a solid boundary (wall) (see Fig. 7.2). A simple case is first considered: a plane jet discharged tangentially to a smooth flat plate in deep still ambient fluid of the same kind. For turbulent plane wall jets with high Reynolds numbers ($R_0 = \frac{U_0 b_0}{\nu} = 10^4 \sim 10^5$, where ν is the kinematic viscosity of the fluid) at the slot exit, the length of the potential core will be $(6.1 \sim 6.7)b_0$ [73], which is in the same range as for

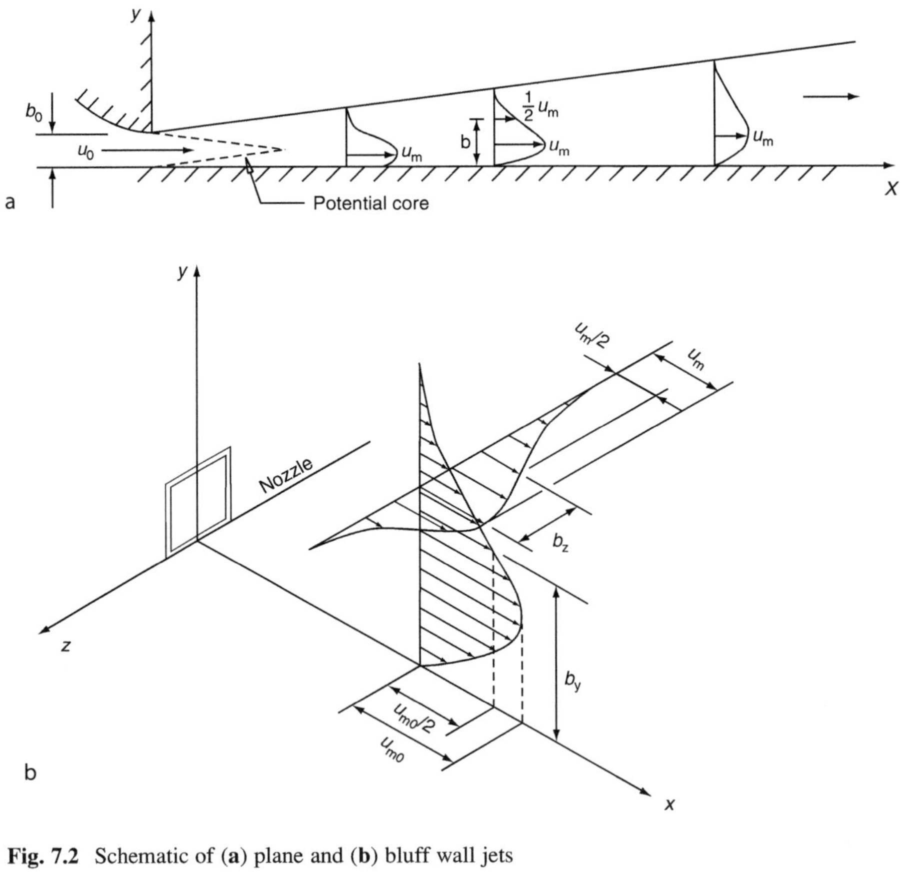

Fig. 7.2 Schematic of (a) plane and (b) bluff wall jets

Table 7.4 Summary of mean properties of plane wall jet and bluff wall jet

Parameter	Plane wall jet	Bluff wall jet
Maximum (centerline) concentration C_m		
Cross-sectional average concentration C_{avg}	$\frac{C_{avg}}{C_0} = \frac{\alpha_1}{\frac{x}{b_0}}$, where α_1 = 4.032 in [73]	
Maximum (centerline) velocity u_m	$\frac{u_m}{U_0} = \frac{\alpha_3}{\sqrt{\frac{x}{b_0}}}$, where α_3 = 3.50 in [78] & [73]	
Velocity half-width b	$b = \alpha_4 x$, where α_4 = 0.068 in [73]; $\frac{db}{dx}$ = 0.073 in [44]	$\frac{b_y}{d_0} = 0.90 + \alpha_{41}\frac{x}{h}$ $\frac{b_z}{B} = \alpha_{42}\frac{x}{B} - 1.25$ where α_{41} =0.045, α_{42} =0.20, B is the nozzle (horizontal) width in [77] & [73]; $\frac{db_y}{dx} = 0.048$ and $\frac{db_z}{dx} = 0.26$ in [44]
Concentration half-width b_C		
Entrainment Coefficient α_e	$\alpha_e = 0.035$ in [73]	

Source: [44, 73, 77, 78].

simple jets. As expected, experiments show that, near the wall, there exists a thin layer (boundary layer) where the jet velocity increases from zero at the wall to a maximum velocity u_m; above the boundary layer (named free mixing region), the et velocity decreases from u_m to zero at some large distance y from the wall. Similarly as for simple jets, the jet width may be defined as where the jet velocity is 50% (or 37%) of u_m and $\frac{\partial u}{\partial y} < 0$ (i.e., in the free mixing region). In the boundary layer region, the boundary layer theories may be used to further divide this region into two or three sub-layers: in the sub-layer very close to the wall, the velocity distribution is linear with y; some distance away from the wall, the velocity distribution can be described by the logarithmic law [87]. For the velocity distributions of the entire wall jet, after some distance (about $20b_0$) from the slot exit, they exhibit self-similarity [23, 99]. Verhoff [99] proposed an empirical equation which agreed well with the experimental data:

$$\frac{u}{u_m} = 1.48\left(\frac{y}{b}\right)^{\frac{1}{7}}\left[1 - \operatorname{erf}\left(0.68\frac{y}{b}\right)\right] \tag{7.14}$$

Using the equations of motion and the integral method, the following results could be obtained for plane wall jets: $u_m \propto x^{-\frac{1}{2}}$; $b \propto x$. The detailed results are listed in Table 7.4. To study the effect of wall roughness on wall jets, readers may refer to Rajaratnam [72], Tachie et al. [95], Dey et al. [19], and Rostamy et al. [83]. To study the jets impinging on walls, readers may refer to Beltaos and Rajaratnam [10], Rajaratnam [73], and Chan et al. [14].

For non-buoyant circular wall jets (herein termed "bluff wall jet" to include semicircular and rectangular wall jets with aspect ratio not very different from unity; the properties of bluff jets are not very different from those of circular jets), a number of studies have been conducted: square wall jets by Sforza and Herbst [88], circular wall jets by Newman et al. [61], bluff (including square, rectangular, circular, elliptic, and equilateral triangular) wall jets by Rajaratnam and Pani [77], square wall jets by Lübcke et al. [53], and circular wall jets by Agelin-Chaab [3]. These experiments show that after some distance from the potential core, the velocity distributions both in the vertical central plane and in the horizontal plane (see Fig. 7.2) are self-similar. From similarity analysis on the equations of motion or from dimensional analysis, the following results can be obtained for bluff wall jets: $u_m \propto x^{-1}$; $b_y \propto x$; $b_z \propto x$ [73]. Experimental results support these predictions, and the results are listed in Table 7.4.

Surface Jets

A surface jet can be produced by discharging a fluid at the surface of an ambient fluid (see Fig. 7.3). One typical example is the surface discharge of heated water from a power plant through either an open-channel or a pipe into an ocean, a lake, or river. Rivers flowing into lakes, reservoirs, and oceans and storm water discharges into rivers may be also viewed as surface jets. In this section, our attention will be limited in the region from the end of the jet potential core to the end of the near field (where the mixing is still dominated by the jet momentum and buoyancy). The length of the near field is in the order of $100\sqrt{A_0}$, where A_0 is the cross-sectional area of the flow at discharge [74]. For the mixing in the far field (where the turbulence in rivers, lakes, or oceans dominates further mixing), readers can refer to Fischer et al. [21] and Rutherford [85].

Non-buoyant plane surface jets in stagnant water are now considered. Equations 7.1–7.8 for plane submerged jets also work for plane surface jets. Using the integral method and some mathematical manipulations, some useful results can be obtained: $u_m \propto x^{-\frac{1}{2}}$; $b \propto x$; $c_m \propto x^{-\frac{1}{2}}$, which are in the same form as those for plane submerged jets (Table 7.1). Experiments on plane surface jets were conducted by Chu and Vanvari [16], Rajaratnam and Humphries [76], and others. The experimental results are listed in Table 7.5. The results confirm that essentially a plane surface jet is quite similar to half of the corresponding plane submerged jets, but with slightly different coefficients. For example, the jet spreading rate of plane surface jets $\frac{db}{dx} = 0.07$, smaller than the value of 0.10 for plane submerged jets.

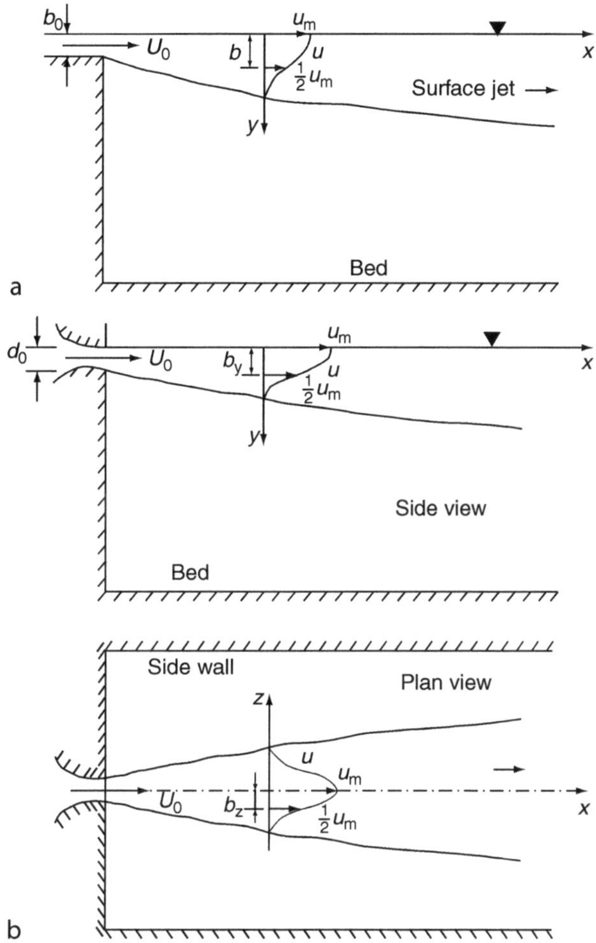

Fig. 7.3 Schematic of (**a**) plane and (**b**) bluff surface jets

Next, non-buoyant bluff surface jets in stagnant water are considered. From the experiments of Rajaratnam and Humphries [76], the Gaussian function describes the velocity distributions well both in vertical (half-Gaussian) and transverse directions, unless there is excess wave generation at the water surface (in this case u_m occurs some distance below the water surface). Using the integral method, it can be shown that: $u_m \propto x^{-1}$;$b_z \propto x$; $b_y \propto x$, which are in the same forms as for circular submerged jets. Rajaratnam and Humphries' experimental results show that the jet spreading rate in the transverse direction $\frac{db_z}{dx} = 0.09$, twice of that in the vertical direction $\frac{db_y}{dx} = 0.044$. The phenomenon of the several times faster transverse spreading has also been observed in the studies of Anthony and Willmarth [6], Gholamreza-kashi et al. [24], and Cuthbertson and Davies [18]. These studies further found that there exists a thin layer (called "surface current") at the free surface, which exhibits even faster transverse spreading compared to that below the

Table 7.5 Summary of mean properties of non-buoyant plane surface jet and bluff surface jet

Parameter	Non-buoyant plane surface jet	Non-buoyant bluff surface jet
Maximum (centerline) concentration C_m		
Cross-sectional average concentration C_{avg}		
Maximum (centerline) velocity u_m	$\frac{u_m}{U_0} = \frac{\alpha_3}{\sqrt{\frac{x}{b_0}}}$, where $\alpha_3 = 3.1$ in [76]	$\frac{u_m}{U_0} = \frac{\alpha_3}{x/d_0}$, where $\alpha_3 = 13$ in [24]
Velocity half-width b	$\frac{db}{dx} = \alpha_4$, where $\alpha_4 = 0.07$ in [74] & [76]	In transverse direction: $db_z/dx = \alpha_{41}$ In vertical direction: $db_y/dx = \alpha_{42}$ where $\alpha_{41} = 0.09$ in [74] & [76], 0.12 (below the free surface) and 0.22 (at the free surface) in [24]; and $\alpha_{42} = 0.044$ in [74] & [76], 0.025 in [24]
Concentration half-width b_C	$b_C/b = \alpha_5$, where $\alpha_5 = 1.15$ in [74]	
Entrainment coefficient α_e	$\alpha_e = 0.037$ in [74]	

Source: [24, 73, 74, 76].

layer. Comparing Table 7.5 with Table 7.4, one may find that surface jets are somehow similar to wall jets, e.g., they both spread faster in the horizontal direction than in the vertical direction due to the boundary constraint in the vertical direction, and they have similar forms of jet equations.

For the surface discharges such as heated water into rivers or wastewater into the oceans, the effect of buoyancy needs to be considered. Experimental results have indicated that the behavior of buoyant surface jets is mainly controlled by three parameters: the Richardson number at the outfall, $Ri_0 = gd_0\Delta\rho_0/(\rho_a U_0{}^2)$; the depth (thickness) of the surface jet, d_0 (or b_0 for a plane jet); and the depth of the surface stratified layer formed at the end of the near field of the surface jet, $\overline{b_\infty}$. For a buoyant surface jet with a fixed Ri_0, depending on the value of $\frac{d_0}{b_\infty}$, there could be four possible hydraulic phenomena: a surface jet, a surface (density) jump at the outfall, a surface jet followed by a surface jump, or a drowned jump. Rajaratnam and Subramanyan [79] presented a graph to distinguish which of the four possibilities may happen for a plane buoyant surface jet. For the case of a pure plane buoyant surface jet (without any jump), the experiments of Rajaratnam and Subramanyan [79] show that: initially the jet spreading rate $\frac{db}{dx}$ follows the equation of the plane non-buoyant surface jet, but after some longitudinal distance, the spreading rate slows down, and eventually the jet thickness approaches a constant. In other words, generally the buoyancy effect constrains the spreading of a plane surface jet. Their results also indicate that $\frac{u}{u_m}$ is self-similar at different x-sections; however, these self-similarities can no longer be described by the Gaussian distribution and seem to be related with Ri_0. The results of Chu and Vanvari [16] reveal that the entrainment coefficient α_e of a plane buoyant surface jet decreases continuously with the increase of bulk Ri (defined as $\frac{U_0 d_0 g \Delta\rho_0}{\rho_a {u_m}^3}$) with x; and α_e equals to zero when Ri increases to 0.2.

For bluff buoyant surface jets, a number of experiments have shown that the jet behavior is strongly affected by Ri_0 at the outfall. For convenience, bluff surface jets may be classified into two classes, the small Ri_0 class ($Ri_0 \leq 0.1$) and the large Ri_0 class ($Ri_0 > 0.1$). From a number of experiments, the common findings for the two classes are that the vertical velocity profile $u(y)$ in the center-plane and the transverse (across the jet) velocity profile $u(z)$ just below the water surface are self-similar, and the self-similarities can be well described by half-Gaussian or Gaussian distribution. Using the similarity analysis of the simplified equations of motion, the following relations can be obtained: for the small $Ri_\mathbf{0}$ class, $u_m \propto x^{-1}$, $C_m \propto x^{-1}$, $b_y \propto x$, $b_z \propto x$; and for the large Ri_0 class, $u_m \propto x^{-\frac{1}{3}}$, $C_m \propto x^{-\frac{2}{3}}$, $b_y =$ constant, $b_z \propto x$. The detailed results are listed in Table 7.6.

Jets and Plumes in Coflow

Similarly as solid bed or free surface of ambient fluid, coflowing or crossflowing ambient fluid itself can be viewed as some sort of boundary affecting jet behaviors. When jets are discharged in the direction of flowing ambient fluids, this is the problem of jets in coflow (Fig. 7.4). Extensive experimental and numerical studies show that beyond the potential core, the jet concentration and the jet excess velocity relative to the ambient velocity exhibit self-similarity. The self-similarity may be described by the Gaussian distribution, exponent function, or cosine expression [45, 73]. In the following, plane jets in uniform coflow will be briefly introduced, followed by circular jets.

For coflowing plane jets, the integral form of equation of motion is:

$$\frac{d}{dx}\int_{\mathbf{0}}^{\infty} \rho u \Delta u dy = 0 \tag{7.15}$$

where $\Delta u = u - U_a$, the jet excess velocity. Equation 7.15 states that the excess momentum flux is conserved in x direction. Using Eq. 7.15 and similarity analysis, Rajaratnam [73] obtained the following asymptotic relations: for the strong jet region (i.e., $\frac{\Delta u_m}{U_a} \gg 1$), $\Delta u_m \propto x^{-\frac{1}{2}}$ and $b \propto x$; and for the weak jet region (i.e., $\frac{\Delta u_m}{U_a} \ll 1$), $\Delta u_m \propto x^{-\frac{1}{2}}$ and $b \propto x^{\frac{1}{2}}$. Based on the experimental results in the literature, Rajaratnam [73] derived that:

Table 7.6 Summary of Mean properties of buoyant plane surface jet and bluff surface jet

Parameter	Buoyant plane surface jet	Buoyant bluff surface jet with $Ri_0 \leq 0.1$	Buoyant bluff surface jet with $Ri_0 > 0.1$
Maximum (centerline) concentration C_m			$\frac{C_m}{C_0} = \frac{\alpha_1}{\left(\frac{x}{\sqrt{A_0}}\right)^{\frac{2}{3}}}$, where $\alpha_1 = 2.83$ in [74]
Cross-sectional average concentration C_{avg}			
Maximum (centerline) velocity u_m	$\frac{u}{u_m}$ exhibits self-similarity, but cannot be well described by Gaussian profile, as $\frac{u}{u_m} \approx 0$ at $\frac{y}{b} \approx 1.4$ [79]	$\frac{u_m}{U_0} = \frac{\alpha_3 Ri_0^{\frac{1}{10}}}{\sqrt{\frac{x}{A_0}}}$, where $\alpha_3 = 15.3$ in [74] & [75]	$\frac{u_m}{U_0} = \frac{\alpha_3}{\left(\frac{x}{\sqrt{A_0}} - 5.0\right)^{1/3}}$, where $\alpha_3 = 1.25$ in [74]
Velocity half-width b	First $\frac{b}{b_0}$ increases linearly as non-buoyant surface jet to some point $\frac{x}{b_0}$ (the location depends on Ri_0); then the increase rate $\frac{db}{dx}$ decreases nonlinearly, finally approaches asymptotically a horizontal line [79].	$\frac{db_y}{dx} = \alpha_{41}$ $\frac{db_z}{dx} = \alpha_{42}$ where α_{41} decreases with Ri_0 ($\alpha_{41} = 0.044$ for $Ri_0 = 0$, 0.02 for $Ri_0 = 0.038$, 0 for $Ri_0 = 0.09$); α_{42} increases with Ri_0 in [74] & [75]	$\frac{b_{y*}}{d_0} = \frac{\alpha_{41}}{Ri_0^{1/8}}$ $\frac{b_z}{\sqrt{A_0}} = \alpha_{42}\left(\frac{x}{\sqrt{A_0}} + 2.0\right)$ where b_{y*} is the average of b_y which changes slightly with y; $\alpha_{41} = 0.29$ in [63], 0.26 in [74]; $\alpha_{42} = 0.54$ in [74]
Concentration half-width b_C		$\frac{b_{yC}}{b_y} = \alpha_{51}$ $\frac{b_{zC}}{b_z} = \alpha_{52}$ where $\alpha_{51} = 1.0$ in [74] & [75] and $\alpha_{52} = 1.15$ in [74]	$\frac{b_{yC}}{b_y} = \alpha_{51}$ $\frac{b_{zC}}{b_z} = \alpha_{52}$ where $\alpha_{51} = 1.12$ in [63] & [74]; and $\alpha_{52} = 1.9$ in [63], 1.6 in [74]
Entrainment coefficient α_e	α_e decreases from about 0.04 to 0 when Ri increases from 0 (non-buoyant plane surface jet) to 0.2 [16]		

Source: [16, 63, 74, 75, 79].

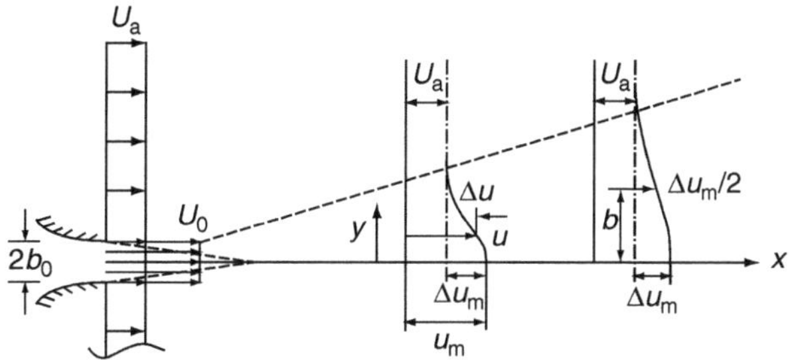

Fig. 7.4 Schematic of plane jets in coflow (for circular jets, replace $2b_0$ by d_0 and y by r)

$$\frac{\Delta u_m}{\sqrt{U_0(U_0 - U_a)}} = \frac{3.41}{\sqrt{\frac{x}{b_0}}} \tag{7.16}$$

$$\frac{b}{b_0} = 0.118\frac{x}{b_0}\frac{1}{1 + \frac{0.41}{\sqrt{\alpha(\alpha-1)}}\sqrt{\frac{x}{b_0}}} \tag{7.17}$$

where $\alpha = \frac{U_o}{U_a}$, the ratio of jet exit velocity to ambient velocity. Note that Rajaratnam [73] also summarized other more complex forms of equations for Δu_m and b, which were derived by Patel [68] and Pande and Rajaratnam [62].

For circular jets in coflow, the integral momentum equation can be derived:

$$\frac{d}{dx}\int_0^\infty \rho 2\pi r u \Delta u dr = 0 \tag{7.18}$$

which says that the jet excess momentum is conserved in axial direction. Using Eq. 7.18 and similarity analysis on the equations of motion, Rajaratnam [73] presented the following asymptotic relations: for the strong jet region (i.e., $\frac{\Delta u_m}{U_a} \gg 1$), $\Delta u_m \propto x^{-1}$ and $b \propto x$; for the weak jet region (i.e., $\frac{\Delta u_m}{U_a} \ll 1$), $\Delta u_m \propto x^{-\frac{2}{3}}$ and $b \propto x^{\frac{1}{3}}$. Pande and Rajaratnam [62] proposed a complex expression for Δu_m. Lee and Chu [45] also derived asymptotic solutions for circular jets in coflow: for the strong jet region, the jet solution is assumed to be the same as in stagnant water (see Table 7.1); for the weak jet region,

$$\frac{\Delta u_m}{U_a} = 2.14\left(\frac{x}{{l_m}^*}\right)^{-\frac{2}{3}} \tag{7.19}$$

$$\frac{b}{{l_m}^*} = 0.385\left(\frac{x}{{l_m}^*}\right)^{\frac{1}{3}} \tag{7.20}$$

where l_m* is the excess momentum length scale defined as $\frac{M_{e0}^{1/2}}{U_a}$; $M_{e0} = (U_0 - U_a)\frac{U_0 \pi d_0^2}{4}$, the jet specific excess momentum at discharge.

To completely model circular jets in coflow, Lee and Chu [45] formulated an integral model based on a Lagrangian jet spreading hypothesis:

$$U^{*2} + U^* - \frac{1}{\pi B^{*2}} = 0 \tag{7.21}$$

$$\frac{dB^*}{dx} = \beta_s \frac{U^*}{1 + U^*} \tag{7.22}$$

where $U^* = \frac{\Delta U}{U_a}$; $B^* = \frac{B}{l_m^*}$; ΔU and B are, respectively, the excess velocity and half of the width of the top-hat profile (instead of the Gaussian profile) of an equivalent jet, which carries the same mass flow and excess momentum flux as the actual jet; and $\beta_s = \frac{dB}{dx}$ in stagnant water. It can be proved that $\Delta U = \frac{\Delta u_m}{2}$ and $B = \sqrt{2b}$, where Δu_m and b are, respectively, the maximum excess velocity and 37% half-width for the Gaussian profile. From Eqs. 7.21 and 7.22, ΔU and B can be solved. The actual jet centerline dilution can be obtained:

$$S_c = \frac{C_0}{C_m} = \frac{\lambda^2 \pi B^2 \left(U_a + \frac{2}{1+\lambda^2} \Delta U\right)}{2Q_0^2} \tag{7.23}$$

where Q_0 and C_0 are the initial jet discharge and concentration, respectively; λ is the ratio of concentration half-width to velocity half-width using the Gaussian profile ($\lambda \approx 1.2$). The modeling results of Lee and Chu [45] reveal that the centerline dilution of a circular jet in coflow is only slightly smaller than that in stagnant water; and the centerline excess velocity decays in a similar way as in stagnant water.

Jets and Plumes in Crossflow

Now consider a non-buoyant circular jet discharged at an oblique angle (not 0 or 180 degree) to a flowing ambient fluid. In fact, most outfalls or diffusers in oceans or rivers discharge effluents as jets in crossflow, as the jet (effluent) dilution can be considerably enhanced even in a weak crossflow [45, 73]. Jets in crossflow have been studied extensively by Abramovich [1], Rajaratnam [73], Fischer et al. [21], Wright [102, 103], Andreopoulos [5], Hodgson and Rajaratnam [29], Margason [54], Smith and Mungal [91], Lee and Chu [45], Huang et al. [31], Kikkert et al. [38], and others. According to these studies, the evolution of jets in crossflow can be divided into three regions: the potential core region, the maximum deflection region, and the vortex region (see Fig. 7.5). The length of the potential core has

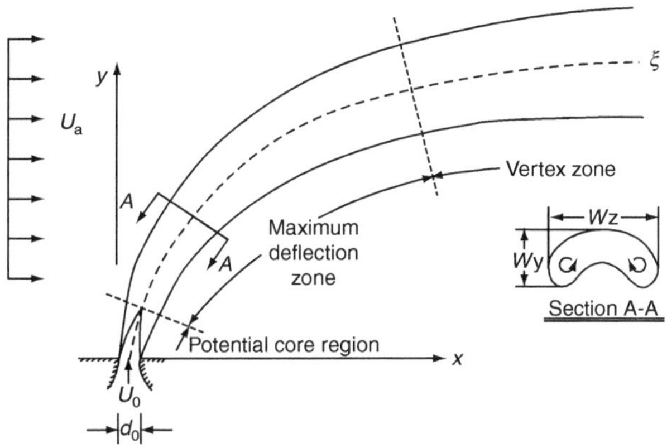

Fig. 7.5 Schematic of jets in crossflow

been found to be mostly controlled by the relative strength of the jet compared to the crossflow ($\alpha = \frac{U_0}{U_a}$), and typically in the range of 2–6d_0 which is smaller than that of a free jet [70].

Beyond the potential core region, the jet would be largely deflected due to the stagnation pressure exerted by the free stream and the entrainment of ambient fluid (and thus horizontal momentum). Jet deflection probably is the most distinct feature in crossflow. After the maximum deflection region, the jet would be gradually parallel to the direction of ambient flow. Laboratory experiments [1, 29, 45] and numerical simulations [45] have found that after some distance beyond the potential core, the jet cross section would be like a kidney shape with a pair of two counter-rotating vortices (see Fig. 7.5). The vortex pair significantly entrains ambient fluid in the form of tornado vortices into the jet [45], which explains the considerable enhancement of jet dilution in crossflow. The concentrations at the centers of the two vortices have been found to be about 1.1–1.6 times of the jet centerline concentration [29, 45].

For non-buoyant jets in crossflow, it is common to analyze them in three regions: the momentum dominated near field (MDNF), the momentum dominated far field (MDFF), and the transition between MDNF and MDFF. If the jet trajectory $\xi \ll l_m$, where l_m is defined as:

$$l_m = \frac{M_{v0}^{\frac{1}{2}}}{U_a} \tag{7.24}$$

and M_{v0} is the vertical momentum at the exit, then the jet is in MDNF, where the effect of jet momentum is much stronger than that of the ambient crossflow. In MDNF, the classic equations for jets in stagnant ambient fluid are approximately valid. If $\xi \gg l_m$, the jet is in MDFF, where the effect of ambient crossflow is dominant over the jet momentum. In MDFF, the jet properties can be studied with physical and numerical models. Fischer et al. [21] used dimensional analysis to find the asymptotic formulas for MDNF and MDFF. Lee and Chu [45] studied the jets in

MDFF based on the analogy to advected line puffs. Using length-scale analysis and numerical models, they proposed the formulas which can represent satisfactorily the experimental results:

$$\frac{y_c}{l_m} = 1.56\left(\frac{x}{l_m}\right)^{\frac{1}{3}} \tag{7.25}$$

$$b_{vc} = 0.28y_c \tag{7.26}$$

$$S_c = 0.46\frac{U_a {y_c}^2}{Q_0} \tag{7.27}$$

where y_c is the vertical location of the centerline concentration, b_{vc} is the vertical centerline half-width defined by 37% of the centerline concentration and S_c is the centerline dilution.

The most common jet discharge angle in crossflow is 90 degree, i.e., jets are discharged at right angle to the ambient crossflow. Rajaratnam [73] summarized the early studies in 1950s to 1970s that mostly focused on jet trajectories. Hodgson and Rajaratnam [29] conducted detailed laboratory experiments on circular jets at right angle to crossflow and proposed the following equations:

$$S_c = 1.09\left(\frac{\alpha x}{d_0}\right)^{0.56} \tag{7.28}$$

$$\frac{y_c}{\alpha d_0} = 1.46\left(\frac{x}{\alpha d_0}\right)^{0.26} \tag{7.29}$$

$$\frac{W_z}{\alpha d_0} = 1.20\left(\frac{x}{\alpha d_0}\right)^{0.29} \tag{7.30}$$

$$\frac{W_y}{\alpha d_0} = 0.78\left(\frac{x}{\alpha d_0}\right)^{0.37} \tag{7.31}$$

where W_z and W_y are the jet width and thickness (see Fig. 7.5). Equations 7.28–7.31 have also been validated by a field experiment in the Lesser Slave River, Canada. Hodgson and Rajaratnam's equations are mainly derived based on the experiments conducted in the range of $\frac{x}{\alpha d_0} = 1 \sim 1{,}000$. It is interesting to note that Hodgson and Rajaratnam's expressions fit the experimental data satisfactorily both in MDNF, MDFF, and the transition between the two.

Now consider a circular plume in crossflow. Similarly as l_m, a length scale l_b needs to be defined to compare the relative strength of plume buoyancy with crossflow:

$$l_b = \frac{B_0}{U_a^3} \tag{7.32}$$

If the jet trajectory $\xi \ll l_b$, then the plume is in the buoyancy dominated near field (BDNF) where the effect of buoyancy is dominant over crossflow; if $\xi \gg l_b$, then plume is in the buoyancy dominated far field (BDFF) where the effect of crossflow is more pronounced than the buoyancy. In BDNF, the plume is essentially vertical and only slightly advected, thus the equations for plumes in stagnant fluid are approximately valid. Similarly as jets in crossflow, after some distance from the nozzle, the plume cross section will become a kidney shape that is made up of a vortex pair, and the concentration at the centers of the vortices have been found to be 1.4–1.7 times of the plume centerline concentration. In BDFF, the plume bends over and finally approaches the ambient flow direction. The analysis on the plume properties in BDFF relies on experiments or numerical models. Based on the equations of motion and the use of similarity solutions, Fischer et al. [21] derived asymptotic formulas for the BDNF and BDFF. As the plume in the BDFF behaves similarly as the advected line thermal, Lee and Chu [45] used numerical models to obtain the plume characteristics. The predictions are comparable to experimental results. The formulas Lee and Chu derived are:

$$\frac{y_c}{l_b} = 1.3\left(\frac{x}{l_b}\right)^{\frac{2}{3}} \tag{7.33}$$

$$b_{vc} = 0.4y_c \tag{7.34}$$

$$S_c = 0.46\frac{U_a {y_c}^2}{Q_0} \tag{7.35}$$

Note Eq. 7.35 for BDFF is exactly in the same form as the equation for MDFF.

Now consider the case of a circular buoyant jet in crossflow. In this case, the relative strengths of buoyancy, momentum, and crossflow need to be considered. If $l_b \ll l_m$, then the jet would in sequence experience in MDNF, MDFF, and BDFF; if $l_b \gg l_m$, then the sequence would be MDNF, BDNF, and BDFF [21, 43, 45]. Equations should be selected carefully according to the studied location of the jet (e.g., in MDNF or MDFF or BDNF or BDFF). The transitions between MDNF and MDFF and between BDNF and BDFF are better treated using numerical models [45].

For jets directed at an oblique angle to crossflow, the reader may refer to Platten and Keffer [69] and Kikkert et al. [38]. For plane jets and plume in crossflow, the reader may refer to Girshovich [25], Jones and Wille [36], Kalita et al. [37], and Huang et al. [31].

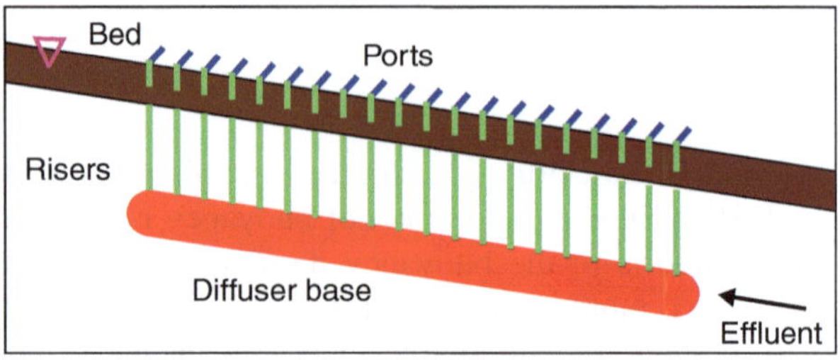

Fig. 7.6 An example of a unidirectional diffuser

Multiple Jets

Effluents may be discharged via single port outfalls or multiport diffusers. Multiport diffusers are commonly used given their fast mixing and diluting ability and thus less adverse impacts on the environment. The jets issuing from the ports of a multiport diffuser are usually viewed as multiple jets. The characteristics of multiple jets are primarily determined by the arrangements of multiport diffusers. Generally, multiport diffusers can be classified into three categories: unidirectional diffuser (where net horizontal momentum flux is imparted perpendicular to diffuser line), staged diffuser (where net horizontal momentum flux is imparted parallel to the diffuser line), and alternating diffuser (where no net horizontal momentum flux is imparted) [20]. Figure 7.6 illustrates one typical example of a unidirectional diffuser. It is expected that different types of diffusers have significantly different jet mixing and spreading properties.

Studies on multiple jets (or diffusers) have been reported extensively in the past decades ([2, 21, 32, 34, 35, 39, 40, 45, 46, 94, 96, 100, 107], etc.). Most of these studies focused on the deep water ambient condition (e.g., in oceans and lakes), and limited studies dealt with the shallow water condition (e.g., in rivers). In this section, only the deep water condition will be considered. Theoretically, multiple jets in sequence experience: the individual free jet zone (where jets has no effect on each other), the jet merging zone (where the interaction between jets are strong), and the two-dimensional zone thereafter [66, 107] (see Fig. 7.7). In practice, multiple jets are usually simplified as one line momentum source, neglecting the interactions between individual jets that are complex and not well understood [35, 46].

In the previous sections, some basic characteristics of three-dimensional free jets and two-dimentional (plane) jets have been introduced; hence, in the following, the studies on the merging process of unidirectional non-buoyant circular jets will be briefly summarized. As yet, only limited studies on jet merging have been reported [30, 40, 64, 71, 100]. To calculate the concentration or velocity field in the jet merging process, the most widely used method is superposition. However, as the momentum equation is nonlinear, simple superposition of individual jets would overestimate the jet velocity. Knystautas [40] studied the velocity field of merging jets in still ambient fluid and showed that the jet velocity can be modeled by superposing the momentum (u^2) of individual jets (based on Reichardt's hypothesis).

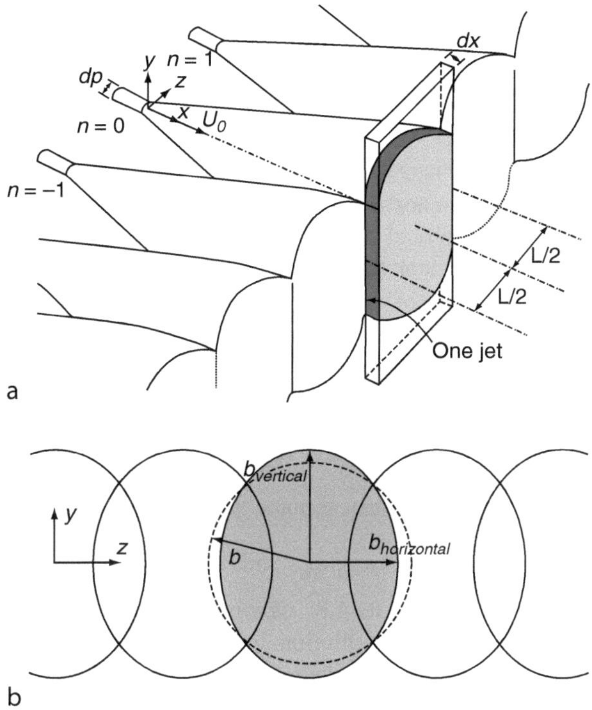

Fig. 7.7 (**a**) Schematic of multiple merging jets, with the indication of (**b**) jet cross-section deformation (Modified from [100])

Hodgson et al. [30] reported the following equation for the velocity field of the merging jets in still water:

$$u^2 = \sum_{i=-n}^{i=+n} {u_i}^2 = {u_m}^2 \sum_{i=-n}^{i=+n} \exp\left[-150\left(\frac{z-iL}{x}\right)^2\right] \tag{7.36}$$

where $(2n + 1)$ is the number of jets; the centerline velocity of each individual jet $u_m = \frac{6.13U_0d_0}{x}$; L is the distance between the centers of neighboring jets; z is the transverse distance from the central jet axis. Hodgson et al. [30] extended the Reichardt's hypothesis on lateral transport of momentum to the lateral transport of pollutant, and showed that uC is additive. After solving the velocity field from Eq. 7.36, the pollutant concentration field can be obtained from Eq. 7.37:

$$uC = \sum_{i=-n}^{i=+n} u_iC_i = u_mC_m \sum_{i=-n}^{i=+n} \exp\left[-130\left(\frac{z-iL}{x}\right)^2\right] \tag{7.37}$$

where the centerline concentration of each individual jet $C_m = \frac{5.34C_0d_0}{x}$. Hodgson's experimental results validated the use of Eqs. 7.36 and 7.37.

Hodgson et al. [30] revealed some basic physics in the jet merging, where the jet spreading rate $\frac{db}{dx}$ and the ratio of the jet concentration half-width to the velocity half-width $\frac{b_C}{b}$ are both assumed to be constant during merging. Wang and Davidson [100] developed a similar model for jet merging in stagnant ambient fluid, but allows the change of $\frac{db}{dx}$ and $\frac{b_C}{b}$ during merging, as well as the change of these parameters in horizontal (the jet merging) plane and vertical (the free entrainment) plane. Theoretical analysis and experimental data indicate that the jet merging in stagnant ambient fluid occurs at $4.5 < \frac{x}{L} < 12$. Note that in Wang and Davidson [100], the start of merging refers to the location where the jet interaction begins to influence the bulk properties of the central jet, which is beyond the location where the physical jet boundaries start intersecting. By studying their experimental data and that of Knystautas [40], Wang and Davidson [100] found that, during the jet merging, the jet spreading rate $\frac{db}{dx}$ (or $\frac{db_C}{dx}$) increases by 30% in the vertical plane, while it decreases by a similar amount in the horizontal plane; the ratio of velocity half-width in the vertical plane to that in the horizontal plane $\frac{b_{\mathrm{Vertical}}}{b_{\mathrm{horizontal}}}$ increases from 1 to 1.5 during merging; and the ratio of concentration half-widths $\frac{b_{C,\mathrm{Vertical}}}{b_{C,\mathrm{horizontal}}}$ increases from 1 to 1.8. Obviously, the jet merging process constrains the jet spreading and thus dilution in the jet merging plane, while accelerates them in the free entrainment plane. This phenomenon is similar to the boundary effects found in wall jets or surface jets.

In recent years, researchers started to study the jet merging in coflow [64, 71]. Pun et al. [71] developed a multiple-point hybrid model for merging jets in coflow, which combines a length-scale model and an Eulerian-integral model. The model allows multiple transition points for each parameter (jet velocity, spread and dilution), instead of a single transition point for all these parameters. The multiple-point hybrid model is shown to be able to significantly reduce transition errors during merging compared to the single-point model, and predicts favorable results compared to the integral solution. Pani et al. [64] developed a model based on Reichardt's hypothesis for multiple coflowing jets. Instead of momentum (u^2) is additive in stagnant water, Pani et al. showed that the excess momentum ($u\Delta u$) is additive in coflow and follows Gaussian distribution. Using the method of superposition and a generalized spreading hypothesis, Pani et al. presented the equations for predicting the velocity field and centerline dilution downstream of multiple circular jets in coflow, which appeared to agree with the experimental data.

Multiphase Jets and Plumes

In this section, two kind of multiphase jets and plumes will be introduced: bubbly jets and plumes and slurry jets and plumes, which have wide engineering applications.

Bubbly Jets and Plumes

Bubbly jets are produced by injecting gas-liquid mixtures into liquids, while bubble plumes are produced by injecting gases into liquids. Bubble plumes and bubbly jets are widely used to achieve artificial aeration, circulation and mixing in confined reactors, aeration tanks, polluted water bodies, ice-covered rivers, and deep stratified lakes and reservoirs [48, 86, 101, 104]. Such kind of gas-liquid two-phase flow is also common in some hydraulic structures, e.g., the super-gas saturation downstream of hydro-power dams. So far, most of the early studies were conducted in confined setups, where the sizes and geometry of the setup further complicates the characteristics of bubbly jets and plumes [51]. In this section, the studies in stagnant water of relatively large setup will be introduced and then the case with flowing ambient fluid will be considered.

For two-phase flows, the dissolving of the gas phase into the liquid phase can be derived from Fick's law of diffusion as [60]:

$$\frac{dC}{dt} = K_L a(C_s - C) \tag{7.38}$$

where C is the dissolved gas (e.g., oxygen) concentration in the liquid, C_S is the saturation dissolved gas concentration, t is the time, K_L is the mass transfer coefficient, and a is the gas-liquid interfacial area per unit liquid volume (also named the specific interfacial area). From Eq. 7.38, the gas transfer rate is mainly controlled by K_L and a, which differ significantly in different setups. Previous studies [9, 59] have shown that these two parameters are greatly influenced by bubble size. Bubble size depends on a number of factors: nozzle sizes and types, initial gas volume fractions, the solubility and mass transfer ability of the gas, turbulence intensity and flow structure of the ambient liquid, impurities and surfactants in the ambient liquid, etc. [17, 49–52]. Lima Neto et al. [50] proposed a criterion to judge the sizes and shapes of the bubbles produced by injecting a mixture of air and water into water: if the nozzle Reynolds number $\mathrm{Re} = \frac{U_{w0} d_0}{\nu_w} < 8000$ (where U_{w0} is the superficial water velocity based on the water discharge at the nozzle and nozzle diameter d_0), then large and irregular bubbles will be produced; if $Re \geq 8{,}000$, smaller and uniform bubbles will be produced. A decrease in gas discharge or an increase in liquid discharge will decrease the bubble size [50, 52, 98].

Now consider the vertical injection of a pure gas into a pure stagnant liquid. The bubbles produced at the orifice will coalescence/breakup and rise, inducing ambient liquid entrained into the bubble core and the dissolving of the gas into the ambient. Lima Neto et al. [49] studied air injection into still water with six different nozzles (single orifice, multiple orifices, and airstone), and found that the water entrained into the bubble core under different initial air discharges Q_a and nozzle types can be described as a function of Q_a and vertical distance from the nozzles.

For a vertical bubbly jet in stagnant liquid, Milgram [56], Brevik and Kristainsen [12], and Lima Neto et al. [50] reported that the bubble area typically only occupies 50–90% of the bubbly jet in the radial direction. Lima Neto et al. [50] studied bubbly jets produced by injecting a mixture of air-water into stagnant water and found that the more uniform and smaller the bubble sizes, the wider the bubble core can spread in the radial direction; within the bubbly jets, the radial distributions of the time-averaged bubble concentration (void fraction) and water velocity of the mean flow can be well described by Gaussian distributions, similarly as for single-phase jets or plumes; and $\mathrm{d}b/\mathrm{d}x$ of the bubbly jets is close to that of the pure water jet.

Although the existence of bubbles seems not to change the Gaussian profiles, the entrainment of the ambient into the bubbly jets is significantly enhanced. Milgram [56], Socolofsky and Adams [92], Brevik and Kristainsen [12], and Lima Neto et al. [50] reported the entrainment coefficient of bubbly jets is in the range of 0.03–0.15, much larger than the values of pure jets or plumes. The additional entrainment probably associates with the bubble wakes [47] and additional liquid turbulence caused by interactions of the bubbles and their wakes [50]. At a specific height of the centerline of a bubbly jet, Lima Neto et al. [50] compared the liquid volume flux Q_w of the bubbly jet with that of a pure water jet with the same nozzle diameter and water flow rate:

$$\frac{Q_w}{Q_{w(C_0=0)}} = 1 + 6.426 \times 10^6 \left(\frac{C_0}{\mathrm{Re}}\right)^{\frac{3}{2}} \tag{7.39}$$

where the initial gas volume fraction $C_0 = \frac{Q_{a0}}{(Q_{a0}+Q_{w0})}$; Q_{a0} and Q_{w0} are the initial volumetric flow rates of air and water, respectively. It is expected that the average dilution of a bubbly jet is larger than that of a pure jet due to the additional entrainment of ambient water as shown in the second term of the the right hand side of Eq. 7.39.

For a bubbly jet injected horizontally into a stagnant liquid, there are only limited experimental studies [22, 52, 57, 98]. Lima Neto et al. [52] studied the injection of air-water mixtures into a water tank of relatively large size and reported that: first, the bubbly jet comes out of the nozzle as a whole quasi-horizontal bubbly jet where bubble breakup/coalescence occurs and only a few bubbles escape from the bubbly jet and rise vertically due to buoyancy; then, there follows a separation zone where the quasi-vertical bubble plume partially separates from the water jet (when the initial gas volume fraction $C_0 < 0.15$, the bubble plume will completely separate from the water jet); finally, the bubble plume continues rising and the water jet impinges the water surface and becomes surface jet. In Lima Neto et al. [52], the length and width of the bubble plume as well as the centerline trajectories of the bubble plume and water jet were also proposed in dimensionless forms.

For bubbly jets, bubble properties and mean liquid flow could be non-dimensionalized as functions of the initial gas volume fraction and nozzle Reynolds number for the vertical injection or as functions of the initial gas volume fraction and nozzle densimetric Froude number for the horizontal injection [50, 52]. The variation of bubble properties and mean liquid flow along the jet centerline and across the jet needs further studies. For bubble plume modeling, the reader may refer to Bravo et al. [11] and others.

Crossflow will exert significant effects on bubbly jets or bubble plumes, of which the most distinguishing one is the possibility of separation of bubble plumes (named generally as dispersed phases) from the entrained ambient fluids (continuous phases) [92]. So far, very limited studies have been reported on bubbly jets or bubble plumes in crossflow. Socolofsky and Adams [92] conducted laboratory experiments on bubble plumes produced by injecting air, air and oil, as well as air and alcohol in uniform crossflow. In weak crossflow, the separation between the bubble plumes and the entrained fluid does not occur before the plumes reach the surface. While in strong crossflow, the separation is significant and the separation height h_s can be given by an empirical relation:

$$h_s = \frac{5.1B_0}{\left(U_a {u_s}^{2.4}\right)^{0.88}} \tag{7.40}$$

where U_a is the horizontal crossflow velocity; u_s is the bubble slip (terminal rise) velocity; $B_0 = Q_0 g'$; Q_0 is the discharge of the plume fluid; $g' = \frac{(\rho_a - \rho_p)g}{\rho_a}$; ρ_a is the ambient density of water; and ρ_p is the density of the plume fluid. Socolofsky and Adams also reported that before h_s, the plumes can be treated as single-phase plumes; after h_s, the bubble plumes follow the trajectory of the vector sum of u_s and U_a (i.e., the bubble plumes rise in a linear line), and the separated entrained fluid behaves like a momentum jet (the momentum is gained by the acceleration of bubbles before the separation).

Slurry Jets and Plumes

Slurry jets are produced by injecting a mixture of liquid phase and solid phase (such as sand or clay particles). Slurry jets have wide applications in pumping industrial (e.g., mining or petroleum) tailings into settling tanks, dredging and land reclamation, discharging storm water and industrial waters that have solid particles, etc. A number of experimental and numerical studies in this area have been reported [7, 13, 26, 33, 55, 67, 90]. Usually, two injection ways were used, vertically upward and vertically downward. In the following, the main focus will be on the vertically downward injection of slurry jets into stagnant ambient fluids. Compared to single-phase jets, the adding of the solid phase will change the properties of the flow [89].

Previous experiments have indicated that the velocity and concentration of the solid phase across slurry jets follow self-similar Gaussian distributions [26, 33, 90]. In the slurry jets with dilute solid particles, Jiang et al. [33] reported that the velocity and concentration of the liquid phase also exhibit self-similar Gaussian profiles. However, this may not be valid for the slurry jets with high concentration solid particles.

The spreading of the solid phase has been found to increase linearly along the axial direction [13, 26, 55]. Brush [13] reported that the spreading rate of the velocity of the solid phase $\frac{db_s}{dx}$ depended on the particle size. Mazurek et al. [55] confirmed this by photographic measurements on sand jets, and further generalized the spreading as a function of the initial densimetric Froude number of the solid particle. Recently, Hall et al. [26] conducted detailed experiments on pure sand jets and sand-water slurry jets in stagnant water. With the densimetric Froude number at the nozzle exit ($\boldsymbol{F_0} = \frac{U_0}{\sqrt{\frac{gd_0(\rho_s-\rho_w)}{\rho_w}}}$ where $\boldsymbol{U_0}$ is the initial velocity of the sand particle from a nozzle of diameter $\boldsymbol{d_0}$ and ρ_s is sand density) in the range of $2 \sim 6$, $\frac{db_s}{dx}$ measured $0.087 \sim 0.109$, not very different from that of single-phase jets in stagnant fluids. The difference lies in the ratio of concentration to velocity spreading rates of the solid phase ($\frac{b_{s,C}}{b_s}$). Hall et al. [26] reported that, for sand jets and slurry jets with high $\boldsymbol{F_0}$ ($\boldsymbol{F_0} \approx 6$), $\frac{b_{s,C}}{b_s} = 0.86 \sim 0.92$, which means the sand concentration spreads slower than the velocity; and for slurry jets with low $\boldsymbol{F_0}$ ($\boldsymbol{F_0} \approx 2$), $\frac{b_{s,C}}{b_s} \approx 1.0$ which means they have almost equal spreading rates. This finding is contrary to the classic single-phase jet theory, which states that the concentration scale spreads faster than the velocity scale ($\frac{b_C}{b} \approx 1.2$).

Similar as single-phase jets, along the axial direction of slurry jets, the velocity and concentration of both solid phase and liquid phase decay rapidly. According to Hall et al. [26], beyond the potential core (about $2.9\, d_0\, F_0^{\frac{2}{5}}$), the axial concentration of the solid phase can be well described by:

$$\frac{C_m}{C_0} = \frac{17.12}{\left(\frac{x}{d_0 F_0^{\frac{2}{5}}}\right)^{\frac{5}{3}} + 11.39} \tag{7.41}$$

In Eq. 7.41, the −5/3 power relation is very similar to that of single-phase plume (Table 7.2), as Eq. 7.41 was established in the region at $x > 5L_M$ where the buoyant slurry jet behaved like a slurry plume [65]. Similar −5/3 power relation was also built for the sand concentration in sand jets. For both sand jets and slurry jets, the axial velocity of the solid phase was found to decrease rapidly and then reach a final plateau region. Generally, before the plateau region, the axial velocity of the solid phase in slurry jets can be well represented by:

$$\frac{u_m F_0}{U_0} = \frac{1.63}{\left(\frac{x}{d_0 F_0}\right)^{\frac{1}{3}} + 0.56} \tag{7.42}$$

Similar as Eq. 7.42, for sand jets, before reaching the velocity plateau region, the axial sand velocity was also found to follow the $-1/3$ power relation, which is very similar to that of single-phase plumes (Table 7.2). The terminal (settling) centerline velocity of the solid phase was found in the range of 0.32 $\sim$ 0.43 m/s, which is larger than the settling velocity of 0.033 m/s for individual solid particles. The larger terminal velocity probably can be attributed to the interactions between solid particles, i.e., the wake of previous solid particles tends to decrease the drag forces for the following particles.

For the studies dealing with particle interactions, the reader may refer to Lain and Garcia [42], Tamburello and Amitay [93], and Yan et al. [106]. For the effect of solid particle size on velocity distribution, concentration profile and turbulent properties, the reader may refer to Azimi et al. [8]. For vertically upward slurry jets, the reader may refer to Jiang et al. [33].

Future Directions

Some of the basic characteristics of a variety of jets and plumes have been reviewed, including simple jets and plumes, buoyant jets, surface jets, wall jets, jets and plumes in coflow and crossflow, multiple jets, bubbly jets and plumes, and slurry jets and plumes. The turbulence and turbulence structures in these flows have not been discussed. Interested readers may refer to the works of Heskestad [27], Wygnanski and Fielder [105], and Launder and Rodi [44]. Also, the behaviors of jets and plumes in stratified environment have not been considered. The readers may refer to the works of Morton et al. [58], Turner [97], and Roberts et al. [80, 81]. To facilitate the applications of jets and plumes theories, software packages have been developed. The USEPA-supported CORMIX is perhaps the most commonly used expert system for dealing with environmental problems involving jets and plumes. Other models are VISJET and Visual Plume. In addition to the simple jets and plumes, all the other varieties of jets and plumes are still currently under active studies. This constitutes the general tone for the future research. Herein, some of key areas are listed as follows:

- Physics and models of bubbly jets, and their application in aeration and mixing of ponds, lakes, and wastewater treatment plants
- Physics and models of slurry jets, and their industrial applications
- Physics and models of three phase jets and plumes, especially oil-water-gas plumes produced by oil spills in oceans
- Physics and models of jets and plumes in stratified environment
- Development, improvement, and validation of computational fluid dynamics (CFD) modeling (such as direct numerical modeling (DNS); large eddy simulation (LES); k-epsilon modeling; and others).

Bibliography

Primary Literature

1. Abramovich GN (1963) The theory of turbulent jets. MIT Press, Cambridge, MA, 671 p
2. Adams EE (1982) Dilution analysis for unidirectional diffusers. ASCE J Hydraul Div 108(HY3):327–342
3. Agelin-Chaab M (2010) Experimental study of three-dimensional offset jets and wall jets. Ph.D. thesis, Department of Mechanical and Manufacturing Engineering, University of Manitoba, Winnipeg
4. Albertson ML, Dai YB, Jensen RA, Rouse H (1950) Diffusion of submerged jets. Trans ASCE 115:639–697
5. Andreopoulos J (1985) On the structure of jets in a crossflow. J Fluid Mech 157:163–197
6. Anthony DG, Willmarth WW (1992) Turbulence measurements in a round jet beneath a free surface. J Fluid Mech 243:699–720
7. Awaya Y, Fujisaki K, Matsunaga K (1985) Transition in the behavior of sediment laden vertical buoyant jet. J Hydrosci Hydr Eng 3(1):63–74
8. Azimi AH, Zhu DZ, Rajaratnam N (2010) Effect of particle size on the characteristics of sand jets in water. J Eng Mech (in press)
9. Barnhart EL (1969) Transfer of oxygen in aqueous solutions. J Sanit Eng Div 95(3):645–661
10. Beltaos S, Rajaratnam N (1972) Plane turbulent impinging jets. J Hydr Res 11:29–59
11. Bravo HR, Gulliver JS, Hondzo M (2007) Development of a commercial code-based two-fluid model for bubble plumes. Environ Modell Softw 22(4):536–547
12. Brevik I, Kristiansen Ø (2002) The flow in and around air bubble plumes. Int J Multiph Flow 28:617–634
13. Brush LMJ (1962) Exploratory study of sediment diffusion. J Geophys Res 67(4):1427–1433
14. Chan T, Zhou Y, Liu M, Leung C (2003) Mean flow and turbulent measurements of the impingement wall jet on a semi-circular convex surface. Exp Fluids 34(1):140–149
15. Chen CJ, Rodi W (1980) Vertical turbulent buoyant jets – a review of experimental data. Pergamon, Oxford. 83 p
16. Chu VH, Vanvari MR (1976) Experimental study of turbulent stratified shearing flow. J Hydraul Div ASCE 102(6):691–706
17. Clift R, Grace JR, Weber ME (1978) Bubbles, drops, and particles. Academic, New York, 380 p
18. Cuthbertson AJS, Davies PA (2008) Deposition from particle-laden, round, turbulent, horizontal, buoyant jets in stationary and coflowing receiving fluids. J Hydraul Eng 134(4):390–402
19. Dey S, Nath TK, Bose SK (2010) Fully rough submerged plane wall-jets. J Hydro Environ Res 4(4):301–316
20. Doneker RL, Jirka GH (2007) CORMIX user manual – a hydrodynamic mixing zone model and decision support system for pollutant discharges into surface waters. U.S. EPA-823-K-07-001, pp 1–236
21. Fischer HB, List EJ, Imberger J, Brooks NH (1979) Mixing in inland and coastal waters. Academic, New York, 483 p
22. Fonade C, Doubrovine N, Maranges C, Morchain J (2001) Influence of a transverse flowrate on the oxygen transfer performance in heterogeneous aeration: case of hydro-ejectors. Water Res 35(14):3429–3435
23. Förthmann E (1936) Turbulent jet expansion. English translation N.A.C.A. TM-789. (Original paper in German, 1934. Ing. Archiv., 5)
24. Gholamreza-kashi S, Martinuzzi RJ, Baddour RE (2007) Mean flow field of a nonbuoyant rectangular surface jet. J Hydraul Eng 133(2):234–239
25. Girshovich TA (1966) Theoretical and experimental study of a plane turbulent jet in a crossflow. Fluid Dyn 9:84–86

26. Hall N, Elenany M, Zhu DZ, Rajaratnam N (2010) Experimental study of sand and slurry jets in water. J Hydraul Eng 136(10):727–738
27. Heskestad G (1965) Hot-wire measurements in a plane turbulent jet. Trans ASME J Appl Mech 32:1–14
28. Hinze JO, Van Der Hegge Zijnen BG (1949) Transfer of heat and matter in the turbulent mixing zone of an axially symmetrical jet. Appl Sci Res A1:435–461
29. Hodgson JE, Rajaratnam N (1992) An experimental study of jet dilution in crossflows. Can J Civ Eng 19:733–743
30. Hodgson JE, Moawad AK, Rajaratnam N (1999) Concentration field of multiple circular turbulent jets. J Hydr Res 37(2):249–256
31. Huang JF, Davidson MJ, Nokes RI (2005) Two-dimensional and line jets in a weak cross-flow. J Hydr Res 43(4):390–398
32. Huang H, Fergen RE, Rroni JR, Tsai JJ (1996) Probabilistic analysis of ocean outfall mixing zones. J Environ Eng 122(5):359–367
33. Jiang JS, Law AWK, Cheng NS (2005) Two-phase analysis of vertical sediment-laden jets. J Eng Mech 131(3):308–318
34. Jirka GH, Akar PJ (1991) Hydrodynamic classification of submerged multiport-diffuser discharges. J Hydr Eng 117(9):1113–1128
35. Jirka GH, Harleman DRF (1973) The mechanics of submerged multiport diffuser for buoyant discharges in shallow water. R.M. Parsons Laboratory Report, Massachusetts Institute of Technology, Boston
36. Jones WP, Wille M (1996) Large-eddy simulation of a plane jet in a cross-flow. Int J Heat Fluid Flow 17:296–306
37. Kalita K, Dewan A, Dass AK (2002) Predication of turbulent plane jet in crossflow. Numer Heat Transfer A 41:101–111
38. Kikkert GA, Davidson MJ, Nokes RI (2009) A jet at an oblique angle to a cross-flow. J Hydro Env Res 3(2):69–76
39. Kim DG, Seo IW (2000) Modeling the mixing of heated water discharged from a submerged multiport diffuser. J Hydraul Res 38(4):259–269
40. Knystautas R (1964) The turbulent jet from a series of holes in Line. Aeronaut Q 15:1–28
41. Kotsovinos NE, List EJ (1977) Plane turbulent buoyant jet. Part I. Integral properties. J Fluid Mech 81:25–44
42. Lain S, Garcia J (2006) Study of four-way coupling on turbulent particle-laden jet flows. Chem Eng Sci 61:6775–6785
43. Lam KM, Lee WY, Chan CHC, Lee JHW (2006) Global behaviors of a round buoyant jet in a counterflow. J Hydr Eng 132(6):589–604
44. Launder BE, Rodi W (1983) The turbulent wall jet - measurement and modeling. Ann Rev Fluid Mech 15:429–459
45. Lee JHW, Chu VH (2003) Turbulent jets and plumes – a lagrangian approach. Kluwer, Norwell, 390 p
46. Lee JHW, Jirka GH (1980) Multiport diffuser as line source of momentum in shallow water. Water Resour Res 16(4):695–708
47. Leitch AM, Baines WD (1989) Liquid volume flux in a weak bubble plume. J Fluid Mech 205:77–98
48. Lima Neto IE, Zhu DZ, Rajaratnam N, Yu T, Spafford M, McEachern P (2007) Dissolved oxygen downstream of an effluent outfall in an ice-cover river: natural and artificial aeration. J Env Eng 133(11):1051–1060
49. Lima Neto IE, Zhu DZ, Rajaratnam N (2008) Air injection in water with different nozzles. J Environ Eng 134(4):283–294
50. Lima Neto IE, Zhu DZ, Rajaratnam N (2008) Bubbly jets in stagnant water. Int J Multiph Flow 34(12):1130–1141
51. Lima Neto IE, Zhu DZ, Rajaratnam N (2008) Effect of tank size and geometry on the flow induced by circular bubble plumes and water jets. J Hydr Eng 134(6):833–842

52. Lima Neto IE, Zhu DZ, Rajaratnam N (2008) Horizontal injection of gas-liquid mixtures in a water tank. J Hydr Eng 134(12):1722–1731
53. Lübcke HM, Rung Th, Thiele F (2003) Predication of the spreading mechanism of 3D turbulent wall jets with explicit Reynolds-stress closures. Int J Heat Fluid Flow 24: 434–443
54. Margason RJ (1993) Fifty years of jet in cross flow research. Computational and experimental assessment of jets in cross flow. AGARD, Report CP-534
55. Mazurek KA, Christison K, Rajaratnam N (2002) Turbulent sand jet in water. J Hydr Res 40(4):527–530
56. Milgram H (1983) Mean flow in round bubble plumes. J Fluid Mech 133:345–376
57. Morchain J, Moranges C, Fonade C (2000) CFD modeling of a two-phase jet aerator under influence of a crossflow. Water Res 34(13):3460–3472
58. Morton BR, Taylor GI, Turner JS (1956) Turbulent gravitational convection from maintained and instantaneous sources. Proc R Soc Lond A 234:1–23
59. Motarjemi M, Jameson GJ (1978) Mass transfer from very small bubbles – the optimum bubble size for aeration. Chem Eng Sci 33(11):1415–1423
60. Mueller JA, Boyle WC, Pöpel HJ (2002) Aeration: principles and practice. CRC Press, New York
61. Newman BG, Patel RP, Savage SB, Tjio HK (1972) Three-dimensional wall jet originating from a circular orifice. Aeronaut Quart 23:188–200
62. Pande BBL, Rajaratnam N (1979) Turbulent jets in co-flowing streams. ASCE J Eng Mech 105:1025–1038
63. Pande BBL, Rajaratnam N (1977) An experimental study of bluff buoyant turbulent surface jets. J Hydr Res 15(3):261–275
64. Pani BS, Lee JHW, Lai ACH (2009) Application of Reichardt's hypothesis for multiple coflowing jets. J Hydro-environ Res 3(3):121–128
65. Papanicolaou PN, List EJ (1988) Investigations of round vertical turbulent buoyant jets. J Fluid Mech 195:341–391
66. Parr AD, Sayre WW (1979) Multiple jets in shallow flowing receiving waters. ASCE J Hydraul Div 105(HY11):1357–1374
67. Parthasarathy RN, Faeth GM (1987) Structure of particle laden turbulent water jets in still water. Int J Multiph Flow 13(5):699–716
68. Patel RP (1971) Turbulent jets and wall jets in uniform streaming flow. Aeronaut Q 22:311–326
69. Platten JL, Keffer JF (1971) Deflected Turbulent Jet Flows. Trans ASME J Appl Mech 38E (4):756–758
70. Pratte BD, Baines WD (1967) Profiles of the round turbulent jets in a cross flow. J Hydr Div ASCE 92:53–64
71. Pun KL, Davidson MJ, Wang HJ (2000) Merging jets in a co-flowing ambient fluid – a hybrid approach. J Hydraul Res 38(2):105–114
72. Rajaratnam N (1967) Plane turbulent wall jets on rough boundaries. Water Power 49:149–242
73. Rajaratnam N (1976) Turbulent jets. Elsevier Scientific, Amsterdam, 304 p
74. Rajaratnam N (1984) Turbulent jets and plumes, Lecture Notes, University of Alberta, Edmonton, Canada
75. Rajaratnam N (1985) An experimental study of bluff surface discharges with small Richardson number. J Hydraul Res 23(1):47–55
76. Rajaratnam N, Humphries JA (1984) Turbulent non-buoyant surface jets. J Hydraul Res 22(2):103–105
77. Rajaratnam N, Pani BS (1974) Three-dimensional turbulent wall jets. Proc ASCE J Hydraul Div 100:69–83
78. Rajaratnam N, Subramanya K (1967) Diffusion of rectangular wall jets in wide channels. J Hydraul Res 5:281–294

79. Rajaratnam N, Subramanyan S (1985) Plane turbulent buoyant surface jets and jumps. J Hydraul Res 23(2):131–146
80. Roberts PJW, Snyder WH, Baumgartner DJ (1989) Ocean outfalls I: Submerged waste field formation. J Hydraul Eng 115(HY1):1–25
81. Roberts PJW, Snyder WH, Baumgartner DJ (1989) Ocean outfalls II: Spatial evolution of submerged waste field. J Hydraul Eng 115(HY1):26–48
82. Rodi W, Spalding DB (1970) A two-parameter model of turbulence and its application to free jets. Wärme Stoffübertragung 3:85–95
83. Rostamy N, Bergstrom D, Sumner D (2010) An experimental study of a turbulent wall jet on smooth and rough surfaces. In: Nickels TB (ed) IUTAM symposium on the physics of wall-bounded turbulent flows on rough walls, Cambridge, UK
84. Rouse H, Yih CS, Humphreys HW (1952) Gravitational convection from a boundary source. Tellus 4:201–210
85. Rutherford JC (1994) River mixing. Wiley, West Sussex, 347 p
86. Schierholz EL, Gulliver JS, Wilhelms SC, Henneman HE (2006) Gas transfer from air diffusers. Water Res 40(5):1018–1026
87. Schlichting H, Gersten K (2000) Boundary layer theory. Springer, Berlin Heidelberg, 801 p
88. Sforza PM, Herbst G (1970) A study of three-dimensional incompressible turbulent wall jets. Am Inst Aeronaut Astron J 8:276–283
89. Sheen HJ, Jou BH, Lee YT (1994) Effect of particle size on a two-phase turbulent jet. Exp Therm Fluid Sci 8:315–327
90. Singansetti SR (1966) Diffusion of sediment in submerged jet. J Hydraul Div 92(2):153–168
91. Smith SH, Mungal MG (1998) Mixing, structure and scaling of the jet in crossflow. J Fluid Mech 357:83–122
92. Socolofsky SA, Adams EE (2002) Multi-phase plumes in uniform and stratified crossflow. J Hydraul Res 40(6):661–672
93. Tamburello DA, Amitay M (2008) Active manipulation of a particle laden jet. Int J Multiph Flow 34:829–851
94. Tang HS, Paik J, Sotiropoulos F, Khangaonkar T (2008) Three-dimensional numerical modeling of initial mixing of thermal discharges at real-life configurations. J Hydraul Eng 134(9):1210–1224
95. Tachie MF, Balachandar R, Bergstrom DJ (2004) Roughness effects on turbulent plane wall jets in an open channel. Exp Fluid 37(2):281–292
96. Tian X, Roberts PJW, Daviero GJ (2004) Marine wastewater discharges from multiport diffuser. II: unstratified flowing water. J Hydraul Eng 130(12):1147–1155
97. Turner JS (1973) Buoyancy effects in fluids. Cambridge University Press, Cambridge, 367 p
98. Varley J (1995) Submerged gas-liquid jets: bubble size prediction. Chem Eng Sci 50(5):901–905
99. Verhoff A (1963) The two-dimensional turbulent wall jet with and without an external stream. Report 626, Princeton University
100. Wang HJ, Davidson MJ (2003) Jet interaction in a still ambient fluid. J Hydraul Eng 129(5):349–357
101. Whipple WJ, Yu SL (1970) Instream aerators for polluted rivers. J Sanit Eng Div 96(5):1153–1165
102. Wright SJ (1977) Mean behavior of buoyant jet in a crossflow. J Hydraul Div ASCE 103(HY5):499–513
103. Wright SJ (1984) Buoyant jets in density-stratified crossflow. J Hydraul Eng ASCE 110(5):643–656
104. Wüest A, Brooks NH, Imboden DM (1992) Bubble plume modeling for lake restoration. Water Resour Res 28(12):3235–3250
105. Wygnanski I, Fielder H (1969) Some measurements in the self-preserving jet. J Fluid Mech 38:577–612

106. Yan J, Luo K, Fan J, Tsuji Y, Cen K (2008) Direct numerical simulation of particle dispersion in a turbulent jet considering inter-particle collisions. Int J Multiph Flow 34: 723–733
107. Zhang W, Zhu DZ (2011) Near-field mixing downstream of a multiport diffuser in a shallow river. J Environ Eng 137(4): 230–240

Books and Reviews

Cheremisinoff NP (ed) (1986) Encyclopedia of fluid mechanics, vol 2, Dynamics of single-fluid flow and mixing. Gulf, Houston, 1505 p

Cheremisinoff NP (ed) (1986) Encyclopedia of fluid mechanics, vol 3, Gas-liquid flow. Gulf, Houston, 1535 p

Cheremisinoff NP (ed) (1993) Encyclopedia of fluid mechanics, vol 2, Advances in multiphase flow. Gulf, Houston, 674 p

Davies PA, Valente Neves MJ (eds) (1994) Recent research advances in the fluid mechanics of turbulent jets and plumes. Kluwer, Dordrecht, 514 p

Fay JA (1973) Buoyant plumes and wakes. Ann Rev Fluid Mech 5:151–160

Jirka GH (2004) Integral model for turbulent buoyant jets in unbounded stratified flow. Part 1: single round jet. Environ Fluid Mech 4:1–56

Jirka GH (2006) Integral model for turbulent buoyant jets in unbounded stratified flow. Part 2: plane jet dynamics resulting from multiport diffuser jets. Environ Fluid Mech 6:43–100

List EJ (1982) Turbulent jets and plumes. Ann Rev Fluid Mech 14:189–212

Philip AA (1997) Jets, plumes and mixing at the coast. In: Allen PA (ed) Earth surface processes. Blackwell, Oxford, pp 241–266, Chapter 7

Rodi W (1982) Turbulent buoyant jets and plumes. Pergamon, Oxford, 184 p

Rubin H, Atkinson J (2001) Environmental fluid mechanics. Marcel Dekker, New York, 728 p

Shen HH, Cheng AHD, Wang KH, Teng MH, Liu CCK (eds) (2002) Environmental fluid mechanics theories and applications. ASCE, Reston, 467 p

Wood AW (2010) Turbulent plumes in nature. Ann Rev Fluid Mech 42:391–412

Wood IR, Bell RG, Wilkinson DL (eds) (1993) Ocean disposal of wastewater. World Scientific, Singapore, 425 p

Chapter 8
Atmosphere-Water Exchange

Bernd Jähne

Glossary

Bulk coefficients	c_i relate the *transfer velocity* k for a species i to the *wind velocity* U_r in a reference height, typically at 10 m above the mean water level: $c_i = k_i / U_r$. From the bulk coefficient, the flux density j_i of a species can be computed as $j_i = c_i(C_r - C_0) U_r$, where C_r and C_0 are the corresponding concentrations at the reference height and right at the water surface, respectively. For momentum density (ρU) the bulk coefficients is also known as the *drag coefficient* c_D. It can also be expressed as $c_D = (u_*/U_r)^2$ with the momentum flux given by $j_m = \rho u_*^2$; u_* is the *friction velocity*.
Friction velocity	u_* is a measure for the tangential force per area applied by the wind at the water surface, the shear stress $\tau = \rho u_*^2$, which is also equal to the vertical momentum flux density j_m.
Mass boundary layer	Thickness of the layers at both sides of the water surface in which transport of mass by turbulence is smaller than by molecular diffusion.

This chapter was originally published as part of the Encyclopedia of Sustainability Science and Technology edited by Robert A. Meyers. DOI:10.1007/978-1-4419-0851-3

B. Jähne (✉)
Institute for Environmental Physics and Heidelberg Collaboratory
for Image Processing (HCI), University of Heidelberg, Speyerer Straße 6,
69115 Heidelberg, Germany
e-mail: Bernd.Jaehne@iwr.uni-heidelberg.de

J.S. Gulliver (ed.), *Transport and Fate of Chemicals in the Environment: Selected Entries from the Encyclopedia of Sustainability Science and Technology*,
DOI 10.1007/978-1-4614-5731-2_8,

Schmidt and Prandtl numbers, Sc and Pr — The Schmidt and Prandtl numbers are the ratio of kinematic viscosity ν (molecular diffusion coefficient for momentum) and the molecular diffusion coefficients for the corresponding chemical species, D, and heat, D_h, respectively. Thus, these numbers express how much slower chemical species and heat, respectively, are transported by molecular processes than momentum. In air, these numbers are in the order of one; in water, the Prandtl number is about 10 and the Schmidt number about 1,000.

Transfer velocity — k is the velocity by which a momentum, heat, and chemical are transported across the surface; because of the concentration discontinuity at the water surface, the transfer velocity on the air side is different from the transfer velocity on the water side.

Viscous boundary layer — Thickness of the layers at both sides of the water surface in which transport of momentum by turbulent is smaller than by molecular friction, resulting in a linear velocity profile in this layer.

Definition of the Subject

Gaseous and volatile chemical species reside not only in the atmosphere. Because they dissolve in water, they are also distributed in the hydrosphere. The by far largest part of the hydrosphere is the ocean. Therefore, the exchange between atmosphere and oceans is the most important process for the fate of gaseous and volatile chemical species (Table 8.1).

For long-lived species, that is, life time by reaction larger than residence times in either the atmosphere or the ocean, two basic factors determine the fate of these species. For time scales longer than the turnover time of ocean mixing, which is much slower than the same process in the atmosphere, the solubility of the species determines where it ends up. On much shorter time scales, the speed by which the species are exchanged between atmosphere and ocean and the concentration difference determines the flux between these two compartments.

While turbulent mixing is quite fast within the ocean and the atmosphere, it becomes more and more inefficient, the closer a species is transported to the surface. The final step toward the interface must be overcome by molecular diffusion, which is a slow process, especially in water. Therefore, very thin mass boundary layers at both sides of the water surface, in which turbulent transport is

Table 8.1 Comparison of size (as depth in m over entire earth surface) and mass of atmosphere and global water storage

System	Depth (m)	Mass (10^{18} kg)
Oceans	2,624	1,382
Ice caps and glaciers	47	25
Total ground water	46	24
Lakes	0.35	0.18
Rivers	0.004	0.002
Atmosphere	8,000	5.3

slower than molecular transport, are the bottleneck of the exchange between atmosphere and ocean.

In this way, tiny layers only about 20–200 μm thick in the water and 200–2,000 μm thick in the air control the exchange process. In consequence, the crucial question is which processes determine the thickness of these layers.

Introduction

Research in air–water gas exchange dates back more than a century. Bohr [1], for example, was one of the first to study the exchange of carbon dioxide in a stirred bottle. With the early observation of increasing carbon dioxide concentration in the atmosphere interest in air-sea gas exchange rose considerably, triggered by the question how much of the excess carbon dioxide would go into the ocean [2, 3]. Since then significant and continuous research work was devoted for a better understanding of air–water mass exchange, but the mechanisms of the exchange processes are still not fully understood even today.

This entry is divided into three major parts and partly based on previous reviews of the author on the same subject [4–6]. Section "Mechanisms of Atmosphere–Water Exchange" describes the basic mechanisms of air–water exchange and the many parameters influencing it. Focusing on field data, section "Experimental Techniques and Results" summarizes the various experimental techniques and gives a synthesis of the available experimental data. Using the information collected in the previous two sections, section "Empiric Parametrization" tries to give the best possible empirical parametrization of the exchange rate and section "Future Directions" points to future directions of research.

Mechanisms of Atmosphere–Water Exchange

Turbulent and Molecular Transport

The transfer of gases and volatile chemical species between the atmosphere and oceans is driven by a concentration difference and the transport by molecular and turbulent motion. Both types of transport processes can be characterized by *diffusion coefficients*, denoted by D and K_c, respectively. The resulting flux area density $\boldsymbol{j}_c$ is proportional to the diffusion coefficient and the concentration gradient:

$$j_c = -(D+K_c(z))\nabla_c. \tag{8.1}$$

In a stationary homogeneous case and without sinks and sources by chemical reactions, the flux density $\boldsymbol{j}$ is in vertical direction and constant. Then integration of Eq. 8.1 yields vertical concentration profiles:

$$C(z_r) - C(0) = j_c \int_0^{z_r} \frac{1}{D + K_c(z)} dz. \tag{8.2}$$

The molecular diffusion coefficient is proportional to the velocity of the molecules and the free length between collisions. The same concept can be applied to turbulent diffusion coefficients. Far away from the interface, the free length (called *mixing length*) is set proportional to the distance from the interface and the turbulent diffusion coefficient K_c for mass transfer is:

$$K_c = \frac{k}{Sc_t} u_* z, \tag{8.3}$$

where $\kappa = 0.41$ is the *von Kármán constant*, u_*, the *friction velocity*, a measure for the velocity fluctuations in a turbulent flow, and $Sc_t = K_m/K_c$ the *turbulent Schmidt number*. Closer to the interface, the turbulent diffusion coefficients are decreasing even faster. Once a critical length scale l is reached, the Reynolds number $Re = u_* l/\nu$ (ν is the kinematic viscosity, the molecular diffusion coefficient for momentum) becomes small enough so that turbulent motion is attenuated by viscosity. The degree of attenuation depends on the properties of the interface. At a smooth solid wall, $K_c \propto z^3$, at a free water interface it could be in the range between $K_c \propto z^3$ and $K_c \propto z^2$ depending on surface conditions.

Viscous and Mass Boundary Layers

Boundary layers are formed on both sides of the interface (Fig. 8.1). When the turbulent diffusivity becomes equal to the kinematic viscosity, the edge of the *viscous boundary layer* is reached. As the name implies, this layer is dominated by viscous dissipation and the velocity profile becomes linear because of a constant diffusivity. The edge of the *mass boundary layer* is reached when the turbulent diffusivity becomes equal to the molecular diffusivity. The relative thickness of both boundary layers depends on the dimensionless ratio $Sc = \nu/D$ (*Schmidt number*).

The viscous and mass boundary layers are of about the same thickness in the air, because values of D for various gaseous species and momentum are about the same (Sc_{air} is 0.56 for H_2O, 0.63 for heat, and 0.83 for CO_2). In the liquid phase, the situation is completely different. With Schmidt numbers in the range from 100 to 3,000 (Fig. 8.2, Table 8.3), molecular diffusion for a dissolved volatile chemical species is two to three orders of magnitude slower than diffusion of momentum (Table 8.2). Thus, the mass boundary layer is significantly thinner than the viscous boundary layer in the liquid phase. This means that the transfer of gases is much slower and almost entirely controlled by the tiny residual turbulence in the small top fraction of the viscous boundary layer. This basic fact makes it difficult to investigate the mechanism of air–water gas transfer both theoretically and experimentally. In addition, the transfer process depends strongly on the water temperature because the Schmidt number decreases by about a factor of 6 from 0°C to 35°C (Fig. 8.2, Table 8.3).

Description of Transport

The amount of species exchanged between the air and water across the interface can be described by a quantity with the units of a velocity. It represents the velocity with which a tracer is pushed by an imaginary piston across the surface. This quantity is known as the *transfer velocity* k (also known as the *piston velocity*, *gas exchange rate* or *transfer coefficient*). It is defined as the flux density divided by the concentration difference between the surface and the bulk at some reference level z_r:

$$k = \frac{j_c}{C_s - C_b}. \tag{8.4}$$

The inverse of the transfer velocity is known as the *transfer resistance* R:

$$R = \frac{C_s - C_b}{j_c}. \tag{8.5}$$

The indices s and b denote the surface and bulk. Both quantities can directly be related to vertical concentration profiles by introducing Eqs. 8.4 and 8.5 into Eq. 8.2:

$$R = \frac{1}{k} = \int_0^{z_r} \frac{1}{D + K_e(z)} dz. \tag{8.6}$$

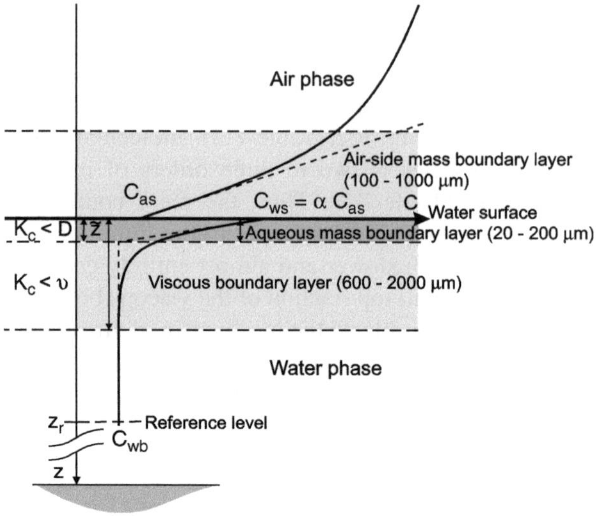

Fig. 8.1 Schematic graph of the mass boundary layers at a gas–liquid interface for a tracer with a solubility $\alpha = 3$

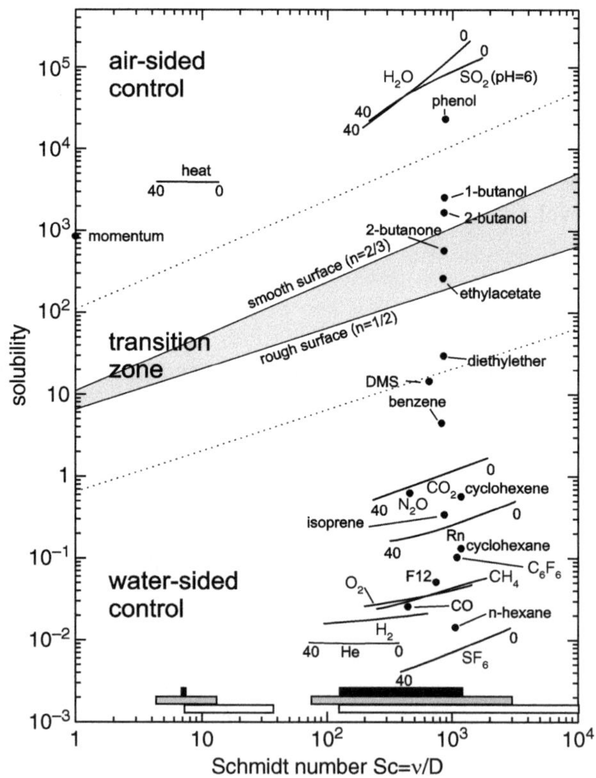

Fig. 8.2 Schmidt number/solubility diagram including various volatile tracers, momentum, and heat for a temperature range (C) as indicated. *Filled circles* refer to only a temperature of 20°C. The regions for air-side, mixed, and water-side control of the transfer process between gas and liquid phase are marked. At the solid lines, the transfer resistance is equal in both phases. The following dimensional transfer resistances were used: $r_a = 31$, $r_w = 12Sc^{2/3}$ (smooth), $r_w = 6.5Sc^{1/2}$ (wavy surface) with $r_a = R_a u_{*a}$ and $r_w = R_w u_{*w}$ (Adapted from Jähne and Haußecker [6])

Table 8.3 Schmidt numbers of various gases and volatile species in the temperature range from 0°C to 40°C

	Schmidt number								
Species	0°C	5°C	10°C	15°C	20°C	25°C	30°C	35°C	40°C
Heat	13.45	11.19	9.46	8.10	7.02	6.14	5.42	4.82	4.32
^3He	329	254	200	160	130	107	88	74	63
^4He	379	293	230	184	149	122	102	85	72
^4He[a]	411	319	252	201	164	135	112	94	79
Ne	768	579	445	347	276	221	180	148	122
Kr	2,045	1,478	1,090	819	625	483	379	301	241
Xe	2,701	1,930	1,409	1,047	791	606	471	370	294
^{222}Rn	3,168	2,235	1,611	1,182	883	669	514	400	314
H_2	633	473	360	278	219	174	140	114	94
H_2[a]	663	502	387	304	242	195	159	131	109
CH_4	1,908	1,400	1,047	797	616	483	383	308	250
CO_2	1,922	1,397	1,036	782	600	466	367	293	236
DMS	2,593	1,905	1,428	1,089	844	662	527	423	344
CH_3Br	2,120	1,545	1,150	870	669	522	412	329	266
F_{12} (CCl_2F_2)	3,639	2,624	1,931	1,447	1,102	851	666	527	422
F_{11}(CCl_3F)	3,521	2,549	1,883	1,416	1,082	839	658	523	420
SF_6	3,033	2,208	1,640	1,239	952	741	585	467	376

[a]In sea water

Thus, the transfer resistances over several height intervals can be added in the same way as electrical resistances that are connected in series. Typical values of the transfer velocity across the water-side mass boundary layer are 10^{-6}–10^{-5} m/s (1–10 m/day). With respect to typical mixed layer depths in the ocean of about 100 m, gas transfer is a very slow process. It takes a time constant $\tau = h/k =$ 10–100 days for the concentration of dissolved gases in the mixed layer to come into equilibrium with the atmosphere.

Boundary Layer Thickness

The *boundary layer thickness* $\tilde{z}$ is defined as the thickness of a fictional layer in which the flux is maintained only by molecular transport: $j = D(C_s - C_b)/\tilde{z}$. Then with Eq. 8.4 the boundary layer thickness is given by:

$$\tilde{z} = \frac{D}{k}. \tag{8.7}$$

Geometrically, $\tilde{z}$ is given as the intercept of the tangent to the concentration profile at the surface and the bulk (Fig. 8.1). With thicknesses between 20 and 200 μm, the mass boundary layer is extremely thin.

Table 8.2 Diffusion coefficients for various gases and volatile chemical species in deionized water and in some cases in sea water collected from Jähne et al. [18] and King et al. [20]. Die data for momentum (kinematic viscosity) are from Sündermann [22]. Column 3 and 4 contain the parameters for the fit of the diffusion coefficient: $D = A \exp[-E_a/(RT)]$, the last four columns the diffusion coefficients for 5°C, 15°C, 25°C, and 35°C

Species	Molecular mass	A (10^{-5} cm²/s)	E_a (kJ/mole)	σ(Fit) %	Diffusion coefficient (10^{-5} cm²/s) 5°C	15°C	25°C	35°C
Momentum	–				1,517	1,139	893.4	724.3
Momentum[a]	–				1,560	1,182	934.5	763.9
Heat	–	379.2	2.375		135.80	140.72	145.48	150.08
^{3}He[b,c]	3.02	941	11.70	2.1	5.97	7.12	8.39	9.77
^{4}He	4.00	818	11.70	2.1	5.10	6.30	7.22	8.48
^{4}He[a]		886	12.02	1.8	4.86	5.88	7.02	8.03
Ne	20.18	1,608	14.84	3.5	2.61	3.28	4.16	4.82
Kr	83.80	6,393	20.20	1.6	1.02	1.41	1.84	2.40
Xe	131.30	9,007	21.61	3.5	0.77	1.12	1.47	1.94
^{222}Rn[b]	222.00	15,877	23.26	11	0.68	0.96	1.34	1.81
H_2	2.02	3,338	16.06	1.6	3.17	4.10	5.13	6.23
H_2[a]		1,981	14.93	4.3	3.05	3.97	4.91	5.70
CH_4	16.04	3,047	18.36	2.7	1.12	1.48	1.84	2.43
CO_2	44.01	5,019	19.51	1.3	1.07	1.45	1.91	2.43
DMS[b]	62.13	2,000	18.10		0.80	1.05	1.35	1.71
CH_3Br[b]	94.94	3,800	19.10		0.98	1.31	1.71	2.20
F_{12}[b] (CCl_2F_2)	120.91	4,100	20.50		0.58	0.79	1.05	1.37
F_{11}[b](CCl_3F)	137.37	3,400	20.00		0.60	0.81	1.07	1.38
SF_6[b]	146.05	2,900	19.30		0.69	0.92	1.20	1.55

[a]In sea water
[b]Values of diffusion coefficients from fit, not measured values
[c]Set 15% higher than ^{4}He

Boundary Layer Time Constant

The *time constant* $\tilde{t}$ for the transport across the mass boundary layer is given by:

$$\tilde{t} = \frac{\tilde{z}}{k} = \frac{D}{k^2}. \tag{8.8}$$

Typical values for $\tilde{t}$ are 0.04–4 s. Any chemical reaction with a time constant larger than $\tilde{t}$ does not significantly affect the transfer process. Therefore, CO_2 can be regarded as an inert gas, but not fast hydrating acid gases such as SO_2.

The definitions of the three parameters k, $\tilde{z}$, $\tilde{t}$ are generally valid and do not depend on any models of the boundary layer turbulence. According to Eqs. 8.7 and 8.8, they are coupled via the molecular diffusion coefficient. Therefore, only one of them needs to be measured to get knowledge of all three parameters provided the molecular diffusion coefficient of the species is known.

Partitioning of Transfer Between Air and Water

Because a mass boundary layer exists on both sides, it is important to determine which one controls the transfer, that is, exhibits the largest transfer resistance (or lowest transfer velocity). At the surface itself, the thermodynamic solubility equilibrium is assumed to be established between the tracer concentrations c_a in the gas phase and c_w in the liquid phase:

$$C_{ws} = \alpha C_{as}, \tag{8.9}$$

where α is the dimensionless solubility (partition coefficient). A solubility $\alpha \neq 1$ causes a concentration jump at the surface (Fig. 8.1). Thus, the resulting total transfer velocity k_t or transfer resistance R_t can either be viewed from the gas phase or the liquid phase. Adding them up, the factor α must be considered to conserve the continuity of the concentration profile:

$$\text{air side:}\ \frac{1}{k_{at}} = \frac{1}{k_a} + \frac{1}{\alpha k_w}, \quad R_{at} = R_a + R_w/\alpha, \tag{8.10}$$

$$\text{water side:}\ \frac{1}{k_{wt}} = \frac{\alpha}{k_a} + \frac{1}{k_w}, \quad R_{wt} = \alpha R_a + R_w.$$

The total transfer velocities in air and water differ by the factor α: $k_{at} = \alpha k_{wt}$. The ratio $\alpha k_w/k_a$ determines which boundary layer controls the transfer process. A high solubility shifts control of the transfer process to the gas-phase boundary layer, and a low solubility to the aqueous layer. The solubility value for a transition from air-sided to water-sided control depends on the ratio of the transfer velocities. Typically k_w is about 100–1,000 times smaller than k_a. Thus, the transfer of even moderately soluble volatile chemical species with solubilities up to 30 is controlled by the water side. Some environmentally important species lie in a transition zone where it is required to consider both transport processes (Fig. 8.2). The transfer of highly soluble volatile and/or chemically reactive gas is controlled by the air-side transfer process and thus analogous to the transfer of water vapor. The following considerations concentrate on the water-side transfer process.

Gas Exchange at Smooth Water Surfaces

At smooth water surfaces, the theory of mass transfer is well established because it is equivalent to mass transfer to a smooth solid wall. The turbulent diffusivity can be described by the classic approach of Reichardt [7] with an initial z^3 increase that smoothly changes to a linear increase in the turbulent layer as in Eq. 8.3.

Then integration of Eq. 8.6 yields the following approximation for Schmidt numbers higher than 100 [12, 16]:

$$k_w = u_{*w} \frac{1}{12.2} Sc^{-2/3} \quad Sc > 60. \tag{8.11}$$

This equation establishes the basic analogy between momentum transfer and gas exchange. The transfer coefficients are proportional to the friction velocity in water, which describes the shear stress (tangential force per unit area) $\tau = \rho_w u_{*w}^2$ applied by the wind field at the water surface. Assuming stress continuity at the water surface, the friction velocity in water is related to the friction velocity in air by:

$$u_{*w} = u_{*a} \left(\frac{\rho_a}{\rho_w} \right)^{1/2}. \tag{8.12}$$

The friction velocity in air, u_{*a}, can further be linked via the drag coefficient to the wind speed U_R at a reference height: $c_D = (u_{*a}/U_R)^2$. Depending on the roughness of the sea surface, the drag coefficient has values between 0.8 and $2.4 \cdot 10^{-3}$. In this way, the gas exchange rate is directly linked to the wind speed. The gas exchange further depends on the chemical species and the water temperature via the Schmidt number.

Influence of Waves

A free water surface is neither solid nor is it smooth as soon as short wind waves are generated. On a free water surface, velocity fluctuations are possible. Thus, there can be convergence or divergence zone at the surface; surface elements may be dilated or contracted. At a clean water surface, dilation or contraction of a surface element does not cause restoring forces, because surface tension only tries to minimize the total free surface area, which is not changed by this process. As a consequence of this hydrodynamic boundary condition, the turbulent diffusivity normal to the interface can now increase with the distance squared from the interface, $K_c \propto z^2$. Then:

$$k_w = u_{*w} \frac{1}{\beta} Sc^{-1/2}, \tag{8.13}$$

where β is a dimensionless constant.

In comparison to the smooth case in Eq. 8.11, the exponent n of the Schmidt number drops from $-2/3$ to $-1/2$. This increases the transfer velocity for a Schmidt number of 600 by about a factor of 3. The total enhancement depends on the value of the constant β.

Wind waves cannot be regarded as static roughness elements for the liquid flow because their characteristic particle velocity is of the same order of magnitude as

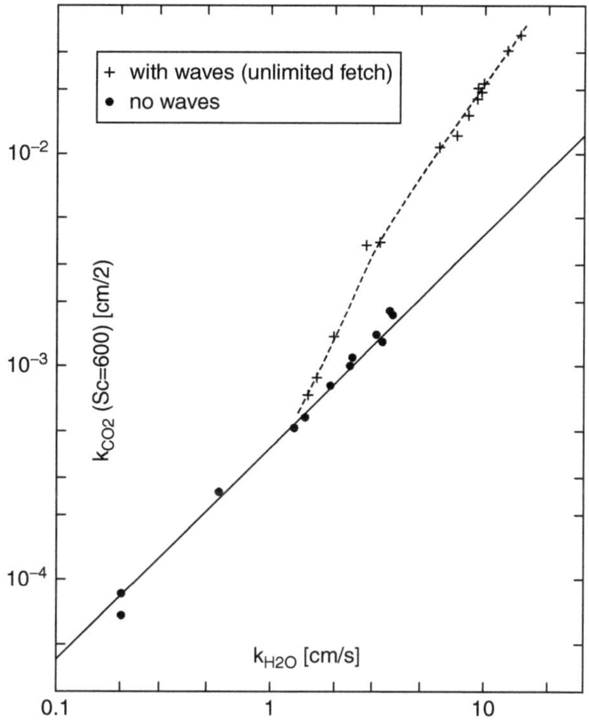

Fig. 8.3 Transfer velocity of carbon dioxide plotted against the transfer velocity of water vapor. Measurements from a small annular wind/wave facility Jähne [15]

the velocity in the shear layer at the surface. This fact causes a basic asymmetry between the turbulent processes on the air and on the water sides of the interface. Therefore, the wave effect on the turbulent transfer in the water is much stronger and of quite different character than in the air. This basic asymmetry can be seen if the transfer velocity for CO_2 is plotted against the transfer velocity for water vapor (Fig. 8.3, [15]). At a smooth water surface, the points fall well on the theoretical curve predicted by the theory for a smooth rigid wall. However, as soon as waves occur at the water surface, the transfer velocity of CO_2 increases significantly beyond the predictions.

Even at high wind speeds, the observed surface increase is well below 20% [24]. When waves are generated by wind, energy is not only transferred via shear stress into the water but a second energy cycle is established. The energy put by the turbulent wind into the wave field is transferred to other wave numbers by nonlinear wave–wave interaction and finally dissipated by wave breaking, viscous dissipation, and turbulence. The turbulent wave dissipation term is the least known term and of most importance for enhanced near-surface turbulence. Evidence for enhanced turbulence levels below wind waves has been reported from field and laboratory measurements. Experimental results also suggest that the gas transfer rate is better correlated with the *mean square slope* of the waves as an integral measure for the nonlinearity of the wind wave field than with the wind speed.

It is not yet clear, however, to what extent *microscale wave breaking* can account for the observed enhanced gas transfer rates. A gravity wave becomes instable and generates a steep train of capillary waves at its leeward face and has a turbulent wake. This phenomenon can be observed even at low wind speeds, as soon as wind waves are generated. At higher wind speeds, the frequency of microscale wave breaking increases.

At high wind speeds, wave breaking with the entrainment of bubbles may enhance gas transfer further. This phenomenon complicates the gas exchange between atmosphere and the oceans considerably [19, 26]. First, bubbles constitute an additional exchange surface. This surface is, however, only effective for gases with low solubility. For gases with high solubility, the gas bubbles quickly come into equilibrium so that a bubble takes place in the exchange only for a fraction of its life time. Thus bubble-mediated gas exchange depends – in contrast to the exchange at the free surface – on the solubility of the gas tracer. Second, bubble-mediated gas transfer shifts the equilibrium value to slight supersaturation due to the enhanced pressure in the bubbles by surface tension and hydrostatic pressure. Third, breaking waves also enhance near-surface turbulence during the breaking event and the resurfacing of submerged bubbles [17].

Experimental data are still too sparse for the size and depth distribution of bubbles and the flux of the bubbles through the interface under various sea states for a sufficiently accurate modeling of bubble-mediated air-sea gas transfer and thus a reliable estimate of the contribution of bubbles to the total gas transfer rate. Some experiments from wind/wave tunnels and the field suggest that significant enhancements can occur, other experiments could not observe a significant influence of bubbles.

Influence of Surface Films

A film on the water surface creates pressure that works against the contraction of surface elements. This is the point at which the physicochemical structure of the surface influences the structure of the near-surface turbulence as well as the generation of waves [13, 21]. As at a rigid wall, a strong film pressure at the surface maintains a two-dimensional continuity at the interface just as at a rigid wall. Therefore, Eq. 8.11 should be valid for a smooth film-covered water surface and has indeed been verified in wind/wave tunnel studies as the lower limit for the transfer velocity. As a consequence, both Eqs. 8.11 and 8.13 can only be regarded as limiting cases. A more general approach is required that has not yet been established. One possibility is a generalization of Eqs. 8.11 and 8.13 to:

$$k_w = u_{*w} \frac{1}{\beta(s)} Sc^{-n(s)}, \qquad (8.14)$$

where both β and n depend on dimensionless parameters describing the surface conditions s. Even films with low film pressure may easily decrease the gas transfer rate to half of its value at clean water surface conditions. But still too few measurements at sea are available to establish the influence of surfactants on gas transfer for oceanic conditions more quantitatively.

Experimental Techniques and Results

Laboratory Facilities

Laboratory facilities play an important role in the investigation of air-sea gas transfer. Only laboratory studies allow a systematic study of the mechanisms and are thus an indispensable complement to field experiments. Almost all basic knowledge about gas transfer has been gained by laboratory experiments in the past.

Geochemical Tracer Techniques

The first oceanic gas exchange measurements were performed using geochemical tracer methods such as the ^{14}C [2], $^{3}He/T$ [23], or $^{222}Rn/^{226}Ra$ [8, 9] methods. The volume and time-average flux density is given by mass balance of the tracer concentration in a volume of water V_w:

$$V_w \dot{c}_w = F_w j \quad or \quad j = h_w \dot{c}_w, \tag{8.15}$$

where F_w, and h_w are the surface area and the effective height V_w / F_w of a well-mixed water body, respectively. The time constant $\tau_w = h_w / k$ is in the order of days to weeks. It is evident that the transfer velocities obtained in this way provide only values integrated over a large horizontal length scales and time scales in the order of τ_w. Thus a parametrization of the transfer velocity is only possible under steady state conditions over extended periods. Moreover, the mass balance contains many other sources and sinks besides air-sea gas exchange and thus may cause severe systematic errors in the estimation of the transfer velocity. Consequently, mass balance methods are only poorly suited for the study of the mechanisms of air–water gas transfer.

Tracer Injection

The pioneering lake studies for tracer injection used sulfur hexafluoride (SF_6). However, the tracer concentration decreases not only by gas exchange across the interface but also by horizontal dispersion of the tracer. This problem can be overcome by the *dual tracer technique* [10] simultaneously releasing two tracers with different diffusivities (e.g., SF_6 and ^3He). When the ratio of the gas transfer velocities of the two tracers is known, the dilution effect by tracer dispersion can be corrected, making it possible to derive gas transfer velocities. But the basic problem of mass balance techniques, that is, their low temporal resolution, remains also with artificial tracer approaches.

Eddy Correlation Flux Measurements

Eddy correlation techniques are used on a routine basis in micrometeorology, that is, for tracers controlled by the boundary layer in air (momentum, heat, and water vapor fluxes). Direct measurements of the air-sea fluxes of gas tracers are very attractive because the flux densities are measured directly and have a much better temporal resolution than the mass balance based techniques. Unfortunately, large experimental difficulties arise when this technique is applied to gas tracers controlled by the aqueous boundary layer [Broecker 1986, 14]. The concentration difference in the air is only a small fraction of the concentration difference across the aqueous mass boundary layer. But after more than 20 years of research has this technique delivered useful results. Some successful measurements under favorable conditions have been reported and it appears that remaining problems can be overcome in the near future.

Passive and Active Thermography

The basic idea of this technique is to determine the concentration difference across the mass boundary layer when the flux density j of the tracer across the interface is known. The local transfer velocity can be determined by simply measuring the concentration difference Δc across the aqueous boundary layer (cold surface skin temperature) according to Eq. 8.4 with a time constant $\tilde{t}$ for the transport across the boundary layer Eq. 8.8. This technique is known as the *controlled flux technique* (CFT).

Heat proves to be an ideal tracer for the CFT. The temperature at the water surface can then be measured with high spatial and temporal resolution using IR thermography. A known and controllable flux density can be applied by using infrared radiation. Infrared radiation is absorbed in the first few ten micrometers at the water surface. Thus a heat source is put right at top of the aqueous viscous boundary layer. Then the CFT directly measures the water-side heat transfer velocity.

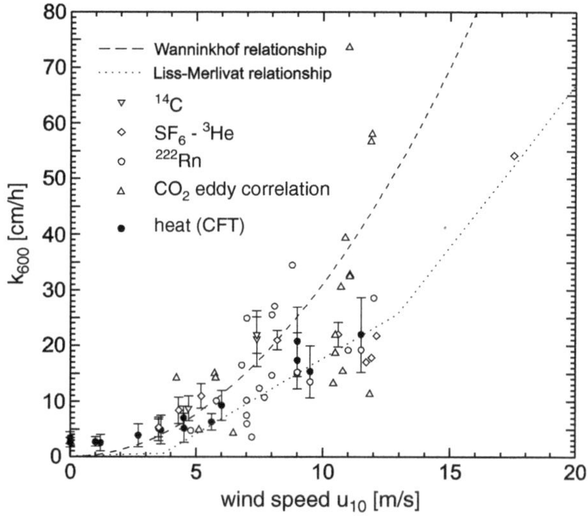

Fig. 8.4 Summary of gas exchange field data normalized to a Schmidt number of 600 and plotted versus wind speed together with the empirical relationships of Liss and Merlivat [11] and ? (After Jähne and Haußecker [6])

A disadvantage of the CFT is that the transfer velocity of gases must be extrapolated from the transfer velocity of heat. The large difference in the Schmidt number (7 for heat, 600 for CO_2) casts some doubt whether the extrapolation to so much higher Schmidt numbers is valid.

Two variants of the technique proved to be successful. Active thermography uses a CO_2 laser to heat a spot of several centimeter in diameter on the water surface. The heat transfer rates are estimated from the temporal decay of the heated spot. Passive thermography uses the naturally occurring heat fluxes caused by latent heat flux j_l, sensible heat flux j_s and longwave emission of radiation j_r. The net heat flux $j_n = j_l + j_s + j_r$ results according to Eq. 8.4 in a temperature difference across the interface of $\Delta T = j_h/(\rho c_p k_h)$. Because of the turbulent nature of the exchange process any mean temperature difference is associated with surface temperature fluctuations which can be observed in thermal images. With this technique, the horizontal structure of the boundary layer turbulence can be observed. Surface renewal is directly observable in the IR image sequences, which show patches of fluid being drawn away from the surface.

With some knowledge about the statistics of the temperature fluctuations, the *temperature difference* ΔT across the interface as well as the time constant $\tilde{t}$ of heat transfer can be computed from the temperature distribution at the surface. Results obtained with this technique are shown in Fig. 8.4.

Summary of Field Data

A collection of field data is shown in Fig. 8.3. Although the data show a clear increase of the transfer velocity with wind speed, there is substantial scatter in the data that can only partly be attributed to uncertainties and systematic errors in the measurements.

Thus, in addition, the field measurements reflect the fact that the gas transfer velocity is not simply a function of the wind speed but depends significantly on other parameters influencing near-surface turbulence, such as the wind-wave field and the viscoelastic properties of the surface film.

Empiric Parametrization

Given the lack of knowledge, all theories about the enhancement of gas transfer by waves are rather speculative and are not yet useful for practical application. Thus, it is still state of the art to use semiempiric or empiric parameterizations of the gas exchange rate with the wind speed. Most widely used is the parametrization of Liss and Merlivat [11]. It identifies three physically well-defined regimes (smooth, wave-influenced, bubble-influenced) and proposes a piece-wise linear relation between the wind speed U and the transfer velocity k:

$$k = 10^{-6} \begin{cases} 0.472U(Sc/600)^{-2/3} & U \leq 3.6\,\mathrm{m/s} \\ 7.971(U-3.39)(Sc/600)^{-1.2} & U > 3.6\,\mathrm{m/s\ and} \\ & U \leq 13\,\mathrm{m/s} \\ 16.39(U-8.36)(Sc/600)^{-1/2} & U > 13\,\mathrm{m/s} \end{cases} \tag{8.16}$$

At the transition between the smooth and wavy regime, a sudden artificial jump in the Schmidt number exponent n from 2/3 to 1/2 occurs. This actually causes a discontinuity in the transfer rate for Schmidt number unequal to 600.

The empiric parametrization of Wanninkhof [25] simply assumes a quadratic increase of the gas transfer rate with the wind speed:

$$k = 0.861 \cdot 10^{-6}(s/m)U^2(Sc/600)^{-1/2}. \tag{8.17}$$

Thus, this model has a constant Schmidt number exponent $n = 1/2$. The two parameterizations differ significantly (see Fig. 8.3). The Wanninkhof parametrization predicts significantly higher values. The discrepancy between the two parameterizations and many others proposed (up to a factor of two) mirrors the current uncertainty in estimating the air-sea gas transfer rate.

Future Directions

In the past, progress toward a better understanding of the mechanisms of air–water gas exchange was hindered by inadequate measuring technology. However, new techniques have become available and will continue to become available that will give a direct insight into the mechanisms under both laboratory and field conditions.

This progress will be achieved by interdisciplinary research integrating such different research areas as oceanography, micrometeorology, hydrodynamics, physical chemistry, applied optics, and image processing.

Optical and image processing techniques will play a key role because only imaging techniques give direct insight to the processes in the viscous, heat, and mass boundary layers on both sides of the air–water interface. Eventually, all key parameters including flow fields, concentration fields, and waves will be captured by imaging techniques with sufficient spatial and temporal resolution. The experimental data gained with such techniques will stimulate new theoretical and modeling approaches.

Bibliography

Primary Literature

1. Bohr C (1899) Definition und methode zur bestimmung der invasions- und evasionscoefficienten bei der auflösung von gasen in flüssigkeiten. Werthe der genannten constanten sowie der absorptionscoefficienten der kohlensäure bei auflösung in wasser und in chlornatriumlösungen. Ann Phys Chem 68:500–525
2. Bolin B (1960) On the exchange of carbon dioxide between the atmosphere and the sea. Tellus 12(3):274–281. ISSN 2153-3490
3. Revelle R, Suess HE (1957) Carbon dioxide exchange between atmosphere and ocean and the question of an increase of atmospheric CO_2 during the past decades. Tellus 9:18–27
4. Jähne B (1982) Dry deposition of gases over water surfaces (gas exchange). In: Flothmann D (ed) Exchange of air pollutants at the air-earth interface (dry deposition). Battelle Institute, Frankfurt
5. Jähne B (2009) Air-sea gas exchange. In: Steele JH, Turekian KK, Thorpe SA (eds) Encyclopedia ocean sciences. Elsevier, Boston, pp 3434–3444
6. Jähne B, Haußecker H (1998) Air-water gas exchange. Annu Rev Fluid Mech 30:443–468
7. Reichardt H (1951) Vollständige darstellung der turbulenten geschwindigkeitsverteilung in glatten leitungen. Z Angew Math Mech 31:208–219
8. Peng TH, Broecker WS, Mathieu GG, Li Y-H, Bainbridge A (1979) Radon evasion rates in the Atlantic and Pacific oceans as determined during the geosecs program. J Geophys Res 84(C5):2471–2487
9. Roether W, Kromer B (1984) Optimum application of the radon deficit method to obtain air–sea gas exchange rates. In: Brutsaert W, Jirka GH (eds) Gas transfer at water surfaces. Reidel, Hingham, pp 447–457
10. Watson AJ, Upstill-Goddard RC, Liss PS (1991) Air-sea exchange in rough and stormy seas measured by a dual tracer technique. Nature 349(6305):145–147
11. Liss PS, Merlivat L (1986) Air-sea gas exchange rates: introduction and synthesis. In: Buat-Menard P (ed) The role of air-sea exchange in geochemical cycling. Reidel, Boston, pp 113–129
12. Deacon EL (1977) Gas transfer to and across an air-water interface. Tellus 29:363–374
13. Frew NM (1997) The role of organic films in air-sea gas exchange. In: Liss PS, Duce RA (eds) The sea surface and global change. Cambridge University Press, Cambridge, pp 121–171
14. Jacobs C, Nightingale P, Upstill-Goddard R, Kjeld JF, Larsen S, Oost W (2002) Comparison of the deliberate tracer method and eddy covariance measurements to determine the air/sea

transfer velocity of CO_2. In: Saltzman E, Donelan M, Drennan W, Wanninkhof R (eds) Gas transfer at water surfaces. Geophysical Monograph, vol 127. American Geophysical Union
15. Jähne B (1980) Zur Parametrisierung des Gasaustauschs mit Hilfe von Laborexperimenten. Dissertation, Institut für Umweltphysik, Fakultät für Physik und Astronomie, Univ. Heidelberg, http://d-nb.info/810123614. IUP D-145
16. Jähne B (1987) Image sequence analysis of complex physical objects: nonlinear small scale water surface waves. In: Proceedings of 1st international conference on computer vision, London, pp 191–200
17. Jähne B (1991) New experimental results on the parameters influencing air-sea gas exchange. In: Wilhelms SC, Gulliver JS (eds) Air-water mass transfer, Selected papers from the 2nd international symposium on gas transfer at water surfaces, Minneapolis, 11–14 Sep 1990. ASCE, pp 582–592
18. Jähne B, Heinz G, Dietrich W (1987) Measurement of the diffusion coefficients of sparingly soluble gases in water. J Geophys Res 92(C10):10,767–10,776
19. Keeling RF (1993) On the role of large bubbles in air-sea gas exchange and supersaturation in the ocean. J Marine Res 51:237–271
20. King DB, Bryun WJD, Zheng M, Saltzman ES (1995) Uncertainties in the molecular diffusion coefficient of gases in water for use in the estimation of air-sea exchange. In: Jähne B, Monahan E (eds) Air-water gas transfer, Selected papers, 3rd international symposium on air-water gas transfer. AEON, Hanau, pp 13–22
21. McKenna SP, Bock EJ (2006) Physicochemical effects of the marine microlayer on air-sea gas transport. In Gade M, Hühnerfuss H, Korenowski GM (eds) Marine surface films: chemical characteristics, influence on air-sea interactions, and remote sensing. Springer, Berlin/ Heidelberg, pp 77–91
22. Sündermann J (ed) (1986) Landolt-Börnstein, vol V 3c, Oceanography. Springer, Heidelberg
23. Torgersen T, Top Z, Clarke WB, Jenkins WJ (1977) A new method for physical limnology-tritium-helium-3 ages results for Lakes Erie, Huron, and Ontario. Limnol Oceanogr 22:181–193
24. Tschiersch J, Jähne B (1980) Gas Exchange trough a rough water surface in a circular windtunnel; Wave characteristics under limited and unlimited fetch. In: Broecker HC, Hasse L (eds) Berichte aus dem Sonderforschungsbereich 94 Meeresforschung – Symposium on capillary waves and gas exchange, Trier, 2–6 July 1979, number 17. Univ. Hamburg, pp 63–70
25. Wanninkhof R (1992) Relationship between wind speed and gas exchange over the ocean. J Geophys Res 97:7373–7382
26. Woolf D, Leifer I, Nightingale P, Rhee T, Bowyer P, Caulliez G, de Leeuw G, Larsen S, Liddicoat M, Baker J, Andreae M (2007) Modelling of bubble-mediated gas transfer: fundamental principles and a laboratory test. J Marine Syst 66:71–91

Books and Reviews

Banner ML (ed) (1999) The wind-driven air-sea interface, electromagnetic and acoustic sensing, wave dynamics and turbulent fluxes, proceedings of the symposium Sydney, Australia, 11–15 January 1999. University of New South Wales, Sydney

Bengtsson LO, Hammer CU (eds) (2001) Geosphere-biosphere interactions and climate. Cambridge University Press, Cambridge

Borges AV, Wanninkhof R (2007) Fifth international symposium on gas transfer at water surfaces. J Mar Syst 66(1–4):1–308

Brasseur GP, Prinn RG, Pszenny AA (eds) (2003) Atmospheric chemistry in a changing world. An integration and synthesis of a decade of tropospheric chemistry research. Springer, Berlin

Brutsaert W, Jirka GH (eds) (1984) Gas transfer at water surfaces. Reidel, Dordrecht

Csanady GT (2001) Air-sea interaction, laws and mechanisms. Cambridge University Press, Cambridge

Cussler EL (2009) Diffusion – mass transfer in fluid systems, 3rd edn. Cambridge University Press, Cambridge

Danckwerts PV (1970) Gas-liquid reactions. MacGraw-Hill, New York

Davies JT (1972) Turbulence phenomena. An introduction to the eddy transfer of momentum, mass, and heat, particularly at interfaces. Academic, New York/London

Davies JT, Rideal EK (1963) Interfacial phenomena, 2nd edn. Acadamic, New York

Dobson F, Hasse L, Davis R (eds) (1980) Air-sea interaction: instruments and methods. Plenum, New York

Donelan MA, Hui WH, Plant WJ (eds) (1996) The air-sea interface, radio and acoustic sensing, turbulence and wave dynamics, proceedings of the symposium, Marseille, France, 24–30 June 1993. Rosenstiel School of Marine and Atmospheric Science, University of Miami, Miami, Florida

Donelan MA, Drennan WM, Saltzman ES, Wanninkhof R (eds) (2002) Gas transfer at water surfaces. American Geophysical Union, Washington, DC

Emerson SR, Hedges J (2008) Chemical oceanography and the marine carbon cycle. Cambridge University Press, Cambridge

Favre A, Hasselmann K (eds) (1978) Turbulent fluxes through the sea surface, wave dynamics, and prediction, proceedings of the symposium, Marseille, 1977. Plenum, New York

Fogg PGT, Sangster J (2003) Chemicals in the atmosphere: solubility, sources, and reactivity. Wiley, Chichester

Gade M, Hühnerfuss H, Korenowski GM (eds) (2005) Marine surface films: chemical characteristics, influence on air-sea interactions and remote sensing. Springer, Berlin

Garbe CS, Handler RA, Jähne B (eds) (2007) Transport at the air sea interface – measurements, models and parameterizations. Springer, Berlin

Gulliver JS (2007) Introduction to chemical transport in the environment. Cambridge University Press, Cambridge

Jähne B, Monahan EC (eds) (1995) Air-water gas transfer – selected papers from the third international symposium on air-water gas transfer. AEON Verlag & Studio Hanau. http://d-nb.info/946682526

Kantha LH, Clayson CA (2000) Small scale processes in geophysical fluid flows, vol 67, International geophysics series. Acadamic, San Diego

Komori S, McGillis W (eds) (2011) Gas transfer at water surfaces, selected papers from the 6th international symposium. Kyoto University Press, Kyoto

Kraus EB, Businger JA (1994) Atmosphere-ocean interaction, vol 27, 2nd edn, Oxford monographs on geology and geophysics. Oxford University Press, New York

Liss PS, Duce RA (eds) (2005) The sea surface and global change. Cambridge University Press, Cambridge

Robinson IS (2010) Discovering the ocean from space – the unique applications of satellite oceanography. Springer, Heidelberg

Sarmiento JL, Gruber N (2006) Ocean biogeochemical dynamics. Princeton University Press, Princeton

Soloviev A, Lukas R (2006) The near-surface layer of the ocean, vol 31, Atmospheric and oceanographic sciences library. Springer, Dordrecht

Wanninkhof R, Asher WE, Ho DT, Sweeney C, McGillis WR (2009) Advances in quantifying air-sea gas exchange and environmental forcing. Annu Rev Mar Sci 1:213–244

Wilhelms SC, Gulliver JS (eds) (1991) Air-water mass transfer – selected papers from the 2nd international symposium on gas transfer at water surfaces, Minneapolis Minnesota, September 11–14, 1990. American Society of Civil Engineers, New York

Zeebe RE, Wolf-Gladrow DA (2001) CO_2 in seawater: equilibrium, kinetics, isotopes, vol 65, Elsevier oceanography series. Elsevier, Amsterdam

Chapter 9
Sediment–Water Interfaces, Chemical Flux at

Louis J. Thibodeaux and Joseph Germano

Glossary

Benthic boundary layer	A slow moving water layer above the sediment.
Bioturbation transport	A chemical mobility process driven by the presence of macrofauna and macroflora residing near the interface.
Chemical flux	The basic term that quantifies chemical mobility across an interface with units of mass per area per time ($kg/m^2/s$)
Chemical mobility	A general term used to denote the idea that chemicals do move from place to place.
Interface	A real or imaginary plane which separates water from sediment.
Mass transfer rate	The chemical flux times the area perpendicular to its direction of movement (kg/s).
Sediment surface layers	A series of distinctive mud layers occupying thickness of several centimeters depth.
Transport model	One of several concepts for describing a chemical mobility process, and the associated formula or algorithm needed to describe it mathematically (a.k.a., the flux expression).

This chapter was originally published as part of the Encyclopedia of Sustainability Science and Technology edited by Robert A. Meyers. DOI:10.1007/978-1-4419-0851-3

L.J. Thibodeaux (✉)
Cain Department Chemical Engineering, Louisiana State University, Baton Rouge, LA 70820, USA
e-mail: thibod@lsu.edu

J. Germano
Germano & Associates, Inc., 12100 SE 46th Place, Bellevue, WA 98006, USA
e-mail: joe@remots.com

J.S. Gulliver (ed.), *Transport and Fate of Chemicals in the Environment: Selected Entries from the Encyclopedia of Sustainability Science and Technology*,
DOI 10.1007/978-1-4614-5731-2_9, © Springer Science+Business Media New York 2012

Definition of the Subject

Numerous individual transport processes which mobilize chemicals on either side of the interface have been studied. However, a consistent theoretical framework connecting the processes across the interface that correctly quantifies the overall flux remains elusive. This occurs because two fundamentally different individual flux relationships are needed to represent the two very different transport mechanisms needed for quantifying the numerous chemical, biological, and physical processes ongoing at this unique locale. The two basic types of transport processes are the chemical potential driven and the media advection driven. Several theoretical modeling approaches exist for combining these, but all have problematic conceptual features, which will be reviewed. By generalizing flux continuity across the interface, which is the fundamental basis for arriving at the well-known and accepted two-resistance theory, the "interface compartment model" is presented and offered as a unifying theory describing advection-driven and potential-driven transport across the sediment–water interface.

Introduction

All theories for chemical transport across the interface originate from the nineteenth century with the Ohm–Kirchhoff laws of electrical currents and potentials. Lewis and Whitman [1] used an analogous electrical flux approach for deriving the chemical potential–driven flux across a gas–liquid interface. It is presumed that most of the significant, individual chemical transport processes on both sides of the sediment–water interface, which influence the flux of geochemicals as well as the anthropogenic ones, have been discovered. Many have been reported and are the subject of several reviews. They are the result of biological, chemical, and geophysical processes, and most have been verified based on observations in the field and/or in the laboratory. Some have been thoroughly studied while others have not. As a consequence, there are well-developed descriptions for several processes as well as many theoretical equations for the flux. A unified theory is proposed for connecting flux across the interface.

Photographs taken of the interface region, obtained using a sediment profile camera [2, 3] are displayed in Figs. 9.1–9.4. (Figures 9.1–9.4 A collection of color images of the sediment–water interface. These are selected photos taken by Joseph Germano over a time period of 28 years. Four categories are presented.) Fig. 9.1 contains images of the interface and a small sampling of the wide range of effects caused by various macrofauna. Figure 9.2 shows images of the interface being acted upon by submerged aquatic plants with leafy parts in the water column above and holdfasts below. Figure 9.3 has images which indicate an interface under the influence of low oxygen and/or chemical pollutant stresses. Finally, Fig. 9.4 shows the particle advection process. These are but a few glimpses of the character

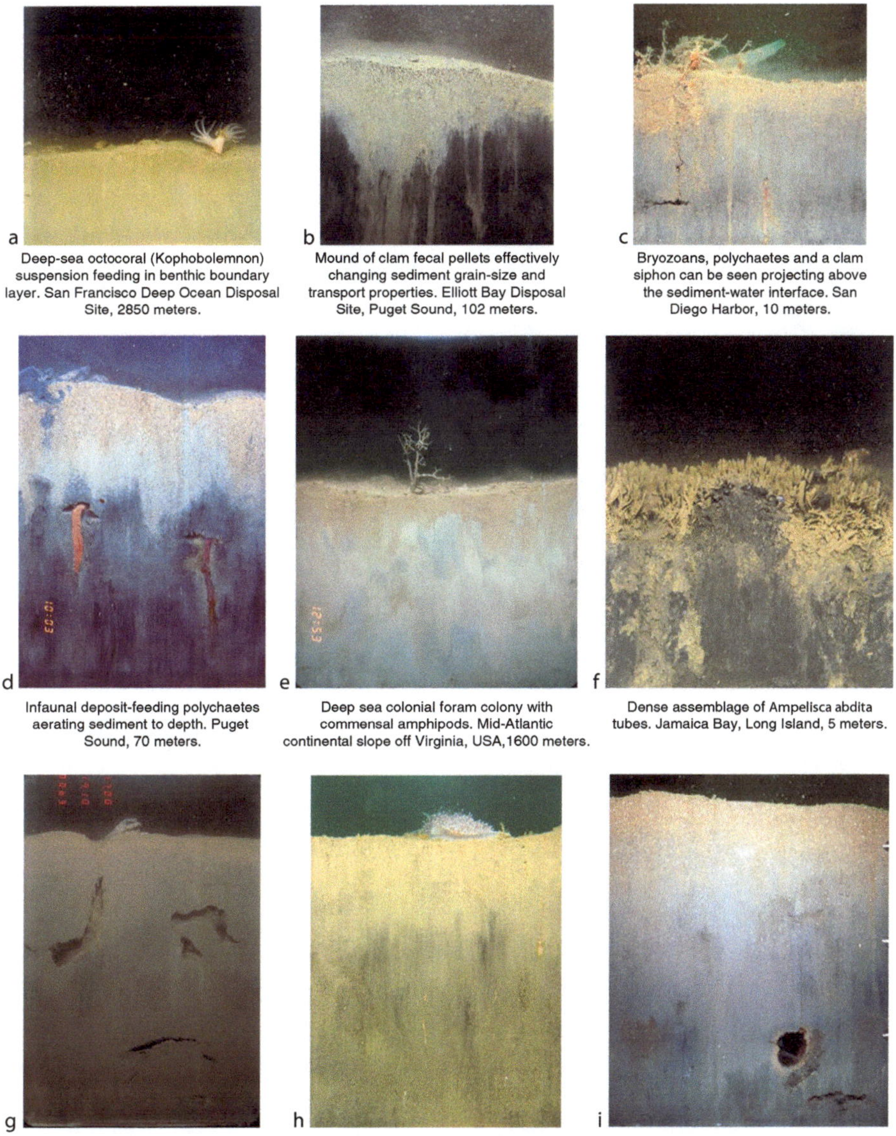

a Deep-sea octocoral (Kophobolemnon) suspension feeding in benthic boundary layer. San Francisco Deep Ocean Disposal Site, 2850 meters.

b Mound of clam fecal pellets effectively changing sediment grain-size and transport properties. Elliott Bay Disposal Site, Puget Sound, 102 meters.

c Bryozoans, polychaetes and a clam siphon can be seen projecting above the sediment-water interface. San Diego Harbor, 10 meters.

d Infaunal deposit-feeding polychaetes aerating sediment to depth. Puget Sound, 70 meters.

e Deep sea colonial foram colony with commensal amphipods. Mid-Atlantic continental slope off Virginia, USA,1600 meters.

f Dense assemblage of Ampelisca abdita tubes. Jamaica Bay, Long Island, 5 meters.

g Subsurface excavations caused by ghost shrimp (Callanassa) visible at sediment surface. Pearl Harbor, Hawaii, 12 meters.

h Deposit-feeding polychaetes are visible in the sediments below the nudibranch at the surface. Nearshore embayment in Unalaska, Aleutian Islands, 36 meters.

i Transected polychaete burrow at depth demonstrates how animals in the "anoxic subsurface sediments" create their own oxygenated mileu. Peconic Estuary, Long Island, NY, 6 meters.

Fig. 9.1 Animal–sediment–fluid relationships (9 photos)

and forms of this important global environmental interface which occupies the largest plane area on Earth. It separates the fluid water–dominated media above and the underlying solid, fluid water–saturated zone below. Figure 9.5 is a conceptual illustration of the interface and its adjoining regions, modified from the original by

a

Subsurface infaunal burrows can be seen beneath the fronds of *Caulerpa prolifera* on the sediment surface. Coastal embayment off Sicily, 14 meters.

b

These sandy clays support a dense assemblage of the invasive *Caulerpa racemosa*. Coastal embayment off Sicily, 8 meters.

c

Mixed assemblage of red algae with a frond of *Laminaria* pushed below the sediment surface by the camera prism. Sinclair Inlet, Puget Sound, 5 meters.

d

Fronds of seagrass can be seen above these silty fine sands. Sinclair Inlet, Puget Sound, 4 meters.

Fig. 9.2 Submerged aquatic vegetation (4 photos)

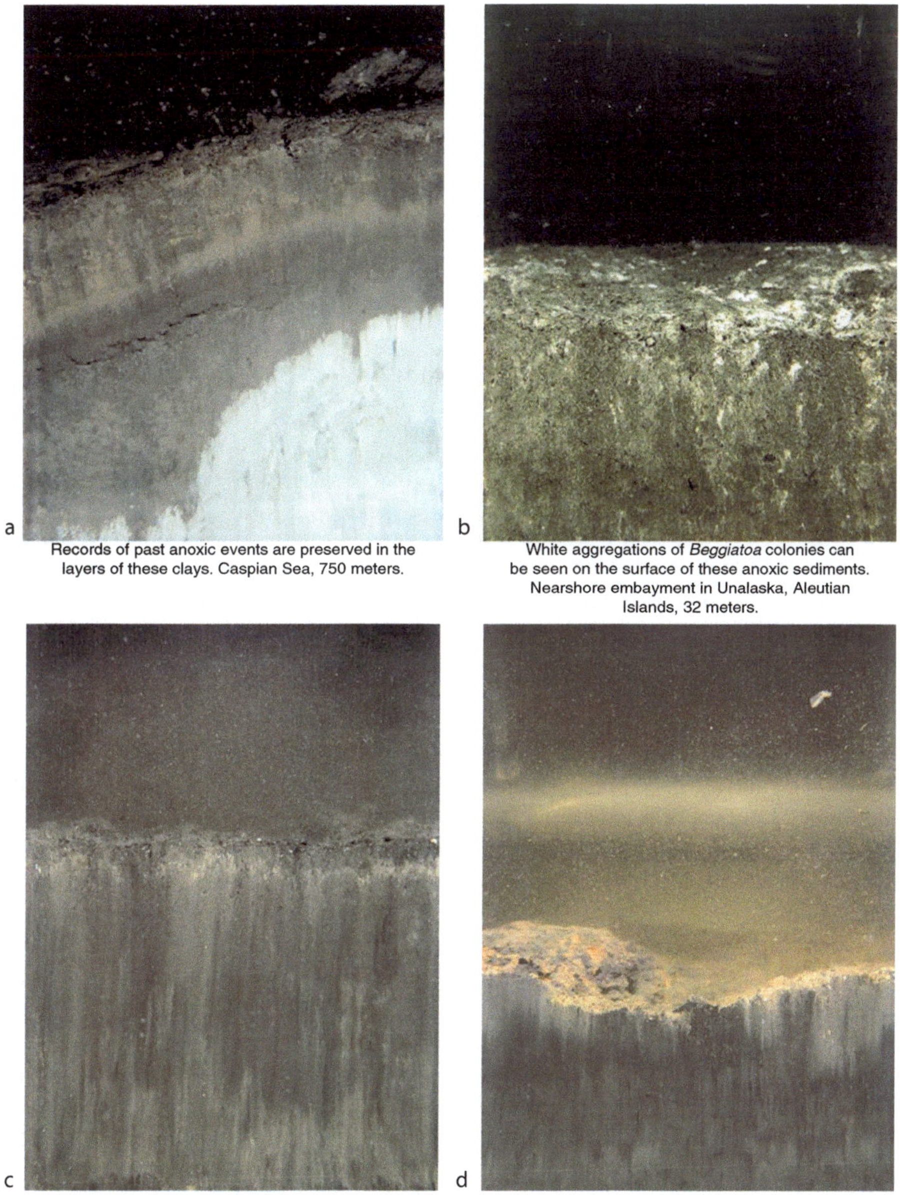

a Records of past anoxic events are preserved in the layers of these clays. Caspian Sea, 750 meters.

b White aggregations of *Beggiatoa* colonies can be seen on the surface of these anoxic sediments. Nearshore embayment in Unalaska, Aleutian Islands, 32 meters.

c No visible RPD can be seen in these organically enriched sediments. Hackensack River, New Jersey, 3 meters.

d A diffusional layer of oxidized sediment can be seen on the surface of these riverine sediments. Hackensack River, New Jersey, 4.5 meters.

Fig. 9.3 Polluted sediments (4 photos)

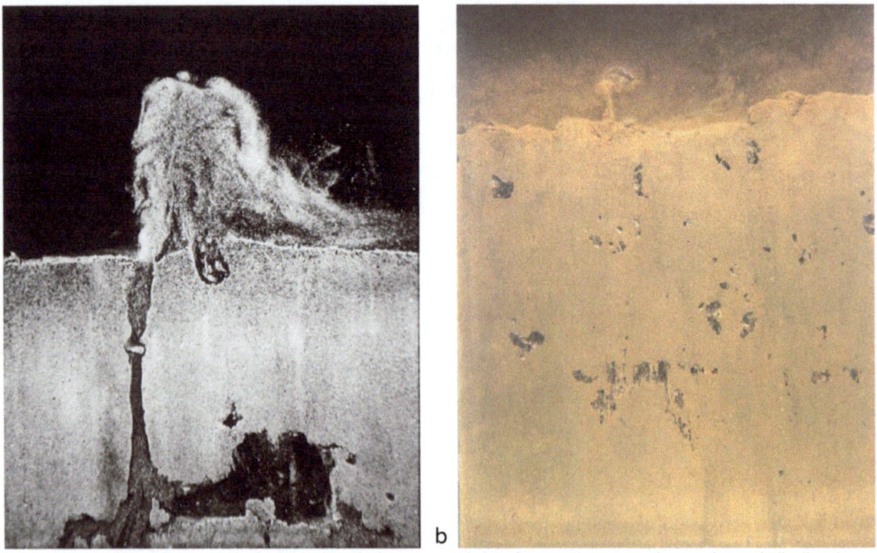

a Deposit-feeding infaunal taxa such as maldanid polychaetes can loosen sediment fabric and eject particles into the benthic boundary layer. Long Island Sound, 15 meters.

b Subsurface methanogenesis can also periodically release gas and sediment particles into the overlying water. Saginaw River , 3 meters.

Fig. 9.4 Examples of particle advection (2 photos)

Santschi et al. [4]. It is an idealized model of the previous images showing the location of the interface plane and some of the processes occurring in the adjoining bulk media phases or compartments.

At any interface locale, it is expected that a combination of individual processes on either side may control the net flux. There may be a dozen or more individual processes, identifying the key ones on either side and coupling them in a logical fashion so as to understand the overall process and quantify the effective flux is an ongoing challenge. As the photographs show, each sediment–water interface has unique characteristics so that the flux is expected to be highly variable from locale to locale. One goal of this chapter is to develop a unified theory for combining the individual processes on the water-side to those on the sediment-side and to obtain the appropriate algorithm for the across-media or interphase flux. The substances of concern are geochemicals such as nitrogen, silica, carbon, and lead and anthropogenic chemicals such as polychlorinated biphenyls, naphthalene, ibuprofen, and caffeine. The development also is applicable to aquasols, nanoparticles, and other identifiable particles moving across the sediment–water interface.

This chapter will first review the available transport processes theories. Several significant individual processes will be listed and summarized. Based on the mechanisms that drive the transport, each will be placed into one of two categories of flux equation types. The theoretical arguments supporting the use of two categories will be covered. Then, a model development section will begin with

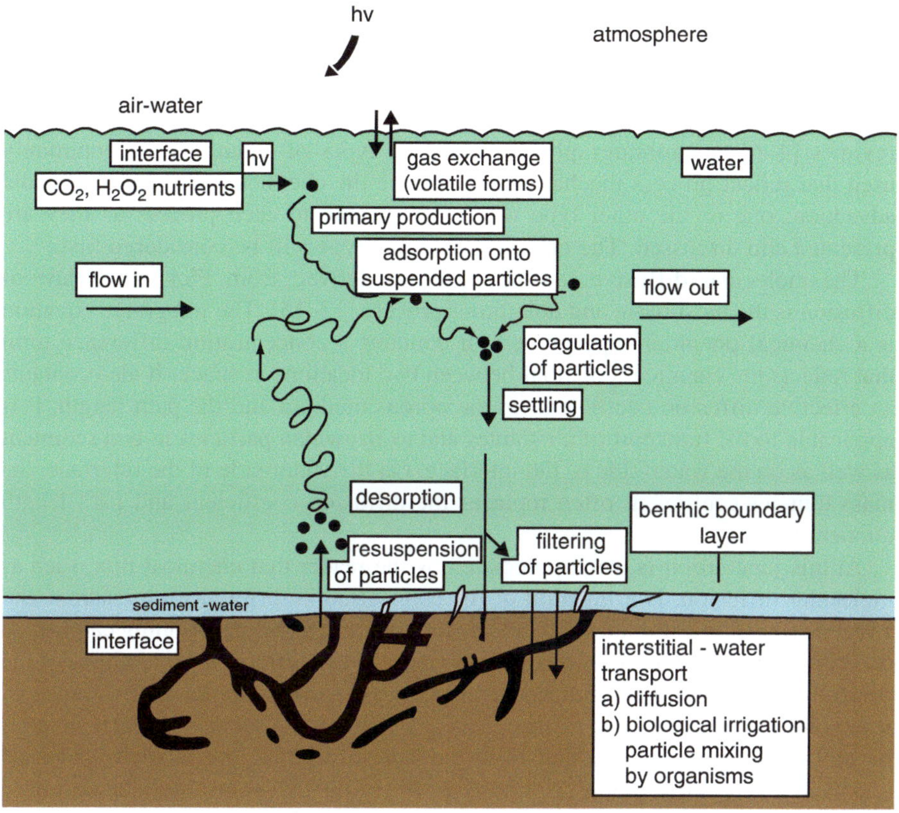

Fig. 9.5 Transport processes near the sediment–water interface. This figure is a slightly modified version of the original produced by Santschi et al. [4]. It is a classical graphical illustration of the interface plane. It is an idealized model of the interface region showing some of the processes occurring in the water column and the bed surface layers

a literature review and summary of the various theoretical approaches proposed for combining the individual processes arriving at the net flux across the interface. The algebraic forms of the flux equations will be given. Finally, the concepts behind the proposed interface compartment (IC) theory will be presented and the flux equation derived. Included will be a discussion of the IC theory in relation to the existing ad hoc protocols in use.

The methods section will describe the modeling approach, the chemicals used in numerical simulations, and the individual processes selected for performing the flux calculations. Including flux calculations in the theoretical section is necessary because it extends and amplifies the IC model theory by providing a layer of reality to accompany the mathematical formulas. In addition, it provides a quantitative means of comparing it to the approaches being used.

Transport Process Theories

There are numerous individual chemical transport processes operating on both sides of the sediment–water interface, and they have been the subject of literature reviews [4–6] and monographs [7–12]. Two types of equations are commonly used that reflect process mechanisms. They are the chemical potential and media advection; one or the other type will be applicable to each process as they are presented and discussed. The chemical potential type will be considered first.

The molecular diffusive transport process derived from Fick's first law of diffusion is the most basic and ubiquitous process [13–15]. The integrated equation is a chemical potential–type process. It contains a concentration difference term that reflects the chemical potential between two locations in space. It also contains an effective diffusion coefficient for the porous medium and the path length. It is applicable to the transport of all solutes and to Brownian particles in bed sediment as well as on the water-side of the interface. On the water-side of the interface, the mass transfer coefficient often replaces the diffusion coefficient and path length quotient [14].

At this juncture, it is well appreciated by the reader that chemical flux, such as molecular diffusion, is a function of the concentration difference or gradient and a kinetic transport parameter. Chemical reactions and phase partitioning can and do occur in the layers on either side of the interface, which affects the magnitude of the respective concentrations and hence the flux. However, applying the transient, reactive-diffusion equation to the chemical species of interest in each layer is beyond the scope of this article. For the sake of clarity, the tactical analytical approach taken in this manuscript portrays the sediment–water interface region so as to isolate and focus only on the transport processes. Therefore, the reader should note that the two layers of interest are assumed to be very thin, void of chemical reactions within (i.e., degradation, oxidation/reduction, polymerization, etc.), and have constant concentration differences across them. In other words, the chemical species are conserved, and a steady-state flux is occurring in the defined sediment–water interface region.

The term Brownian diffusion is conventionally applied to very small particles that respond to the kinetic motions of the surrounding solvent molecules. The transport of colloids, also termed aquasols, in the water on either side of the interface is quantified by a chemical potential–type transport process. These are present in the form of dissolved organic carbon (DOC) particles or inorganic particles with sorbed chemical fractions. The bioturbation transport process of particles in the bed as well as in the adjoining porewater is driven by the presence of macrofauna. It is consistent with the chemical potential–type transport mechanism in that the randomness of a collection of macrofauna-driven particle and fluid motions mimics the molecular kinetic mechanism of diffusion on a larger physical scale. For this reason, it is termed a biodiffusion process and is treated as such mathematically [9, 14, 16–19]. Bioturbation has also been depicted as a convective transport process. This typically involves use of bed turnover rate and nonlocal particle movement rates. However, the potential-based biodiffusion also has theoretical merit in that the range of macrofauna sizes and transport lengths tends to approach a Gaussian distribution, indicating that the process can be described by a diffusion type of equation. In addition, the biodiffusion model has an extensive database of field- and laboratory-measured biodiffusion transport coefficients that convective transport lacks for this process [19].

The two generic forms of the chemical potential–type flux equations are

$$F = (D/h)\left(C_S - C_i\right) \tag{9.1}$$

and

$$F = K\left(C_i - C_W\right), \tag{9.2}$$

where F (kg/m^2/s) is the flux, D is the diffusion coefficient (m^2/s), h is the path length (m), K is the convective mass transfer coefficient (m/s), and C (kg/m^3) represents the concentration in water for the sediment bed (s), the interface (i), and the water column (w) beyond the benthic boundary layer, respectively.

The media convective–type rate equation for chemical flux reflects a transport mechanism driven by the directed motion of a bulk media. Several types operate across the interface region. In-bed porewater convection moves solutes and fine particles in both directions. The porewater flow direction responds to hydraulic pressure differences across the bed layers. These can be long-range pressure differences such as in-bank and water column head differences or localized pressure differences generated by the flowing water column as it encounters local bottom roughness such as sand waves, mounds, etc. [20]. In either case, the chemical flux is the product of the effective Darcian porewater velocity and its aqueous concentration.

Solid particles moving through the water-side benthic boundary layer are also a media convective–type flux. The primary ones are particle deposition onto the bed from the water column and resuspension from the bed surface. These transport mechanisms are initiated and maintained by the action of the flowing water. The fluid-generated shear stress at the bed surface drives particle movements into suspension as well as change the particle deposition probability [21, 22]. However, particles and so-called marine snow are also formed in the water column [23], and others originate as wind-blown dust on the sea surface, etc.; all types are deposited onto the bed. The physics of cohesive and noncohesive sediment transport employs complex algorithms for their estimation [24, 25]. For the purpose of this manuscript, a deposition velocity and a resuspension velocity will be used to characterize the respective processes. In each case, the flux is equal to a velocity–concentration product so the generic forms of the media convective–type equation are

$$F = v_w C_w \tag{9.3}$$

and

$$F = v_p C_p, \tag{9.4}$$

where v_w and v_p (m/s) are the effective media velocities of water and particles perpendicular to the interface plane, and C_w and C_p are the media concentrations.

The above review covers the most well-known, characterized, and quantified individual processes. Several other processes have been observed and described. One termed the "benthic cannon" is dramatic and appears in Fig. 9.4. As shown, the organism responsible for this phenomenon (a maldanid polychaete) is seen injecting a spray of fine particles from its burrow into the lower portion of the water column. In a similar mechanism, gas bubbles generated within the bed can also move to the interface and emerge, likewise injecting fine particles into the water-side boundary layer (Fig. 9.4). The role of the nepheloid layer, made up of submerged aquatic vegetation and other macrofauna activities that enhance and attenuate chemical transport processes, remains to be studied and quantified. Due to the lack of sufficient information on these and other individual processes, they cannot be included in numerical simulations at this time.

Several computational studies have been performed aimed at comparing aspects of various individual processes. The most comprehensive of these is a study of trichlorobiphenyl (TCP) for nine in-bed transport mechanisms [26]. Individual processes were ranked by characteristic times-of-recovery of TCP in freshwater riverine bed sediment. It was concluded that in high-energy environments, sediment transport was likely the dominant sediment-side TCP transport process, while in low-energy environments, bioturbation was likely to dominate the movement rate of TCP in the upper layer of the bed. Singh et al. [21] developed a framework for a comprehensive mathematical model for fine and cohesive sediment transport to be combined with contaminant transport models in rivers, lakes, and estuaries; they observed that much research needs to be done before truly realistic chemical exchange models for the sediment–water interface will be available for practical use. They further noted that sediment transport and chemical transport must be meshed in development of a comprehensive model. In assessing the soluble release process of polychlorinated biphenyls (PCBs) from bed sediment in three North American rivers, Thibodeaux et al. [27] evaluated five individual transport processes by comparing the magnitudes of the mass transfer coefficients. However, such individual process studies do not completely address the interconnections that result in across-media transport at the sediment–water interface. Comprehensive studies of the interconnections of transport processes are lacking. A study limited to solute transport of polychlorinated biphenyls in the Hudson River highlights the importance of connecting processes across the interface plane [28, 29]. It was found that during active in-bed bioturbation, the transport resistance on the water-side benthic boundary layer is significant and that both of these processes regulate the PCB flux from the bed.

The number and complexity of the biological, chemical, and geophysical processes cooperating to drive chemical flux across the interface is daunting, and sorting out the cause-and-effect factors is confusing without the aid of theoretical guidance. Ad hoc approaches are in use, but all have theoretical shortcomings and may not, therefore, extend into areas outside of the data set. A robust theory that can accommodate the types of various individual chemical transport processes on either side of the interface and connect them in a logical and transparent procedure is needed for several reasons. In the first place, there appear to be none available. Second, having one will lead to much better understanding of the overall situation related to transport and will provide a hypothesis for interpretation of flux data from both laboratory and field measurements. Third, modelers of aquatic chemodynamic processes need a theory-based procedure for connecting chemical movement between the adjoining bulk-phase compartments based on first principles.

Theories and Model Development

G. S. Ohm (1787–1854) found that the electric current (I) is directly proportional to the difference in voltaic potential between the ends of a conductor (V) and the proportionality constant is R^{-1}, where R is the resistance. G. R. Kirchhoff (1824–1887) extended Ohm's law. By analogy, the Ohm–Kirchhoff laws were applied by Whitman [30] and Lewis and Whitman [1] to chemical flux (F) across a gas–liquid interface plane. The emf potential, V, was replaced by chemical concentration potential such as ($C_s - C_i$), and the result is the well-known resistance-in-series (RIS) law for chemical mass transfer. When applied to the solute at the sediment–water interface the flux equation is

$$F = (C_s - C_w)/\left[\frac{1}{B} + \frac{1}{D}\right], \tag{9.5}$$

where B and D (m/s) are the water-side and sediment-side mass transport coefficients, respectively. Because of chemical-to-water partitioning, with coefficient K_d (L/kg), the $C_s = w_s/K_d$, where w_s (g/kg) is chemical loading on sediment solids and $D = D_b\, \rho\, K_d/h$, where D_b (m^2/s) is the biodiffusion transport coefficient, ρ (kg/L) particle density, and h (m) bed depth. This result has been verified using field data [29, 31].

Resistance-in-series is applicable only to potential-driven processes. It is often time referred to as Ohm's law [32, 33] and has become a shortcut modeling approach. Misapplication occurs, for example, when media velocity is treated as a potential-type transport coefficient and likewise added in the resistance fashion to obtain the overall resistance. It has been misapplied for atmospheric deposition of gases and particles where both potential and convective processes resistances are summed [34, 35]. Hybrid applications that involve converting the chemical

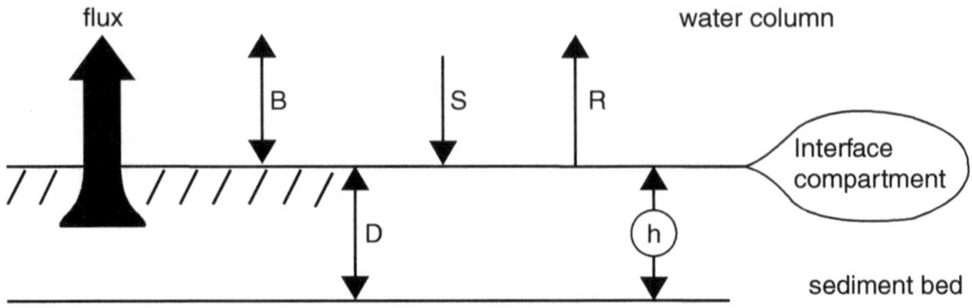

Fig. 9.6 The essence of the interface compartment. The bubble indicates the location of the IC. It is a plane surface so the line depicts one edge. The large arrow to the left depicts the chemical flux and direction. Transport begins in the sediment bed, crosses the interface, and emerges into the water column. This net flux is a result of four individual processes. The B process is solute transport through the water-side benthic boundary layer. The S process is particle deposition (i.e., settling) onto the bed surface and R is particle resuspension from the bed surface. On the sediment side, the D denotes a diffusive process across a distance h. This illustration is used in deriving the IC model equation (i.e., Eq. 9.8)

potential resistance to an equivalent media convective type for use with the adjoining bulk-phase concentrations appear in multimedia compartment (MMC) models [33, 36] for the sediment–water interface. As presented in the next section, a return to first principles embodied in the Ohm–Kirchhoff and Lewis–Whitman laws will mend the problem and produce a unified theoretical construct that accommodates both types individually and in unison.

Environmental models contain multiple phases so interfaces are necessary. The sediment–water interface is depicted as an idealized concept in Fig. 9.5. It is an imaginary plane within the complex transition zones separating bulk water phase from bulk sediment phase as depicted in the Figs. 9.1 to 9.4. The "interfacial compartment" is defined based on the following assumptions and concepts: (a) it is a two-dimensional surface containing no mass that separates the adjoining bulk phases, (b) the chemical flux direction is perpendicular to the surface plane, (c) the net entering and departing fluxes are equal, (d) a hypothetical aqueous chemical concentrations with mass per volume units (mg/m^3) is assumed, and (e) because solid (i.e., particle) phases exist on one side, chemical equilibrium is assumed to exist at the interface plane for estimating the equivalent aqueous concentration. A mass balance on the interface plane is performed. Because of assumption, a steady-state equation results and yields a simple algebraic relationship for the interface compartment concentration. It in turn yields a single flux equation containing the individual process mass transfer coefficients and the bulk chemical concentrations in the adjoining media compartments.

Although there are many individual processes occurring in the region of the interface, for demonstration purpose, only four will be used in the following derivation of the interfacial compartment model. This approach simplifies the mathematics of the theoretical procedure while maintaining the essence of the concept. The combination of chemical potential–type and media convective–type transport mechanisms used is shown in Fig. 9.6. The double-tipped arrows on either side of the interface are the chemical potential type. They represent respectively the solute transport across the benthic boundary layer with coefficient B (m/s) and diffusive transport within the upper sediment layers with coefficient D/h (m/s), where the layer thickness is h (m). The single-tipped arrows depict the media convective–type flux equations. They represent particle deposition or settling with coefficient S (m/s) and particle resuspension with transport coefficient R (m/s). The assumed chemical movement pathway and direction of flux F ($kg/m^2/s$) is depicted by the large arrow. The sediment bed porewater concentration, C_S (kg/m^3), in equilibrium with the bed load fraction, w_a (mg/kg), is at position h, and the bulk water column concentration is C_W (kg/m^3) and is located at the edge of the benthic boundary layer. The overall chemical potential driving force is the concentration difference, $C_S - C_W$.

A steady-state Lavoisier mass balance for the interface compartment requires that the flux from the sediment-side to the interface, F_{si}, equals that departing the interface on the water-side, F_{iw}. It is

$$F_{si} = (D/h)(C_s - C_i) = B(C_i - C_w) + RC_i - SC_w = F_{iw} \tag{9.6}$$

This result yields the concentration in the interface compartment:

$$C_i = ((D/h)C_s + (S + B)C_w)/(D/h + B + R) \tag{9.7}$$

Combining the two equations yields flux between the compartments in terms of the bulk-phase concentrations:

$$F_{IC} = \frac{C_s(1 + R/B) - C_W(1 + S/B)}{h/D + 1/B + Rh/BD} \tag{9.8}$$

This result is the interface compartment model (IC) flux; it is consistent with the traditional potential flux in that it takes the form of Eq. 9.5 when the convective parameters S and R are set to zero. It is the opinion of the authors that the above procedure is the correct one. However, alternative flux relationships have been proposed for the across-interface flux based on various assumptions and methodologies. Two commonly used approaches appear below.

Invoking the RIS concept directly by mimicking the form of Equation 9.5 is one approach used [32]. In its derivation, the water-side conductances are summed and then inverted to obtain the overall water-side resistance which is then added to the sediment-side resistance. This procedure yields

$$F_{RIS} = \frac{C_s - C_W}{h/D + 1/(B + R - S)} \tag{9.9}$$

for the RIS flux equation. The multimedia compartment or box model approach is a hybrid [33]. Its derivation starts by decomposing chemical potential flux into two individual convective-type flux components. It then uses the Lavoisier mass balance for summing the individual fluxes but assumes all are driven by the bulk compartment concentrations. The multimedia compartment (MMC) flux equation is

$$F_{MC} = \frac{C_s - C_w}{h/D + 1/B} + (RC_s - SC_w) \tag{9.10}$$

The flux equation for the RIS and MMC models reduce to Eq. 9.5 when S and R are set to zero.

At this juncture, it is clear that three very different algebraic algorithms based on as many approaches are available for estimating the flux across the sediment–water interface. In all cases, the fluxes are linear in relation to the bulk media concentrations and contain the appropriate conductance. Only in the case of the RIS model must the bulk media concentrations be equal for a zero flux. For both the IC model and the MMC model equations, a simple algebraic proof shows that positive, nonzero bulk media concentrations can yield a zero flux. This is a more realistic outcome because in nature it is possible to have a situation where a mix of conductance produces a zero net flux and the bulk media concentration is unequal. Numerical simulations using the theoretical models in flux calculations provide a realistic and quantitative means demonstrating these and other outcomes. The methods used and the results of the numerical flux calculations are presented and discussed in the next section.

Simulation Methods and Results

Although only four individual transport processes were used in the development of the three theoretical models presented in the previous section, nine individual transport processes will be used in the numerical simulation. This is done in order to realistically mimic and highlight the most significant process typically present in the sediment–water interface region. The four on the water-side include solute transport in the benthic boundary layer, particle resuspension from the bed surface, particle deposition from the water column onto the surface, and colloid Brownian transport through the benthic boundary layer. The five on the bed-side are colloid Brownian diffusion in the porous bed, Darcian water advection into and out the bed, solute molecular diffusion in porewater, particle biodiffusion, and porewater biodiffusion. Altogether, there are five chemical potential–type flux expressions and five media advection type. The types and categories of the processes plus the base case numerical values of the parameters are summarized in Table 9.1. The top four lines represent the water-side transport coefficients and the remaining six represent the bed-side transport coefficients. The large numerical difference in the PCB versus BZ transport coefficients are due to partitioning for the particle-associated processes. For details on how the bed and water column transport

Table 9.1 Transport coefficients (m/day)

Name, location, and type[a]	Benzene	PCB
Solute, water-side mass transfer coefficient, cp.	0.32	0.30
Particle resuspension, ma.	5.72E-5	1.24
Particle deposition, ma.	2.0E-9	4.32E-5
Colloid, water-side, ma.	2.31E-7	4.99E-3
Colloid, bed-side, cp.	2.31E-4	2.31E-4
Darcian velocity into bed (−), ma.	4.35E-4	3.99E-4
Darcian velocity from bed (+), ma.	1.18E-3	1.08E-3
Solute diffusion in bed, cp.	7.48E-4	6.85E-4
Particle biodiffusion, cp.	1.14E-5	0.25
Porewater biodiffusion, cp.	2.0E-4	2.0E-4

[a]*cp* chemical potential, *ma* media advection–type flux

coefficients are calculated, see Chapters 10 through 13 in the *Handbook of Chemical Mass Transport in the Environment* [37].

Data available in the literature on several North American rivers and lakes with bed sediment and water column contaminated with organic chemicals were used. Several studies [27, 29, 38–41] provide the necessary physical, chemical, and biological data and information needed for estimating the bed and water column transport coefficients used in the calculations. Typical bed and water column characteristics at 25°C and 3 m water column depth were used. These characteristics were bed porosity, 75%; bulk density, 572 kg/m^3; sediment layer thickness for active transport, 0.05 m; fraction organic carbon in bed solids, 50 g/kg; dissolved organic carbon (DOC) in the porewater, 50 g/m^3; particle biodiffusion coefficient, 2E-6 m^2/day; porewater biodiffusion coefficient, 2E-5 m^2/day; colloid Brownian diffusivity, 1.61E-5 m^2/day; Darcian water convection into bed, 4E-4 m/day, and out, 1.1E-3 m/day; Peclet number = 1; water column suspended particle concentration, 0.005 kg/m^3; particle deposition velocity, 4E-4 m/day; particle resuspension velocity, 1E-4 m/day; and colloid benthic boundary layer transport coefficient, 2.3E-4 m/day.

The porewater chemical concentrations are separated into dissolved and particle-bound DOC fractions and the fractions transported separately, the dissolved as solute molecular diffusion and the DOC as Brownian particle diffusion. The chemical equilibrium phase distribution partition coefficient is used to relate the solute and DOC-bound concentrations. For characterizing the physical properties of the bed, the New Bedford Harbor estuary site was used [41]. The chemical 2,4,2’,4’-tetrachlorobiphenyl was used as the PCB. Its partition coefficient was 21.6 m^3/kg and that used for benzene (BZ) was 0.001 m^3/kg. The tabulated molecular diffusivity in water for each was used [14]. For the benthic boundary layer, solute transport coefficients were based on those from the Hudson River; they were 0.32 and 0.30 m/day for BZ and PCB, respectively. These two chemicals represent the extremes of hydrophobic properties typically encountered in contaminated bed sediments. In addition, they also represent the extremes of numerous soluble and particle-phase geochemicals.

Table 9.2 PCB flux (g/m^2·d) increasingly active processes

Active processes C_s =.065, C_w =.020	C_i(g/m^3)	IC	RIS	MMC
1. Mol. diff., bed porosity 0.1%	.020	4.3E-9	4.3E-9	5.0E-8
2. Mol. diff., bed 78% porosity	.021	3.1E-5	3.1E-5	3.18-5
3. Water advection into bed, Pe = −10	.0196	−1.3E-4	−2.8E-3	−4.1E-4
4. Mild resuspension, Pe = +10, DOC on.	.0195	4.5E-4	3.3E-4	2.4E-3
5. Aggressive resuspension	.00418	4.5E-4	3.4E-4	8.1E-2
6. "Storm event" resuspension	.00013	4.5E-4	3.4E-4	3.2
7. Mild resusp., mild in-bed biodiff.	.0468	9.4E-3	9.0E-3	1.1E-2
8. Mild resusp., aggressive biodiff.	.0648	1.5E-2	1.5E-2	1.6E-2
9. "Storm" resusp., aggressive biodiff.	.0413	2.1	1.4	3.2

The data appearing above were used with the appropriate algorithms and formulations to estimate the numerical values of the nine transport coefficients [14]. Water (porewater in the bed and in the column above) is the continuous phase across the interface. The transport coefficients use chemical concentrations in water for flux calculations with both the chemical potential–type and media advection–type equations. A summary of typical numerical values of the nine transport coefficients appear in Table 9.1 for BZ and PCB. Several of the base-case transport parameters were perturbed in doing numerical simulations to cover their expected range of variation.

The results of the first numerical study appear in Table 9.2. Calculated fluxes for the PCB using the three theoretical models, the interface compartment (IC) model, resistance-in-series (RIS) model, and multimedia compartment (MMC) model, appear. In addition, the IC model interface concentration, C_i (g/m^3), is given. This simulation uses a porewater concentration of 0.065 g/m^3 and water column 0.020 for a chemical potential gradient driving the PCB flux from the bed to the water column for positive flux values with unit g/m^2·d. The negative values denote fluxes directed into the bed. The first numerical study was to assess the role of transport aggressiveness or intensity on the flux.

The first three simulations (i.e., 1, 2, and 3) represent molecular diffusion and in-bed water advection. The flux varies from low positive to negative. Low bed porosity will produce a low flux; all models have essentially the same values for simulation 2. Porewater advection in the opposite direction is sufficient to reverse flux direction as shown in simulation 3. This reverse in flux behavior is present in all three models, but flux numbers are different for each. This is expected because the algebraic forms are different (see Eqs. 9.8, 9.9, and 9.10).

These IC model simulations reveal an interconnection between the media convective–type erosion process on the water-side and the chemical potential–type diffusive process in the bed (see simulations 4, 5, and 6). In the absence of biodiffusion, which is the case for these three simulations, the flux for the IC and RIS models remains unchanged with increasing particle resuspension aggressiveness. This occurs because the in-bed transport processes are slow; they are combined porewater advection and molecular diffusion. The resistance-in-series law is operating properly in the IC and RIS models, where the slowest process controls. However, an equivalent resistance-in-series functionality is absent in the MMC model, and the flux tracks particle resuspension directly, resulting in substantially higher fluxes. The generally higher MMC model flux values are due to the porewater solute concentration because it, rather than the interface concentration, is the driver and is always numerically larger.

More interesting process interconnections are revealed in simulations 7, 8, and 9. In these cases, in-bed particle and porewater biodiffusion are active as well as particle resuspension. The flux increases for both the IC and RIS model simulations as they track the level of biodiffusion; however, the MMC model does not change much because it is already high. This behavior by the IC and RIS models shows that an active bed-side process must be present to provide chemical mass readily available and in the upmost layer in order for erosion to be an effective transport process. The rapid biodiffusion provides the mass while the slower molecular diffusion and water advection processes cannot. Finally, all three models give high and approximately equal flux for the most aggressive resuspension and biodiffusion transport coefficients (see simulation 9).

To summarize, the simulation study starts with mild in-bed passive molecular diffusion and a very low-porosity sediment layer. The flux is very low. As one moves down the table, level of transport aggressiveness increases. The final one is for aggressive particle biodiffusion and hydraulic flows that result in aggressive particle resuspension. The flux is very high in this case. The variation in flux from low and high is approximately 10^9. The flux numbers for the IC and RIS models are similar in magnitude while the MMC is consistently higher and, in some cases, much higher. Its flux behavior seems to track the particle resuspension process in aggressiveness. Clearly, Eq. 9.10 supports this behavior.

The flux behavior of BZ is somewhat different; no tabular data are provided. The behavior for simulations 1, 2, and 3 are similar to the PCB ones. Low porosity yields low flux, and into bed water advection can reverse the weak diffusion-driven flux. Being less hydrophobic, BZ displays limited sorption to their surfaces. For this reason the in-bed particle transport is not a significant chemical mobilization process for BZ. For simulations 2 through 9, all three models give essentially the same numerical flux values. All models reflect no particle process dependence, and all display only chemical potential flux–driven behavior patterns. Theoretically, in the absence of particle processes, all three models are equal and become identical to Eq. 9.5.

The second numerical study was on the in-bed concentration gradient difference polarity and the effect on the direction of the flux. Equations 9.8, 9.9, and 9.10 show each model has a different mathematical dependence on the porewater and water column concentration. In the above simulations, the concentration difference was set to simulate chemical transport from the bed, so for PCB, the concentration differences as C_s were =.065 and 0.02 g/m^3, and for BZ, they were 20 and 2.0 g/m^3. So as to further test the models for realistic flux behavior characteristics, the

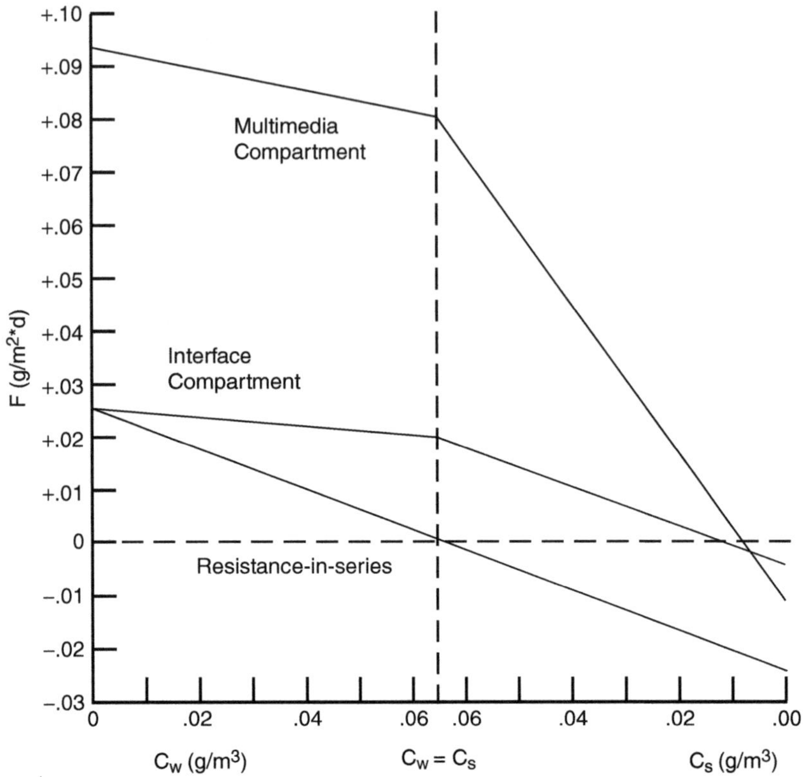

Fig. 9.7 PCB flux across the sediment–water interface. A positive flux, F, is from the bed to the water column and negative is from water to the bed. The concentration differences between bed and water column used in the calculations are displayed on the ordinate. The progression of concentration difference values was devised to force the chemical potential out the bed, on the left, and into the bed on the right so as to drive the flux accordingly. At the vertical dotted line, bed and water concentrations are equal. Only for the RIS model is the flux zero. The three lines depict the theoretical behavior of the PCB flux for the three models: IC, RIS, and MMC

chemical potential gradient range and the direction were reversed; the flux results for reversing the gradient direction appear in Figs. 9.7 and 9.8. In each figure, the flux is on the vertical axis and the imposed concentration difference on the horizontal axis. Each line in Fig. 9.7 represents a model from top to bottom; they are MMC, IC, and RIS.

Figure 9.7 presents aspects of the flux behavior. The vertical axis displays the flux. Positive numbers represent the PCB moving from the bed to the water column. Negative numbers represent the PCB moving from the water column to the bed. The zero flux indicates no net chemical movement in either direction. The horizontal axis displays chemical concentrations in the water column and the porewater. It is an unusual axis in that it has zero on each end and maximum in the center. The horizontal axis consists of two sections divided vertically by a dotted line.

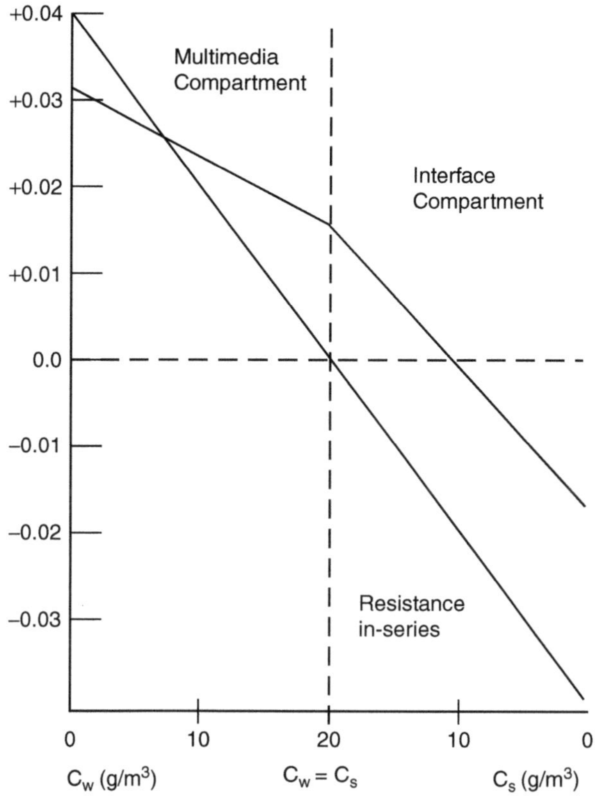

Fig. 9.8 Benzene flux across the sediment–water interface. A positive flux, F, is from the bed to the water column and negative is from water into the bed. The concentration differences between bed and water column used in the calculations are displayed on the ordinate. The progression of concentration values were devised to force the chemical potential out the bed, on the left, and into the bed on the right, so as to drive the flux accordingly. The two lines depict the theoretical behavior of the BZ flux for the three models: IC, RIS, and MMC. The IC and MMC flux values overlap

The concentration is a maximum at the position of the dotted line. To the left side of the vertical line, the porewater concentration in the bed, C_S, is held constant at a value of 0.065 g/m^3 and the water column concentration, C_W, is varied from 0.0 to 0.065 g/m^3 as shown. Under these conditions, the concentration difference tends to move the PCB from the bed to the water column and produce a positive flux. It does for all three models. Only for the RIS model does the flux equal zero for the condition where $C_S = C_W = 0.065$. Due to the particle resuspension process, the PCB flux result for both the IC and MMC models remain positive and nonzero when moving from the bed to the water column. In the next simulation, concentration levels in bed porewater versus the water column concentration will be reversed.

Consider concentration conditions to the right of the dotted line in Fig. 9.7. As shown on the horizontal axis, the water column concentration, C_W, is held constant at 0.065 g/m^3 and the porewater concentration, C_S, is reduced from 0.065 down to 0.0 g/m^3. The flux for the RIS model starts at zero and goes negative as the PCB moves into the bed. The other model results display a positive flux at the start and both trends downward. The MMC model shows a steep decline and reaches zero flux at C_S of about 0.01 g/m^3. The IC flux declines to the same value as well, and this occurs because the imposed concentration gradient encourages PCB movement from the water column to bed.

The point of the above flux versus concentration study for PCBs was to show the different behavior patterns produced by the three models. The horizontal concentration axis was contrived to force the flux to range in magnitude from positive to negative and therefore cover all possible conditions to be encountered in nature. The IC model is theory-based, and so it characterizes the correct PCB flux behavior. Hopefully, future experiments will be able to verify or refute this behavior. The RIS model predicts lower flux values while those for the MMC model are much higher. All three models behave according to the algebraic forms of 9.8, 9.9, and Eqs. 9.10 for the IC, RIS, and MMC, respectively. Both the IC and the MMC models have a flux inflection; this suggests they are better quantitative representations of the overall flux process. Crossing the $C_w = C_s$ line and interchanging concentration represents a switching of processes; it is in the algebra (see Eqs. 9.8 and 9.10).

A somewhat different behavior occurs for BZ. The results are displayed in Fig. 9.8 where the graph is constructed similar to that for PCB. The RIS and MMC flux values are identical and appear as a single diagonal line that starts positive, goes to zero, and then ends negative. The IC model starts at a slightly lower positive flux. It decreases then goes through a slope change at $C_S = C_W$. This is an algebra-driven inflection point in the IC model; it occurs at $C_S = C_W = 20$ g/m^3 (see Eq. 9.8). As C_S decreases toward zero, the flux goes into a steep decline with increasing negative values. It parallels the behavior of the RIS and MC models but with slightly higher flux values. Beyond the IC model, flux is higher than the others. The particle transport process is low, and the MMC model takes the same algebraic form as the RIS. However, the IC model reflects the correct theoretical approach and displays a very different behavior pattern. Being water soluble, BZ has minimal particle association. However, the particle processes enter as ratios (see Eq. 9.8). This has the effect of delaying the zero flux. It occurs at C_S of 11.0 g/m^3.

The flux results for the IC versus RIS and MMC models using this specific simulation with BZ and PCB are different numerically. They also display different behavior patterns as the imposed concentration gradient condition is changed to encourage chemical movement from the bed to a condition of chemical movement into the bed. Presumably, the IC model result is the correct one since it is theory-based. However, at this juncture, it is a hypothesis in need of testing against laboratory and field experimental measurements.

Significance for Aquatic Environments

There are numerous individual transport processes on both sides of the interface that are driven by biological, chemical and geophysical phenomena. Some processes work in parallel while others work in series, forming a connected network of processes moving anthropogenic substances and geochemicals across the interface. The interface compartment (IC) concept and associated mathematical model is developed and presented as the appropriate theoretical approach for understanding the overall process and quantifying the resulting net flux. For aquatic researchers, it is a tool with several uses. It provides a testable hypothesis and a means of interpreting rate data based from measurements in the laboratory or field. It provides a mathematical rate equation for individuals making numerical flux estimates. Finally, the derivation provides a protocol for obtaining one additional model equation for use by chemical fate modeler's connecting mobility across the interface that separates the adjoining bulk sediment and water compartments.

Future Directions

The sediment–water interface is the largest plane surface on Earth. Understanding and quantifying chemical and particle mobility across this semipermeable interface is relevant to the work of a broad community of aquatic researchers. The interface compartment model provides the basic theory for connecting chemical flux across the sediment–water interface. Correctly quantifying the net chemical, particle, and aquasol exchange rates across this plane is a key factor for understanding the fate of numerous natural and anthropogenic substances on Earth and aid in assessing the ecological significance. The potential impact of the IC model's further development and use in the environmental and geosciences fields may be a key contributing factor. Life-forms residing on both sides, in the water column or the surface sediment layers, depend on oxygen and nutrient fluxes. The bed is a source or a sink of soluble and particulate carbon compounds depending on the chemodynamics of the specific locale. The bed is a sink for chemical pollutants entering the aquatic system but, later as conditions improve in the water column, it becomes the source. These are just a few examples of the types of possible uses which require theoretically sound and verified science-based tools. As outlined above, further work is needed on the interface compartment concept before it is accepted as a unified theory and modeling protocol.

Bibliography

1. Lewis WK, Whitman WG (1924) Principles of gas absorption. Indust Eng Chem 16:1215–1220
2. Rhoads DC, Germano JD (1982) Characterization of benthic processes using sediment profile imaging: An efficient method of remote ecological monitoring of the seafloor (REMOTS™ System). Mar Ecol Prog Ser 8:115–128

3. Rhoads DC, Germano JD (1986) Interpreting long-term changes in benthic community structure: a new protocol. Hydrobiologia 142:291–308
4. Santschi P, Hohener P, Benoit G, Brink MB (1990) Chemical processes at the sediment-water interface. Mar Chem 30:269–315
5. Duursma EK, Smies M (1982) Sediments and transfer at and in the bottom interfacial layer. In: Kullenberg G (ed) Pollutant transfer and transport in the sea, vol II. CRC Press, Boca Raton, pp 101–137
6. Krantzberg G (1985) The influence of bioturbation on physical, chemical and biological parameters in aquatic environment – a review. Environ Pollut (Ser A) 39:99–122
7. DiToro DM (2001) Sediment flux modeling. Wiley, New York
8. Kullenberg G (ed) (1976) Pollutant transfer and transport in the Sea-II. CRC Press, Boca Raton
9. Boudreau BP, Jorgensen BB (eds) (2001) The benthic boundary layer. Oxford University Press, New York
10. McCave IN (ed) (1976) The benthic boundary layer. Plenum, New York
11. Windom HL, Duce RA (1976) Marine pollutant transport. Lexington, Lexington
12. Tenore KR, Coull BC (eds) (1980) Marine benthic dynamics. University of South Carolina Press, Columbia
13. Fanning KA, Manheim FT (eds) (1982) The dynamic environment of the ocean floor. Lexington Books, Lexington
14. Thibodeaux LJ (1996) Environmental chemodynamics. Wiley, New York
15. Lerman A (1979) Geochemical processes water and sediment environments. Wiley, New York
16. Boudreau BP (1997) Diagenetic models and their implementation. Springer, Berlin
17. Schink DR, Guinasso NL Jr (1975) Modeling the influence of bioturbation and other processes of $CaCO_3$ dissolution at the sea floor. In: Andersen NR, Malahoff A (eds) The fate of fossil fuel CO_2 in the oceans. Plenum, New York, pp 375–399
18. Berner RA (1980) Early diagenesis. Princeton University Press, Princeton
19. Thibodeaux LJ, Matisoff G, Reible DD (2010) Bioturbation and other sorbed-phase transport processes in surface soils and sediment. In: Thibodeaux LJ, Mackay D (eds) Handbook of chemical mass transport in the environment. CRC Press, Boca Raton (Chap 13)
20. Thibodeaux LJ, Wolfe JR, Dekker TJ (2010) Advective porewater flux and chemical transport in bed-sediment. In: Thibodeaux LJ, Mackay D (eds) Handbook of chemical mass transport in the environment. CRC Press, Boca Raton (Chap 11)
21. Singh VP, Reible DD, Thibodeaux LJ (1988) Mathematical modeling of fine sediment transport. Hydrol J IAH 11:1–3
22. Lick W (2009) Sediment and contaminant transport in surface waters. CRC Press, Boca Raton
23. Lohmann R, Dachs J (2010) Deposition of dissolved and particle-bound chemicals from surface ocean. In: Thibodeaux LJ, Mackay D (eds) Handbook of chemical mass transport in the environment. CRC Press, Boca Raton (Chap 17)
24. Yang CT (2003) Sediment transport. Krieger, Malabar
25. DePinto JV, McCulloch RD, Redder TM, Wolfe JR, Dekker TJ (2010) Deposition and resuspension of particles and associated chemical transport across the sediment-water interface. In: Thibodeaux LJ, Mackay D (eds) Handbook of chemical mass transport in the environment. CRC Press, Boca Raton (Chap 10)
26. Reible DD, Valsaraj KT, Thibodeaux LJ (1991) Chemodynamic models for transport of contaminants from sediment beds. In: Hutzinger O (ed) The handbook of environmental chemistry, part F, vol 2. Springer, Berlin, pp 186–228
27. Thibodeaux LJ, Reible DD, Valsaraj KT (2002) Non-particle resuspension chemical transport from stream beds. In: Lipnick RL, Mason RP, Phillips ML, Pittman CU Jr (eds) Chemicals in the environment, vol 806, ACS symposium series. American Chemical Society, Washington, DC, pp 130–149
28. Thibodeaux LJ, Valsaraj KT, Reible DD (2001) Bioturbation–driven transport of hydrophobic organic contaminants from bed sediment. Environ Eng Sci 18:215–223

29. Erickson MJ, Turner CL, Thibodeaux LJ (2005) Field observation and modeling of dissolved fraction-sediment-water exchange coefficients for PCBs in the Hudson River. Environ Sci Technol 39:549–555
30. Whitman WG (1923) The two-film theory of gas absorption. Chem Metall Eng 29:146–148
31. Thibodeaux LJ, Bierman VJ (2003) The bioturbation-driven chemical release process. Environ Sci Technol 1:253A–258A
32. Monteith JL, Unsworth M (1990) Principles of environmental physics. Butterworth-Heinemann, Oxford
33. Mackay D (2001) Multimedia environmental models. Lewis, Boca Raton
34. Slinn WGN (1978) 4-Wet and dry removal processes. In: NRC (ed) The tropospheric transport of pollutants and other substances to the oceans. National Academy of Sciences, National Academy Press, Washington, DC
35. Trapp S, Matthies M (1998) Chemodynamics and environmental modeling. Springer, Berlin
36. Van de Meent D (1993) Simple box: a generic multimedia fate evaluation model. RIVM Report No. 6727200001. Bilthoven
37. Thibodeaux LJ, Mackay D (eds) (2010) Chapters 10,11,12,13 &17. In: Handbook of chemical mass transport in the environment. CRC Press, Boca Raton
38. Thoma GJ, Koulermos AC, Valsaraj KT, Reible DD, Thibodeaux LJ (1991) The effect of pore-water colloids on the transport of hydrophobic organic compounds from bed sediments. In: Baker RA (ed) Organic substances in water, vol 1, Humics and soils. Lewis, Boca Raton
39. Valsaraj KT, Thibodeaux LJ, Reible DD (1997) A quasi-steady-state pollutant flux methodology for determining sediment quality criteria. Environ Toxicol Chem 16:391–396
40. Savant SA, Reible DD, Thibodeaux LJ (1987) Convective transport within stable river sediments. Water Resour Res 23:1763–1768
41. Thibodeaux LJ, Reible DD, Bosworth WS, Sarapas LC (1990) A theoretical evaluation of the effectiveness of capping PCB contaminated New Bedford Harbor bed sediments. Final Report. HSRC. Middelton Library, Louisiana State University, Baton Rouge

Chapter 10
River Fate and Transport

Zhen-Gang Ji

Glossary

Advection	The horizontal transport by flows that move patches of material around but do not significantly distort or dilute them.
Biodegradation	The breakdown of a compound by enzyme-mediated transformation primarily due to bacteria, and to a lesser extent, fungi.
Dispersion	The mixing of water properties in rivers.
Henry's law	A law which states that at a given temperature, the solubility of a gas is proportional to the pressure of the gas directly above the water.
Hydrograph	A graph showing time variation in flow rate or stage (depth) of water in a river.
Hydrolysis	The reaction of a chemical with water in which splitting of a molecular bond occurs in the chemical and there is formation of a new bond with either the hydrogen component (H^+) or the hydroxyl component (OH^-) of a water molecule.
Manning equation	An empirical formulation relating velocity (or flow rate) depth, slope, and a channel roughness coefficient in a river.
Mineralization	The process by which a dissolved organic substance is converted to dissolved inorganic form.

This chapter was originally published as part of the Encyclopedia of Sustainability Science and Technology edited by Robert A. Meyers. DOI:10.1007/978-1-4419-0851-3

Z.-G. Ji (✉)
Minerals Management Service, 381 Elden Street, Herndon, VA 20170, USA
e-mail: Jeff.Ji@boemre.gov

J.S. Gulliver (ed.), *Transport and Fate of Chemicals in the Environment: Selected Entries from the Encyclopedia of Sustainability Science and Technology*,
DOI 10.1007/978-1-4614-5731-2_10, © Springer Science+Business Media New York 2012

Nonpoint SOURCE	A pollution source that cannot be traced to a specific spot.
Photolysis	The transformation of a compound that results directly from the adsorption of light energy.
Point source	A pollution source that comes from a specific identifiable source such as a pipe.
Residence time	The time required by a particle to cross a river reach.
River	A naturally flowing waterbody.
Volatilization	The process representing a chemical substance entering the atmosphere by evaporation from water.

Definition of the Subject and Its Importance

Rivers are naturally flowing waterbodies. Small rivers are also called streams or brooks. Rivers are a watershed's self-formed gutter system and usually empty into an ocean, lake, or another river. This chapter describes the characteristics of rivers and the fate and transport in rivers. The mathematical description of river processes and the modeling of rivers are also described here.

Rivers are complex and dynamic. A river often acts as a sink for contaminants discharged along the river, such as effluents from wastewater treatment plants that discharge nutrients, heavy metals, and/or pathogens into the river. Rivers may also act as sources of contaminants in the watershed, depending on the time of the year or the section of the river. The health of a river is directly linked to the health of the surrounding watershed. The water quality in a river will deteriorate, if the watershed condition deteriorates. Via rivers, pollutants can travel hundreds or even thousands of kilometers and cause environmental problems in a waterbody that is located far away from the sources. The common designated uses of a river include aquatic life support, water supply, and recreation activities (such as swimming, fishing, and boating).

Introduction

The rivers and their tributaries, normally occupying less than a few percent of the total drainage basin, are the conduits of the river basin. They are like a gutter system and transport water, nutrients, sediment, and toxicants downstream (often to an estuary or a large lake). Compared with lakes and estuaries, the most distinct characteristic of a river is its natural downstream flow. Lakes typically have much smaller flow velocities than rivers. Flow velocities in estuaries, though their magnitudes can be comparable to the ones in rivers, are tidally driven and can be in either direction (downstream or upstream).

The origins of contaminants can be divided into point and nonpoint sources. Point source pollution comes from a specific, identifiable source such as a pipe.

Nonpoint source pollution cannot be traced to a specific spot. Point sources include wastewater treatment plants, overflows from combined sanitary and storm sewers, and industry discharges. Nonpoint sources include runoffs from urban, agriculture, and mining areas. Point and nonpoint sources have caused a wide range of water quality problems and the deterioration of the ecological state in rivers. Leading pollutants and stressors in the USA include [1]:

1. Pathogens (bacteria)
2. Siltation
3. Habitat alterations
4. Oxygen-depleting substances
5. Nutrients
6. Thermal modifications
7. Toxic metals
8. Flow alterations

Pathogens are the most common pollutant affecting rivers and streams in the USA. Pathogen pollution is a major public health problem especially in the use of river water for water supply and the consumption of fish and shellfish harvested in rivers and estuaries. Bacteria commonly enter surface waters in inadequately treated sewage, fecal material from wildlife, and runoff from pastures, feedlots, and urban areas.

Sediment siltation is one of the leading environmental problems in rivers. The filling of river channels, harbors, and estuaries by sediments brings a high cost to society. The condition of a river's watershed greatly affects the amount of sediment delivered into the river. The sediment sources vary among rivers, and even within a particular river, from year to year. Extreme events, such as hurricanes, can produce dramatic changes in the amounts and types of sediments that are delivered into a river. The vulnerability of a river to sediments and contamination reflects a complex combination of upstream flows, land use, and land-management practices. The vast majority of river sediments is discharged during only 10% of the year (36 days), and 90% of the year represents a very small amount of the sediment load [2]. Low flow rates usually result in net deposition conditions. High flow rates may cause net erosion in upstream reaches and net deposition in downstream reaches or in the estuary into which the river flows.

Often water quality is defined in terms of concentrations of the various dissolved and suspended substances in the water, for example, temperature, salinity, dissolved oxygen, nutrients, phytoplankton, bacteria, heavy metals, etc. The distribution of these substances can be calculated by a mathematical model. Based on the principle of conservation of mass, the concentration change can be represented simply in a one-dimensional form [3]:

$$\frac{\partial C}{\partial t} = -U\frac{\partial C}{\partial x} + \frac{\partial}{\partial x}\left(D\frac{\partial C}{\partial x}\right) + S + R + Q \qquad (10.1)$$

where
C = substance concentration
t = time

x = distance
U = advection velocity in x direction
D = mixing and dispersion coefficient
S = sources and sinks due to settling and resuspension
R = reactivity of chemical and biological processes
Q = external loadings to the aquatic system from point and nonpoint sources

The changes of concentration C in Eq. 10.1 are determined by the following:

1. The hydrodynamic processes control the water depth (D), the advection (represented by the U term), and mixing (represented by the D term).
2. The size and properties of sediment (or particular organic matter) affect the settling and resuspension (represented by the S term).
3. The chemical and biological reactions of pathogens, toxics, and/or nutrients are represented by the R term.
4. External loadings from point and nonpoint sources are included by the Q term.

Fate and Decay

Contaminants in rivers include nutrients, organic toxicants, heavy metals, and pathogens. If no degradation reactions occurred in Nature, every single contaminant discharged in the past would still be polluting the environment. Fortunately, natural purification processes dilute, transport, remove, and degrade contaminants. It is essential to understand the kinetics of reactants and to describe them mathematically. This section summarizes the fate and decay of contaminants and their mathematical formulations.

The fate and transport of contaminants are controlled by two factors: their reactivity and their hydrodynamic transport. Reactivity includes:

1. Chemical processes
2. Biological processes
3. Bio-uptakes

Transport in a river, which will be discussed in the next section, includes three mass transport processes:

1. Advection of water current
2. Diffusion and turbulent mixing within the water column
3. Deposition and resuspension on the water-sediment bed interface

Mathematical Formulations

How long contaminants remain in a waterbody depends on the nature of the compound. Most chemicals undergo chemical or biological decay. Some chemicals are conservative and do not undergo these types of reactions, even though it is very difficult to find a truly conservative chemical in Nature. The fate and decay of a contaminant represent the gradual decrease in the amount of a substance in a river, as the result of various sink processes, including chemical and biological transformation, or dissipation/deposition to other environmental systems.

Although reaction kinetics in aquatic systems can be described in numerous ways, the form for a single reactant is generally expressed as:

$$\frac{dC}{dt} = R = -kC^m \tag{10.2}$$

where
m = the order of reaction
k = rate constant of the m-order reaction

In natural waters, the commonly used forms of Eq. 10.2 are with m = 0, 1, and 2.

Zero-order reactions: A zero-order reaction (m = 0) represents irreversible degradation of a reactant that is independent of the reactant concentration. The solution to Eq. 10.2 is:

$$C = C_0 - kt \tag{10.3}$$

where C_0 = the initial concentration at t = 0. In this case, a plot of concentration versus time should yield a straight line with a slope of k, as shown in the left panel of Fig. 10.1. Zero-order reactions have their reaction rates determined by some factor other than the concentration of the reacting materials.

First-order reactions: First-order reactions (m = 1) have their reaction rates proportional to the concentration of the reactant and are most commonly used in describing chemical and biological reactions. For first-order reactions, the solution to Eq. 10.2 is:

$$C = C_0 e^{-kt} \tag{10.4}$$

Equation 10.4 indicates that for first-order reactions, reactant concentration decreases exponentially with time. In this case, a plot of logarithm concentration versus time should yield a straight line with a slope of k, as shown in the middle panel of Fig. 10.1. Most of the reactions found in the environment can be conveniently expressed by a first-order approximation without much error. Examples of first-order reactions include biochemical oxygen demand in surface waters, death and respiration rates for bacteria, and production reaction of algae.

Second-order reactions: For second-order reactions (m = 2), the solution to Eq. 10.2 is:

$$\frac{1}{C} = \frac{1}{C_0} + kt \tag{10.5}$$

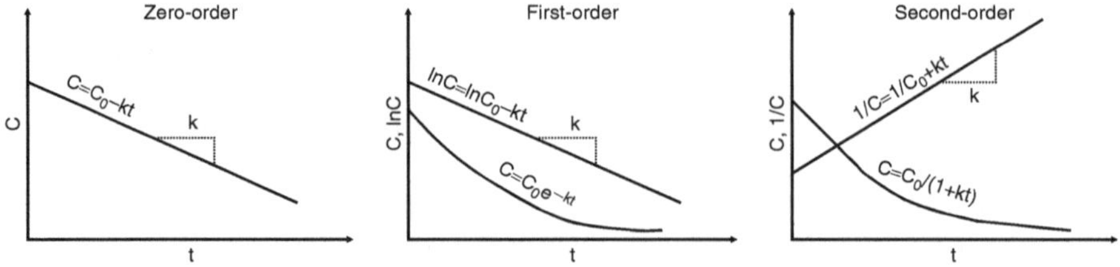

Fig. 10.1 *Left panel*: concentration versus time for zero-order reaction. *Middle panel*: concentration and logarithm concentration versus time for first-order reaction. *Right panel*: concentration and inverse concentration versus time for second-order reaction

Therefore, if a reaction is indeed second-order, a plot of inverse concentration of C (1/C) with time should yield a straight line with a slope of k (the right panel of Fig. 10.1). Equation 10.5 can also be expressed as

$$C = \frac{C_0}{1 + kC_0t} \tag{10.6}$$

which reveals that, similar to the first-order reaction, the resulting concentration of a second-order reaction also decreases and approaches zero as time increases.

Processes Affecting Fate and Decay

The fate and decay of contaminants can result from physical, chemical, and/or biological reactions. In addition to sorption and desorption, processes that can significantly affect the fate and decay processes include:

1. Mineralization and decomposition
2. Hydrolysis
3. Photolysis
4. Biodegradation
5. Bioconcentration
6. Volatilization

Most decay processes are expressed as first-order reactions. The first-order decay coefficients for individual processes are additive and can be linearly superimposed to form a net decay coefficient:

$$k_d = k_m + k_h + k_p + k_{bd} + k_{bc} + k_v \tag{10.7}$$

where
k_d = net decay coefficient

k_m = mineralization coefficient
k_h = hydrolysis coefficient
k_p = photolysis coefficient
k_{bd} = biodegradation coefficient
k_{bc} = bioconcentration coefficient
k_v = volatilization coefficient

In modeling studies, either the net degradation coefficient or the individual coefficients can be specified.

Mineralization and Decomposition

Mineralization is the process by which a dissolved organic substance is converted to dissolved inorganic form. Mineralization makes nutrients, such as nitrogen and phosphorus, available for a fresh cycle of plant growth. Bacteria decompose organic material to obtain energy for growth. Plant residue is broken down into glucose that is then converted to energy:

$$C_6H_{12}O_6 + O_2 \xrightarrow{\textit{energy released}} CO_2 + H_2O \tag{10.8}$$

In water quality models, the term "mineralization" often represents the process by which dissolved organic matter is converted to dissolved inorganic form, and thus includes both heterotrophic respiration of dissolved organic carbon and mineralization of dissolved organic phosphorus and nitrogen.

Hydrolysis

Hydrolysis is the reaction of a chemical with water, in which splitting of a molecular bond occurs in the chemical and there is formation of a new bond with either the hydrogen component (H^+) or the hydroxyl component (OH^-) of a water molecule. This involves ionization of the water as well as splitting of the compound hydrolyzed:

$$RX + H_2O \rightarrow ROH + HX \tag{10.9}$$

Essentially, water enters a polar location on a molecule and inserts itself, with an H^+ component going to one part of the parent molecule and an OH^- component going to the other. The two components then separate. The concentration of hydrogen and hydroxide ions, and therefore pH, is often an important factor in assessing the rate of a hydrolysis reaction. Hydrolysis is a major pathway for the degradation of many toxic organics.

Photolysis

Photolysis is the transformation of a compound that results directly from the adsorption of light energy. Compounds that absorb sunlight may gain sufficient energy to initiate a chemical reaction. Some of these photochemical reactions result in the decomposition or transformation of a substance.

The energy of light varies inversely with its wavelength. Longwave light lacks sufficient energy to break chemical bonds. Short wave light (x-rays and gamma rays) is very destructive. Fortunately for life on earth, this type of radiation largely is removed by the upper atmosphere. Light near the visible spectrum reaches the earth's surface and can break the bonds of many organic compounds, which can be important in the decay of organic chemicals in a water system.

The basic characteristics of photolysis are:

1. Photolysis has two types of energy absorption: direct photolysis and indirect photolysis. The direct photolysis is the result of direct absorption of sunlight by the toxic chemical molecule. Indirect photolysis is the result of energy transfer to the toxic chemical from some other molecule that has absorbed the sunlight.
2. Photolysis is the destruction of a compound activated by the light energy and is an irreversible decay process.
3. Products of photolysis may remain toxic and the photolysis process does not necessarily lead to detoxification of the system.
4. The photolysis coefficient in Eq. 10.7 is usually a function of the quantity and wavelength distribution of incident light, the light adsorption characteristics of the compound, and the efficiency at which absorbed light produces a chemical reaction.

Biodegradation

Biodegradation is the breakdown of a compound by enzyme-mediated transformation, primarily due to bacteria, and to a lesser extent, fungi. Although these types of microbial transformations can detoxify and mineralize toxics, they can also activate potential toxics. The rate of biodegradation can be very rapid, which means that biodegradation is often one of the most important transformation processes in rivers.

Even though the biodegradation process is largely mediated by bacteria, the growth kinetics of the bacteria is complicated and is not well understood. As a result, toxic models often assume constant decay rates rather than modeling the bacteria activity directly. The first-order decay rate is commonly used. Biodegradation rate is influenced by water temperature and can be represented by an Arrhenius function:

$$k_b = k_{b20}\,\theta^{(T-20)} \tag{10.10}$$

where
k_b = biodegradation rate
k_{b20} = biodegradation rate at 20°C
T = water temperature in °C
θ = temperature correction factor

The effect of the Arrhenius function is that a higher temperature will cause a faster chemical reaction rate. It gives a quantitative relationship between the reaction rate and its temperature.

Biodegradation rate is also related to the contaminant concentration and can be expressed by a typical Michaelis–Menten formulation:

$$k_b = k_{bmax} \frac{c}{c + c_{1/2}} \tag{10.11}$$

where
k_{bmax} = the maximum biodegradation rate
c = the contaminant concentration
$c_{1/2}$ = half saturation (Michaelis) constant.

The combination of the above two formulations yields

$$k_b = k_{max}\ \theta^{(T-20)} \frac{c}{c + c_{1/2}} \tag{10.12}$$

where k_{max} = maximum decay rate due to biodegradation. Equation 10.12 combines the effects of contaminant concentration and water temperature on the biodegradation process.

Volatilization

Volatilization represents a chemical substance entering the atmosphere by evaporation from water. Volatilization is often treated as an irreversible decay process, because of its mathematical similarities to these decay processes. However, volatilization is actually a reversible transfer, in which the dissolved concentration in water attempts to equilibrate with the gas phase concentration in the overlying atmosphere. Equilibrium occurs when the partial pressure exerted by the chemical in water equals the partial pressure of the chemical in the atmosphere.

Henry's law states that, at a given temperature, the solubility of a gas is proportional to the pressure of the gas directly above the water. Volatilization is often treated similarly to surface oxygen exchange, where the volatilization flux is proportional to the difference between the chemical concentration in water and the saturation concentration, as:

$$F_V = k_V(c_W - c_{WS}) \tag{10.13}$$

where
F_v = volatilization flux
k_v = transfer rate
c_w = dissolved concentration of the chemical in water
c_{ws} = saturation dissolved concentration of the chemical in water

Equation 10.13 indicates that the chemical enters the water when the chemical in the water is unsaturated ($c_w < c_{ws}$) and the chemical leaves (volatizes from) the water when the chemical in the water is oversaturated ($c_w > c_{ws}$). The saturation dissolved concentration is dependent upon the atmospheric partial pressure and Henry's law constant for the chemical. The transfer rate, k_v, depends on the properties of the chemical as well as the characteristics of the waterbody and the atmosphere, including the molecular diffusion coefficient of the chemical in the water and in the atmosphere, the temperature, the wind speed, the current velocity, and the water depth.

Transport in a River

Rivers have distinct hydrodynamic characteristics that are different from those of lakes or estuaries. This section focuses on the following:

1. River flow and the Manning equation
2. Advection and dispersion processes in rivers

River Flow and the Manning Equation

The flow rate of a river is the volume of water that passes a cross section of the river in a unit of time, which is usually expressed in cubic meters per second (cms) or cubic feet per second (cfs) and is calculated as:

$$Q = A\,V \tag{10.14}$$

where
Q = Flow rate in cms or cfs
A = Area through which the water is flowing in m^2 or ft^2
V = Average velocity in the downstream direction in m/s or ft/s

The river flow can generally be separated into two components:

1. Base flow
2. Storm flow

Base flow is composed largely of groundwater effluent and sustains river flow during dry weather periods. Storm flow is from the runoff during or shortly after a precipitation event. The water from base flow is the precipitation that percolates into the ground and flows slowly through a long path before reaching the river, whereas the water from storm flow is the precipitation that reaches the river shortly after precipitation through runoff. In addition to base flow from the groundwater and the storm flow from the runoff, point sources, such as wastewater treatment plant discharges and tributaries to the river, also contribute to a river flow.

A hydrograph is a graph showing time variation in flow rate or stage (depth) of water in a river. As sketched in Fig. 10.2, the river flow is composed of the storm flow and the base flow. After the beginning of a rainfall, the storm flow from runoff starts to increase and reaches its peak some time after the peak rainfall. There is a time lag between the two peaks. The rising limb is the portion of the hydrograph to the left of the peak of the storm flow, which shows how long the river takes to reach its peak flow rate after a rainfall event. The receding limb is the portion of the hydrograph to the right of the peak, which shows how long the river takes to return to the base flow.

In addition to flood events, low flow conditions are also important characteristics of a river. When there is no precipitation contributing to the storm flow, and the base flow from groundwater is low, the river experiences low flow conditions. Low flow results in less water available for dilution of pollutants from point sources, causing high pollutant concentrations in the river. Therefore, point source discharges during low flow conditions have the most significant impact on the river's water quality, since the discharge may constitute a larger percentage of river flow. For instance, wastewater discharges to the Blackstone River can account for up to 80% of the total river flow in summer [4].

A hydrodynamic model based on momentum and continuity equations is often used to calculate flow velocity, flow rate, and water depth in a waterbody. A simpler approach to calculate these parameters is to use the Manning equation, which is an empirical formulation relating velocity (or flow rate), depth, slope, and a channel roughness coefficient in a river. The Manning equation was derived by curve-fitting data measured in rivers and channels. The equation is:

$$V = \frac{Q}{A} = \frac{R^{\frac{2}{3}} S^{\frac{1}{2}}}{n} \qquad (10.15)$$

where
V = mean flow velocity in m/s
Q = flow rate in m^3/s
A = cross-sectional area in m^2
R = hydraulic radius in m
S = slope of the channel bed in m/m
n = Manning roughness coefficient

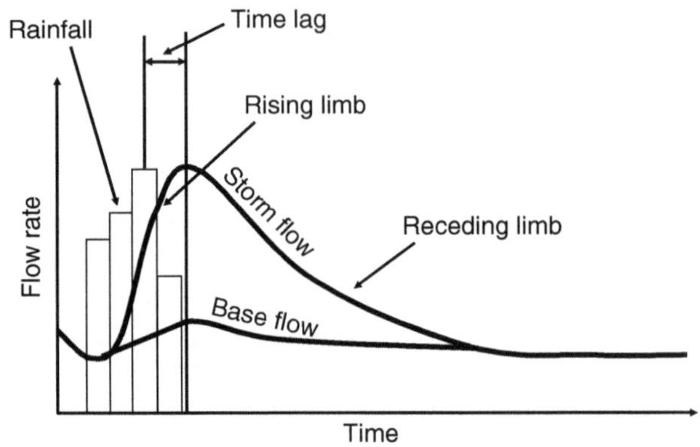

Fig. 10.2 A storm hydrograph of a river

The hydraulic radius is defined as:

$$R = \frac{A}{P} \quad (10.16)$$

where P is the wetted perimeter in m, which is the length of contact of the water with the channel in m, measured in a direction normal to the flow. The Manning roughness coefficient, n, represents the channel roughness that contributes to the dissipation of flow energy. Table 10.1 shows a range of n values for various channels and rivers.

Originally developed in the 1880s, the Manning equation is still widely used in hydraulic calculations with reasonable accuracy today. In hydrodynamic modeling, the Manning equation may serve the purpose of giving a quick estimation of flow conditions in a river. However, the Manning equation is an empirical formulation that may not reflect actual conditions of a river.

Advection and Dispersion in Rivers

Advection refers to horizontal transport by flows that move patches of material around but do not significantly distort or dilute them. In rivers, advection often represents the primary transport process of pollutant in the longitudinal direction. Dispersion is the mixing of water properties. In rivers, a prominent feature is the longitudinal dispersion: the transport and spreading of pollutants downstream from a point source. When a tracer is released into a river, two distinct processes control the tracer transport:

1. Flow advection carries the tracer away from the releasing point.
2. Turbulence dispersion spreads out and dilutes the tracer concentration.

Table 10.1 Values of the Manning roughness coefficient, n, for various channels and rivers [5]

Type of channel	Manning roughness coefficient (n)
Smooth concrete	0.012
Ordinary concrete lining	0.013
Earth channels in best condition	0.017
Straight unlined earth canals in good condition	0.020
Natural rivers and canals	0.020–0.035
Mountain streams with rocky beds and rivers with variable sections and some vegetation along banks	0.040–0.050
Alluvial channels without vegetation	0.011–0.035

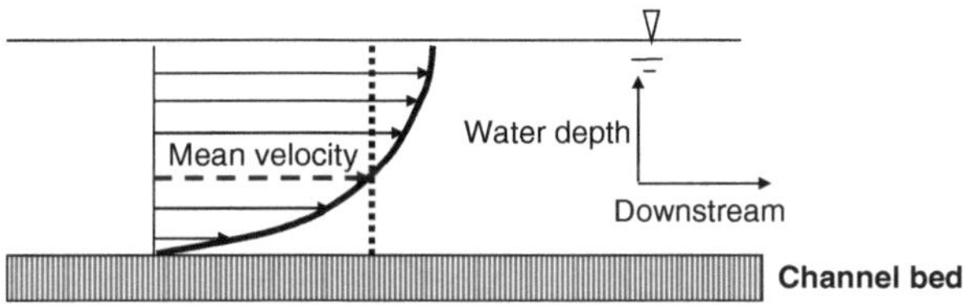

Fig. 10.3 Velocity vertical profile in a channel

Mathematically, the above two processes are represented by the first and second terms on the right-hand side of Eq. 10.1, respectively. Advection results in the pollutant's moving downstream, while longitudinal mixing leads to spreading or smearing in the longitudinal dimension. Lateral and vertical mixing processes determine how long it takes for a pollutant to be completely mixed across a river. The dominant transport process in rivers is the advection due to river flow. Flow velocity controls a river's residence time, the time required by a particle to cross a river reach. The dispersion process in rivers is often less important in the transport of pollutants. The effect of dispersion may be ignored in analyzing a continuous pollutant load to a river. Fig. 10.3 is a velocity vertical profile in a channel. In small rivers, however, the turbulence generated by bed friction is strong, and the depth is generally small, resulting in rivers that are often well mixed vertically.

To illustrate the longitudinal dispersion in a river, an idealized dye release experiment is shown in Fig. 10.4, in which Panel A gives the plain view of the dye transport in the river and Panel B presents the lateral-averaged dye concentration along the river. In the river, a line source of constant concentration is instantaneously released at time $t = 0$, and the longitudinal velocity has parabolic variation across the river. As shown in Panel A, the advection process transports the dye downstream, and the dispersion process spreads the dye and reduces the maximum concentration. Dye travels downstream faster in the middle of the river than near the banks. As a result, the line source released at $t = 0$ becomes approximately a parabolic shape at $t = t_1$ and $t = t_2$. The concentration profiles at t_1 and t_2 in Panel A also reflect the random fluctuations of turbulence activities in the river. Because of variations in flow velocity across the

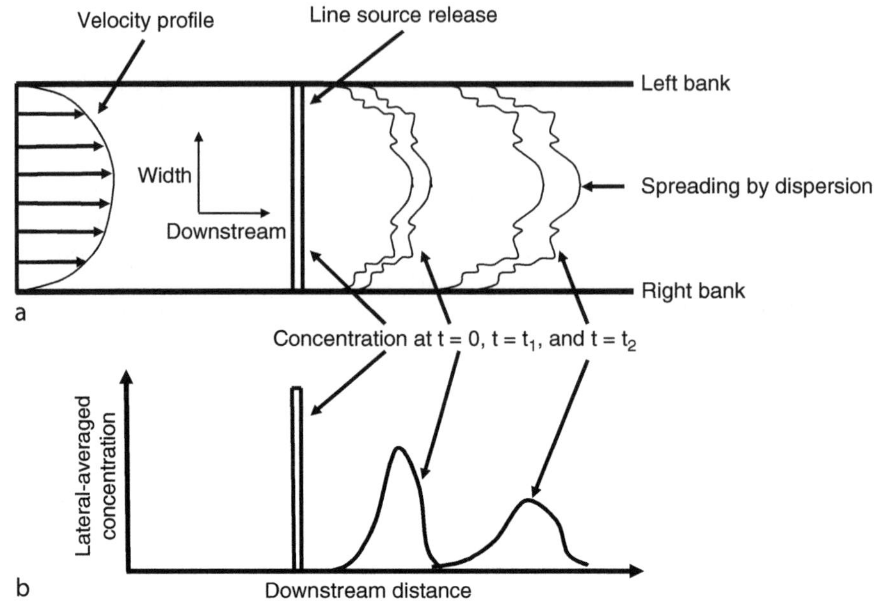

Fig. 10.4 Advection and dispersion processes in a river. *Panel A* gives the plain view of dye transport in the river. *Panel B* presents the lateral-averaged dye concentration along the river

river, dye spreads both along and across the river by dispersion. The laterally averaged dye concentration in Panel B also indicates that the velocity shear and turbulent dispersion contribute to the concentration spreading along the river.

Impacts of River Flow on Water Quality

Water quality processes can be highly dependent on river flow conditions. The time that a pollutant remains within a section of a river is called residence time. The flow velocity and the length of the river section determine the residence time. River flow affects water quality in a river in several ways:

1. *Dilution.* A large volume of flow dilutes concentrations of pollutants that are discharged into the river.
2. *Residence time.* High flow velocity reduces the residence time and affects the amount of material that can be produced or degraded in the river section.
3. *Mixing.* High flow velocity increases mixing in the river, enhances the assimilative capability of the river, and reduces pollutant concentration gradients.
4. *Erosion.* High flow can erode bed material and destabilize the benthic environment.

The impact of pollutant loadings to a river is largely determined by the magnitudes of the loadings and the flow rate. Rapid transport of pollutants by high flow results in a short residence time and often causes minimal water quality problems. Conversely, slow transport of pollutants by low flow results in a long residence time and can lead to water quality problems, such as oxygen depletion and eutrophication. Channel alteration and watershed disturbance can lead to abnormally high flow rates for a given amount of rain and amplify the impact of floods. Watershed disturbance can also increase sedimentation and harm aquatic biota in a river.

In temperate regions, seasonally high flow typically occurs during the periods of snowmelt in early spring and spring rains, whereas seasonally low flow normally occurs in summer and early fall. The river flow affects the concentration and distribution of water quality variables. Generally, point sources have a larger impact on a river during low flow (dry weather) conditions due to less water diluting the pollutants. Low DO concentrations and high algal growth in a river often occur during low flow periods and hot weather conditions. The combination of low flow, minimum dilution, and high temperature often makes summer and early fall the critical periods for evaluating the impact of point sources (such as wastewater treatment plants).

In contrast, nonpoint sources can bring large amounts of pollutants from the watershed into a river during high flow (wet weather) conditions. It is important to examine both point and nonpoint sources in both high and low flow conditions. Point sources of nutrients often cause algal blooms in rivers during low flow conditions, while nonpoint sources may increase nutrient concentrations and turbidity following periods of wet weather events. Municipal discharges, agriculture runoff, and urban runoff are among the most common sources of impairment to rivers.

In the study of the Blackstone River, for example, Ji et al. [2] reported that a discharge from a wastewater treatment plant was the dominant point source of contaminants and had significant impact on the sediment contamination in the river. However, this point source alone is still insufficient to account for the total metal concentrations in the river. Nonpoint sources and the processes of sediment deposition and resuspension are also important factors that control the concentrations of sediment and toxic metals.

Dissolved oxygen is essential to river ecosystems. Processes controlling DO spatial distribution in a river include:

1. Oxidation of the biochemical oxygen demand (BOD): BOD is used to represent all sinks of dissolved oxygen, such as the oxidation of carbonaceous and nitrogenous organic matter, the benthic oxygen demand, and the oxygen utilized by algal respiration.
2. Reaeration of DO from the atmosphere: In addition to atmospheric reaeration, DO produced by photosynthesis and DO contained in incoming flows are also major oxygen sources.
3. Transport due to the river flow: Advection and diffusion processes enhance DO mixing and reaeration within a river.

The pioneering work by Streeter and Phelps [6], who developed the first water quality model to describe the oxygen depletion in the Ohio River, is useful for understanding DO processes in a river. It can be described in a first-order reaction equation:

$$U\frac{dC}{dx} = -k_d B + k_a (C_S - C) \tag{10.17}$$

where x = distance
U = advection velocity in x direction
C = DO concentration
B = BOD concentration
C_s = saturated dissolved oxygen concentration
k_d = deoxygenation rate constant of BOD
k_a = first-order reaeration rate constant of DO

By assuming that BOD has a first-order degradation reaction with a decay rate constant of k_r, the solution to Eq. 10.17 is the famous Streeter-Phelps equation:

$$C = C_S - \frac{k_d L_0}{k_a - k_r}\left(e^{-k_r x/U} - e^{-k_a x/U}\right) - (C_S - C_0)e^{-k_a x/U} \tag{10.18}$$

A schematic representation of the Streeter-Phelps equation is shown in Fig. 10.5, describing a DO sag curve in a river. The DO sag curve gives DO longitudinal variation as the result of oxygen depletion and recovery, after a BOD load is discharged into a receiving river. Between the discharge point (x = 0) and the critical distance ($x = x_c$), oxidation exceeds reaeration (i.e., $k_d B > k_a(C_S - C)$ in Eq. 10.17) because of high BOD concentrations and a small DO deficit ($= C_s - C$). Oxygen in the river is consumed faster than it is resupplied. The DO concentration decreases to a minimum C_{min} at a critical distance x_c (or critical time $t_c = x_c/U$). This position is the critical location where the lowest DO concentration occurs, and the oxidation rate and reaeration rate are equal. After passing the critical location, reaeration exceeds oxidation (i.e., $k_d B < k_a(C_S - C)$ in Eq. 10.17) because of a low BOD concentration and a high DO deficit. Thus, oxygen in a river increases gradually. Further downstream, the rate of supply exceeds the utilization rate, resulting in a full recovery of the DO concentration.

River Modeling

The two primary reasons to conduct river modeling are:

1. To better understand physical, chemical, and biological processes

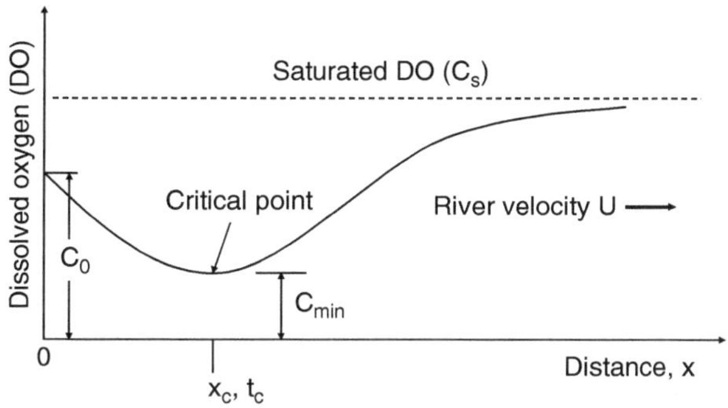

Fig. 10.5 DO sag curve in a river

2. To develop models capable of realistically representing rivers, so that the models can be used to support water quality management and decision making

Water quality management needs to understand key processes affecting environmental problems in order to evaluate management alternatives. Examples of such environmental problems are:

1. Thermal pollution due to power plant discharges
2. Sedimentation in harbors causing siltation and high dredging costs
3. Eutrophication due to excessive nutrient loadings
4. Low dissolved oxygen conditions caused by wastewater discharges
5. Accumulation of toxic materials in the sediment bed

Models play a critical role in advancing the state-of-the-art of hydrodynamics, sediment transport, and water quality, and of water resources management. Because of their requirements for precise and accurate data, models also ultimately contribute to the design of field data collection and serve to identify data gaps in characterizing waterbodies. Models are used to analyze the impact of different management alternatives and to select the ones that result in the least adverse impact to the environment.

Models are often used to improve the scientific basis for theory development, to make and test predictions, and to clarify cause-and-effect relationships between pollutant loadings and the receiving waterbody. Models are often used to evaluate and test potentially expensive water quality management alternatives prior to their implementation. The cost of a hydrodynamic and water quality modeling study is usually a small fraction of the implementation cost. Models can simulate changes in an ecosystem due to changes in internal and/or external conditions, such as water elevation variations or increased external pollutants. These simulations predict positive or negative changes within the river ecosystem due to the management actions, such as improved sewage treatment or reduced agricultural runoff. These simulations are obviously far more cost-effective than testing expensive management actions on a trial-and-error basis, thus making models a useful tool for water quality management.

In the past decades, hydrodynamic and water quality models have evolved from simplified one-dimensional, steady-state models, such as the legendary QUAL2E model [7], to complex three-dimensional, time-dependant models of hydrodynamics, sediment, toxics, and eutrophication. Three-dimensional modeling has matured from a research subject to a practical engineering tool. Over this same period, computational requirements for realistic three-dimensional modeling have changed from supercomputers, to high-end workstations, and then to personal computers.

These advanced three-dimensional and time-dependant models, which can also be readily applied for one- and two-dimensional problem settings, provide a powerful computational tool for sediment transport, water quality, eutrophication, and toxic chemical fate and transport modeling studies. These advanced models often include several coupled submodels for different physical, chemical, biological processes in surface waters, such as:

1. Hydrodynamic model
2. Wind wave model
3. Sediment model
4. Toxic model
5. Eutrophication model
6. Sediment diagenesis model
7. Submerged aquatic vegetation (SAV) model

As an example, Fig. 10.6 illustrates the major components of the Environmental Fluid Dynamics Code (EFDC) model [8]. In addition to computational modules, these advanced models tend to evolve into complex software systems, comprising many tools and sources of information. They may contain components for grid generation, data analysis, preprocessing, postprocessing, statistical analysis, graphics, and other utilities.

Transport in rivers is often dominated by the processes of advection and dispersion. One-, two-, and three-dimensional models have been developed to describe these processes. Study objectives, river characteristics, and data availability are key factors determining model applicability. In river studies, 1D and steady-state models are commonly used, such as the QUAL2E model [7]. If a river is wide enough to have significant lateral variations or deep enough to develop vertical stratifications, 2D (and even 3D) models may be needed to simulate sediment and toxicant transport in the river. For instance, sediment transport within a meandering river is very complex. The velocities are faster at the outer bank and slower at the inner bank. The lateral velocity difference directly influences the sediment transport. There might be erosion occurring along the outer bank and deposition occurring on the inner bank. Using a 1D model to represent the river is equivalent to treating sediment transport as being uniform across the river, eliminating the effect of river meandering on sediment transport and vertical stratifications. A 1D model represents the entire cross section of the river as being either net depositional or net erosional.

One-dimensional models, such as the widely used QUAL2E model [7], are traditionally applied to river modeling. For most small and shallow rivers, these

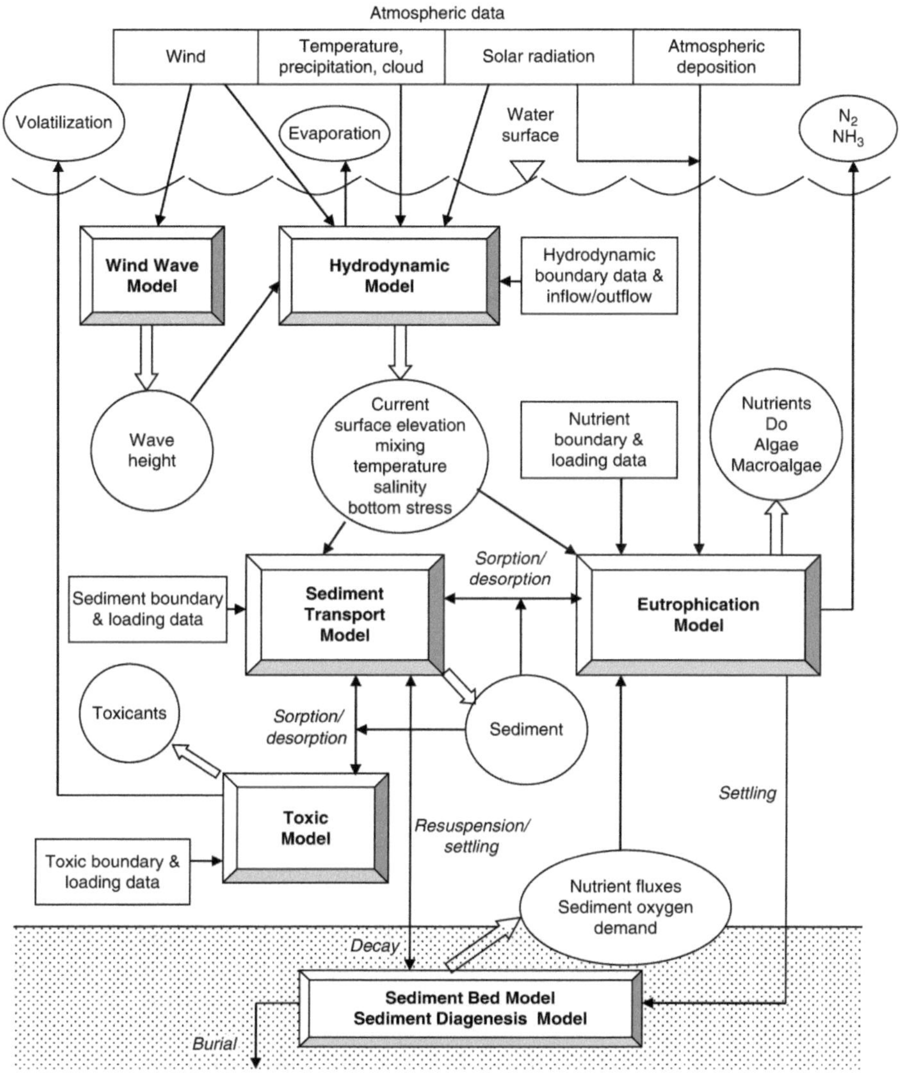

Fig. 10.6 Major components (submodels) of the EFDC model

1D models are often adequate to simulate hydrodynamic and water quality processes. In 1D models, water surface elevation, velocity, and discharge vary only in the longitudinal (along-the-river) direction and are constants in the lateral (across-the-river) direction. This approach provides a simplified mathematical description of river flows.

Rivers with a steep bottom slope often have a relatively high velocity and a shallow water depth, and are characterized by gravel, cobbles, and rocks in the riverbed. Coarse sands and finer particles are washed out by the high velocity.

The dominant gradient of water quality constituents is along the river in the direction of flow. A 1D laterally and verticallyaveraged model is thus appropriate for describing water flow and the transport of sediment and toxic chemicals. Rivers with a moderate bottom slope result in a low-velocity waterway, often characterized by a sediment bed consisting of a mixture of fine-grained cohesive particles and fine sands. The dominant gradient of water quality constituents in this kind of river is in the direction of the flow and a 1D model may still be adequate. One-dimensional models are limited in their ability to capture the complexity of natural rivers. The assumption that the characteristics of the river are uniform both vertically and laterally may not be valid for wide, deep rivers. In this case, the 1D approach may fall short of describing the river processes. Transport in these rivers can have significant gradients either laterally or vertically. In this case, a 2D or 3D model is needed to provide a better representation of the river. Ji et al. [4] gave an example of modeling hydrodynamics, sediment transport, and toxics in a small, shallow river.

Future Directions

The fate and transport of contaminants in rivers are complicated processes that include physical transport and chemical and biological kinetics. Contaminants in a river may be the result of either past or present disposal practices. Shutting off the sources does not always solve the problem (e.g., DDT persists many years). Consequently, it is essential that mathematical models for assessing contaminants are accurate and reliable. In the past decades, significant progress has been made in numerical model development, data collection, and computer software and hardware. These developments have helped mathematical models to become reliable tools for environmental management and engineering applications.

"Modeling is a little like art in the words of Pablo Picasso. It is never completely realistic; it is never the truth. But it contains enough of the truth, hopefully, and enough realism to gain understanding about environment systems" [9]. Water quality management increasingly depends upon accurate modeling. This dependency is further amplified by the adoption of the watershed-based approach to pollution control. Models enable decision-makers to select better, more scientifically defensible choices among alternatives for river water quality management.

In many cases, the models are used to evaluate which alternative will be most effective in solving a long-term water quality problem. The management decisions require the consideration of existing conditions, as well as the projection of anticipated future changes of the water system. In these applications, the river models not only need to represent the existing conditions, but also have to be predictive and give conditions which do not yet exist. Models are also used to provide a basis for economic analysis, so that decision makers can use the model results to evaluate the environmental significance of a project as well as the cost-benefit ratio.

Three key factors have contributed to the great progress in the modeling of rivers:

1. Better understanding and mathematical descriptions of physical, chemical, and biological processes in rivers
2. Availability of fast and efficient numerical schemes
3. Progress in computer technology

The powerful, yet affordable computers in combination with fast numerical algorithms have enabled the development of sophisticated 3D hydrodynamic and water quality models. These advanced models contain very few simplifying approximations to the governing equations. Personal computers (PCs) have evolved rapidly to become the standard platform for most engineering applications (with the exception of very large scale problems). PCs represent the most widely used computer platform today. Models developed on a PC can be transformed to other PCs without much difficulty. The relatively low prices of PCs also make modeling more cost-effective. Due to the rapid advances in computer technology, PCs are now widely used in river modeling studies.

Bibliography

Primary Literature

1. USEPA (2000) National water quality inventory: 1998 report to Congress. EPA 841-R-00-001. US Environmental Protection Agency, Office of Water, Washington, DC
2. CSCRMDE (1987) Sedimentation control to reduce maintenance dredging of navigational facilities in estuaries. Report and symposium proceedings. Committee on sedimentation control to reduce maintenance dredging in estuaries, National Academy Press, Washington, DC
3. Ji Z-G (2008) Hydrodynamics and water quality: modeling rivers, lakes, and estuaries. Wiley, Hoboken, 676 pp
4. Ji Z-G, Hamrick JH, Pagenkopf J (2002) Sediment and metals modeling in shallow river. J Environ Eng 128:105–119
5. Chow V (1964) Handbook of applied hydrology, a comparison of water-resources technology. McGraw Hill, New York
6. Streeter HW, Phelps EB (1925) A study of the pollution and natural purification of the Ohio river. III: factors concerned in the phenomena of oxidation and reaeration. Bulletin Number 146, US Public Health Service
7. Brown LC, Barnwell TO (1987) The enhanced stream water quality models QUAL2E and QUAL2E-UNCAS: documentation and user manual. EPA/600/3-87-007. US Environmental Protection Agency, Athens, Georgia
8. Hamrick JM (1992) A three-dimensional environmental fluid dynamics computer code: theoretical and computational aspects. The College of William and Mary, Virginia Institute of Marine Science, Special Report 317, p 63
9. Schnoor JL (1996) Environmental modeling: fate and transport of pollutants in water, air, and soil. Wiley, New York

Books and Reviews

Blumberg AF, Mellor GL (1987) A description of a three-dimensional coastal ocean circulation model. In: Heaps NS (ed) Three-dimensional coastal ocean models, coastal and estuarine science, vol 4. American Geophysical Union, Washington, DC, pp 1–19
Bowie GL, Mills WB, Porcella DB, Campbell CL, Pagenkopf JR, Rupp GL, Johnson KM, Chan PWH, Gherini SA (1985) Rates, constants, and kinetics formulations in surface water quality modeling, 2nd edn. USEPA, Environmental Research Laboratory, Athens. EPA/600/3-85/040
Casulli V, Cheng RT (1992) Semi-implicit finite difference methods for three-dimensional shallow water flow. Int J Numer Meth Fl 15:629–648
Cerco CF (1999) Eutrophication models of the future. J Environ Eng 125(3):209–210
Chapra SC (1997) Surface water-quality modeling. McGraw-Hill, New York, 844 pp
Chapra SC, Canale RP (1998) Numerical methods for engineers, with programming and scientific applications. McGraw-Hill, New York, 839 pp
Di Toro DM (2001) Sediment flux modeling. Wiley, New York
Fischer HB, List EJ, Imberger J, Brooks NH (1979) Mixing in inland and coastal waters. Academic, New York, 483 pp
Gill AE (1982) Atmosphere-ocean dynamics. Academic, New York, 662 pp
Hutchinson GE (1957) A treatise on Limnology. In: Geography, physics and chemistry, vol I. Wiley, New York, p 1015
Ji Z-G (2004) Use of physical sciences in support of environmental management. Environ Manage 34(2):159–169
Ji Z-G (2005) Water quality models: chemical principles. In: Water encyclopedia, vol 2, Water quality and resources development. Wiley, New Jersey, pp 269–273
Ji Z-G (2005) Water quality modeling-case studies. In: Water encyclopedia, vol 2, Water quality and resources development. Wiley, New Jersey, pp 255–263
Martin JL, McCutcheon SC (1999) Hydrodynamics and transport for water quality modeling. Lewis, Boca Raton
Morel F (1983) Principles of aquatic chemistry. Wiley, New York, 446 pp
Park K, Kuo AY, Shen J, Hamrick JM (1995) A three-dimensional hydrodynamic-eutrophication model (HEM3D): description of water quality and sediment processes submodels. The College of William and Mary, Virginia Institute of Marine Science. Special Report 327, 113 pp
Schumm SA (1977) The fluvial system. Wiley, New York
Thomann RV, Mueller JA (1987) Principles of surface water quality modeling and control. Harper and Row, New York
USEPA (1994) Water quality standards handbook, 2nd edn. US Environmental Protection Agency, Office of Water, Washington, DC, EPA-823-B-94-005b
USEPA (1998) Bacteria water quality standard status report. US Environmental Protection Agency, Office of Water, Washington, DC
USEPA (2000) Nutrient criteria technical guidance manual: rivers and streams. EPA-822-B-00-002. Office of Water, Office of Science and Technology, Washington, DC
Wezernak CT, Gannon JJ (1968) Evaluation of nitrification in streams. J Sanit Eng Div ASCE 94(SA5):883–895
Wool AT, Ambrose RB, Martin JL, Corner EA (2003) Water quality analysis simulation program (WASP), Version 6: Draft users manual. Available at: http://www.epa.gov/athens/wwqtsc/html/wasp.html
Ziegler CK, Nesbitt B (1994) Fine-grained sediment transport in Pawtuxet river, Rhode Island. J Hydraul Eng 120:561–576
Ziegler CK, Nesbitt B (1995) Long-term simulation of fine-grained sediment transport in large reservoir. J Hydraul Eng 121:773–781

Chapter 11
Lake and Reservoir Fate and Transport of Chemicals

Heinz G. Stefan, Xing Fang, and John S. Gulliver

Glossary

Dimictic lake	A lake that has two complete mixing (circulation) periods per year (one in spring after the ice melts and another in fall before the ice forms).
Hyporheic flow	Flow in a region beneath and along a stream bed. It is characterized as mixing of shallow groundwater and surface water which is important to sedimentary oxygen uptake.
Lentic	"Stagnant" waterbody as opposed to "flowing".
Limnology	Science or the study of inland waters, e.g., lakes and reservoirs.
Oxythermal parameter	Variable to define fish habitat in inland waters using dissolved oxygen (DO) and water temperature limits, e.g., TDO3 – temperature at DO = 3 mg/L.
Particulate	Tiny subdivisions of solid matter suspended in a gas or liquid, also known as particulate matter (PM) or fine particles.
Residence time	The mean amount of time that water or a substance would stay or "reside" in a lake or reservoir. Hydraulic residence time is equal to lake volume divided by outflow rate. The residence time of

This chapter was originally published as part of the Encyclopedia of Sustainability Science and Technology edited by Robert A. Meyers. DOI:10.1007/978-1-4419-0851-3

H.G. Stefan (✉)
St. Anthony Falls Laboratory, Department of Civil Engineering,
University of Minnesota,
55414 Minneapolis, MN, USA
e-mail: stefa001@umn.edu

X. Fang
Department of Civil Engineering, Auburn University, 36849-5337 Auburn, AL, USA

J.S. Gulliver
Department of Civil Engineering, University of Minnesota, Minneapolis, MN, USA

J.S. Gulliver (ed.), *Transport and Fate of Chemicals in the Environment: Selected Entries from the Encyclopedia of Sustainability Science and Technology*,
DOI 10.1007/978-1-4614-5731-2_11,

	a substance is equal to the quantity of a substance in volume divided by the change of a substance in volume over time through various lake's removal mechanisms (outflow or flushing, settling, and chemical and biological reactions).
Sedimentation	The process or tendency for particles in suspension to settle out of the fluid in which they are entrained and come to rest against a barrier (e.g., lake or river bed).
Solute	Material dissolved in water of a lake.
Stratification	The formation of horizontal layers (strata) in which water temperature and concentration of substances are different along depth of a lake and reservoir.

Definition of the Subject

"Eutrophication" is originally used to describe aging process whereby a lake is transformed from a lake to a marsh to a meadow (fill the lake with sediments). "Cultural eutrophication" occurs when the lake aging process is quickened or accelerated by excess nutrients from human activities [1]. Understanding of the fate and transport of water quality constituents in lakes and reservoirs is essential to sustaining water quality and fish habitat in these inland waters. Constituent is used generically and does not necessarily mean a polluting substance, e.g., dissolved oxygen (DO) is a relatively benign variable. The fate of a constitute typically depends on its transport (movement) through an inland water system (lake or reservoir) and on sources, sinks, chemical and biological reactions, and other decay mechanisms (e.g., settling). When sediment input is more than sediment outflow or nutrients are more than demands of aquatic plants, a lake or reservoir becomes not sustainable and the aging process of a lake is accelerated. The study of fate and transport of a substance in a lake or reservoir is to qualitatively and quantitatively account for mass balance of the substance through boundaries of and within the waterbody. Using the principle of the conservation of mass to investigate mass balance is not a new topic, but closely examining mass balance of various water quality constituents in inland waters was only started a few decades ago, and sustainability of aquatic systems is a relative new topic to researchers, water resources managers, and the public. Due to waterborne pathogens as one of the prime causes of disease, civil engineers began to plan, design, and construct urban water and wastewater systems in the late nineteenth century, and then the water quality management processes or models from streams to lakes and reservoirs emerged. In the United States (U.S.), Rivers and Harbors Act in 1899 to Federal Water Pollution Control Act in 1972 (subsequently amended and called the Act Clean Water Act) promoted studies on water quality in receiving waters. Sustainability is the capacity to endure. For humans, sustainability is the potential for long-term maintenance of well-being, which has environmental, economic, and social dimensions. Sustainability in lakes and reservoirs involves how biological systems remain diverse and productive over time and how designated uses (e.g., water supply, recreation, fish and wildlife, etc.) endure over time. Lake and reservoir fate and transport involve understanding and maintaining healthy aquatic ecosystems and environments that provide vital goods and services to humans and other organisms.

Inland waterbody provides suitable habitat for survival and reproduction of desirable fish, shellfish, and other aquatic organisms. Sustaining fish habitat in lakes and reservoirs is a major resource management issue. Freshwater fish habitat is constrained by physical, chemical, and biological attributes that relate to water quality, food supply, and human interference [2]. In lakes, water temperature and DO concentration are two of the most significant water quality parameters affecting survival and growth of fishes [2, 3] and sustainability of lake aquatic ecosystem. An increase of atmospheric greenhouse gases is projected to cause climate warming, which would alter water temperature and DO characteristics in lakes. These changes are in turn expected to have an effect on indigenous fish populations. Understanding impacts of future climate changes on water quality and fish habitat in lakes and reservoirs is essential to promoting sustainability of valuable but limited water resources for humanity.

Introduction

"Sustainability" of lakes and reservoirs relates/depends on the transport and transformation of materials into, within, and out of these "lentic" waterbodies. Typically, lakes are formed by natural, i.e., geophysical and hydrogeological processes, whereas reservoirs or impoundments are typically man-made. There is a vast amount of literature on physical, chemical, and biological structure and function of lakes. There is a lot of literature on the engineering of dams and other hydraulic structures, e.g., by Zipparro and Hasen [4], USBR [5], Novak [6], and Press [7], but less on water quality processes in the impoundments they create [8–10]. Decisions on water releases from reservoirs and associated water quality downstream are difficult to make because of overlapping and contradictory objectives such as water storage, flood protection, hydropower generation, sediment transport, recreational water use, and water quality [11].

The study of lakes probably goes back to the scientists of the antiquity with a keen sense of observation such as Plinius. Thieneman in Germany and Birge and Juday in the USA (Wisconsin) were early scientific explorers of the physics, chemistry, and biology of lakes. "Limnology" is the discipline under which early investigators of lakes conducted their research, but geophysicists, biologists, engineers, and people with other diverse backgrounds have contributed significantly. The two volume book by Hutchinson [12] was an early and thorough compilation of the state of scientific knowledge on lakes. An overview of North American Limnology was compiled by Frey [13]. Books, e.g., by Lerman [14], Wetzel [15], Horne and Goldman [16], Lampert and Sommer [17]; conference proceedings of the International Society of Theoretical and Applied Limnology

(SIL); and journal articles, e.g., in Limnology and Oceanography, Hydrobiologia, Water Resources Research, and Aquatic Sciences, have summarized the progressively larger body of knowledge. Physical processes in lakes and their quantitative analysis are described by Henderson-Sellers [18]. Ecosystems in lakes and reservoirs and their modeling are described by Straskraba and Gnauck [19].

Reservoirs were created by man-made dams thousands of years ago in the Middle East and in China to provide water for irrigation of agricultural lands [20]. This practice provided the food source that sustained major cultures. Reservoirs connected by aqueducts to major cities supplied drinking water even in relatively dry regions. Eventually, sediment accumulation in reservoirs as well as natural lakes made it painfully obvious that water-filled depressions in the landscape that receive inflows from a watershed are not sustainable forever. Sedimentary rocks give testimony to this fact on geological timescales.

While sediment transport into reservoirs received some early attention, e.g., in China [21], the limnology of reservoirs has become of interest recently because of fishery issues in the reservoirs (e.g., Columbia River) and water quality and ecological issues in the tailwaters below dams (e.g., Colorado River) [11, 22]. The need to better understand reservoir processes and reservoir operations on the environment became obvious. Information is typically presented at meetings of the major water-resources-oriented societies such as American Geophysical Union AGU, the American Society of Civil Engineers (ASCE), American Water Resources Association (AWRA), International Water Resources Association (IWRA), American Society of Limnology and Oceanography (ASLO), SIL, and International Commission of Large Dams (ICOLD). Symposium proceedings were published by AGU [9] and ASCE [10], and a book on reservoir limnology was written by Thornton et al. [8]. Introductions to the modeling of lakes and reservoirs were given by Imberger in Fischer et al. [23] and Gray [24].

Although often lumped together, lakes and reservoirs differ significantly in several respects: (a) the water levels in a reservoir can be controlled by an outlet structure and, therefore, often varies more widely and at shorter timescales as water demand dictates, whereas the water level of a lake depends on natural outflow, evaporation, and precipitation on the water surface, and groundwater; (b) a reservoirs has typically a shorter hydraulic residence time than a natural lake because a river or stream is flowing through it; (c) horizontal water quality gradients in a reservoir can be far greater than in natural lakes (except the very largest lakes) because of the flow through a reservoir.

Even though entire cultures collapsed because water reservoirs for the irrigation of farm fields could not be sustained, construction of large dams for the purpose of creating new reservoirs has continued to the present (Three Gorges on the Yangtze River in China). Reservoirs have been created now not only for water storage but also to provide hydropower, flood protection, and sufficient water depth in navigable waterways. Small retention and flow retarding structures have been built in large numbers in cities and agricultural areas and in suburban developments for erosion control, to retain stormwater runoff for infiltration and groundwater recharge, and for urban stormwater management, i.e., flow reduction and water quality improvement [25, 26].

Sustainability of lakes and reservoirs is a big issue. Topographic depressions that lead to the formation of lakes in the landscape have been created naturally by a variety of geophysical processes. Large numbers of lakes were created in the northern USA during the last major glacial period only 10,000 year ago. Processes, resulting in over 70 lake types, are reviewed in Hutchinson's 1957 book [12]. However, all lakes and reservoirs are eventually filled by sediments because they are at a low point of a landscape. Prolonging their lifespan while maintaining high water quality and water use potential is the goal of many efforts.

Inputs, Outputs, and Transformations of Materials in a Lake or Reservoir

The fate and transport of a water quality constitute in a lake and reservoir is the study of mass balance of a substance. Mechanistic water quality studies and models are based on the conservation of mass; that is, within a finite volume of water, mass is neither created nor destroyed [1]. A mass balance equation quantitatively accounts for all transfers (fluxes) of matter across the system's boundaries (e.g., water surface and lake bottom) and all transformations occurring within the system (a lake or reservoir) during a finite period.

$$\text{Accumulation} = \text{loading} \pm \text{transport} \pm \text{reactions}$$

The movement of substances through the volume, along with water flow (surface water and groundwater), is termed transport. Mass of substances is gained or lost by transformations or reactions of the substances within the volume. Chemical and biological reactions either add mass by changing another constituent into the substance being studied or modeled or remove mass by transforming the substance into another constitute [1].

The concept of water and material residence times is intimately linked to the "sustainability" of lakes and reservoirs. As lentic waterbodies, they experience periods of low flow and "flushing." In those times, "deposition" of inorganic or organic particulate matter becomes a crucial and defining process. Material fluxes that can be used to develop material budgets for lakes and reservoirs are crucial for the understanding of how lake chemistry and biology function. Material fluxes and budgets can be for total suspended solids (TSS), total dissolved solids (TDS), specific nutrients such as phosphate (PO_4), ammonia and nitrate (NO_3), silica (SiO_2), sulfate (SO_4), organic matter in terms of total (C) or cell count, dissolved oxygen (DO), or subcategories of nutrients such as total Kjeldahl N (TKN) or dissolved reactive phosphorus or orthophosphorus (P).

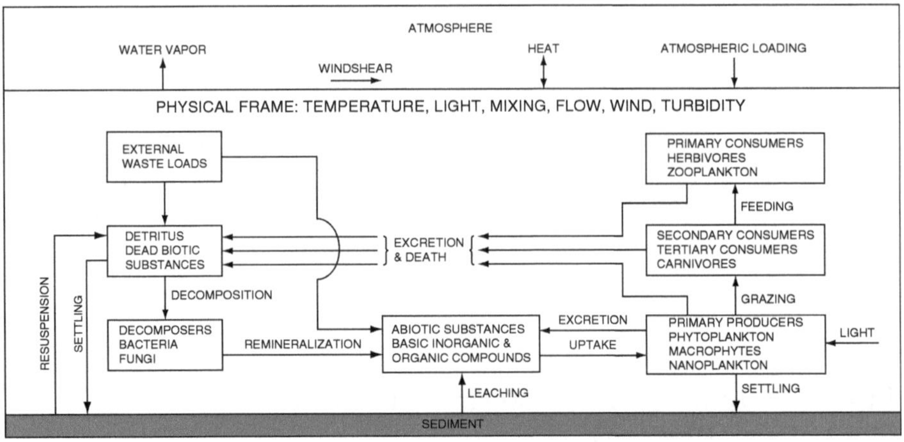

Fig. 11.1 Material flow in a lake or reservoir with emphasis on biological processes (From [27])

If a material budget for a substance is developed, and the control volume is the entire lake or reservoir, the material fluxes can be divided into the following components:

1. External material inputs to a lake or reservoir by runoff from the watershed, i.e., by rivers, small streams, storm sewers, drainage ditches, and "overland" flow from surrounding watersheds
2. External material inputs from the atmosphere through the lake or reservoir surface in form of wet (rain) or dry deposition
3. Material flux to or from the lake or reservoir sediments by organic or inorganic particle deposition or resuspension
4. Material flux of solutes (dissolved substances) into or from the pores of lake or reservoir sediments
5. Material transformations by chemical and biological processes (kinetics) within the various layers of a lake or reservoir

The "overland" flow is storm runoff flowing over the ground surface toward a channel or a waterbody when rainfall has exceeded infiltration capacity and depression storage capacity. Figure 11.1 gives a schematic of fluxes across the upper and lower boundaries of a lake or reservoir and internal transfer processes with emphasis on fundamental biological processes, and Fig. 11.2 is an example of biochemical interactions for various phosphorus components within a lake or reservoir.

External Material Inputs to a Lake or Reservoir by Runoff from the Watershed

Besides the water, material inputs to lakes and reservoirs come in particulate or dissolved form as TSS or TDS. Besides suspended particles, there can also be

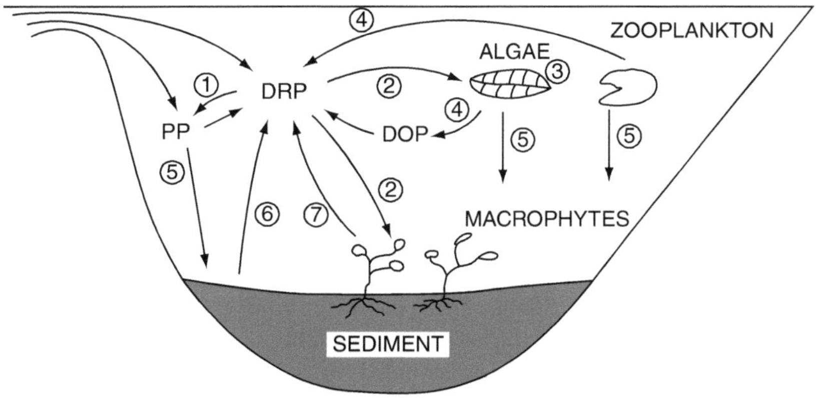

Fig. 11.2 Schematic of phosphorus components and interactions in a lake (From [28]). *DRP* dissolved reactive phosphorus, *DOC* dissolved organic phosphorus, *PP* particulate phosphorus. (1) Adsorption/desorption, (2) assimilation, (3) grazing by zooplankton, (4) regeneration, (5) sedimentation, (6) sediment release/recycling, (7) regeneration (From [27])

a significant bed load component in rivers and streams. These inputs are often detrimental to water quality. Suspended inorganic sediments cause undesirable turbidity and fill lakes and reservoirs when they settle out, reducing the useful life expectancy of reservoirs. This is one of the main arguments against the construction of dams and impoundments. Since there are no known economical methods to remove sediment from reservoirs, water storage in impoundments becomes unsustainable, until incoming and outgoing sediment loads balance. This is a huge problem for reservoirs on sediment-laden rivers. On the Yellow River in China, the life expectancy of a reservoir can be just a few decades. The Colorado River also has significant sedimentation problems. Bed load carried by inflowing rivers contributes to the loss of storage capacity, mostly by the formation of deltas in reservoirs. Diverting clay-laden turbid inflow from Lake Chico, AK, to the nearby Mississippi River increased light penetration and photosynthesis sufficiently to restore a dysfunctional ecosystem and bass fishery [29].

External materials (substances) that are fully dissolved in water and are not reactive are not likely to accumulate long-term in lakes or reservoirs. An example is road salt which is not only a conservative substance but also highly soluble [30]. Snowmelt water from roads in northern cities of the USA will first flow to and accumulate near the bottom of a lake because the salinity makes the water slightly denser than freshwater, some of it will flow into the pores of the lake sediments and the groundwater table underneath, and some will get mixed into the lake and flushed out if there is an outflow of water from the lake. Depending on the annual salt loading and the hydraulic residence times, each lake receiving road salt runoff will reach an equilibrium concentration, which may be high if the loading rate and the residence times are high. Novotny and Stefan [30] give examples for Twin Cities' lakes (7–104 t of Cl input per year and 3–14 years hydraulic residence time). The associated seasonal maximum Cl concentrations at the end of the winter for these lakes were projected to be 107 and 387 mg/L, compared to a chronic water quality standard of 230 mg/L. Road salt accumulation is therefore considered to be fully reversible in lakes but not in groundwater because hydraulic residence times can be several order of magnitudes larger.

Nutrients such as phosphate, ammonium, or nitrate also come from watersheds, especially where fertilizers are used for agriculture. They are dissolved in lake or reservoir water in their original form, but they are reactive, and once they are incorporated in plant mass, they are particulates that precipitate to the lake or reservoir bottom. There have been literally thousands of studies of water quality in runoff from watersheds. Studies have progressed from estimates of annual yields of sediment, phosphorus, or nitrate from agricultural and urban watersheds to computer simulations using fairly detailed models such as SWAT [31] and SWMM [32, 33]. Heat transport in urban runoff has been simulated in models such as SNTEMP [34] for natural watersheds and MINUHET [35] in urban watersheds. Stormwater management is being studied to improve the quality of the runoff into lakes and reservoirs. Reducing the inflow of undesirable nutrients or contaminants to lakes or reservoirs is the goal of many TMDL (Total Maximum Daily Load) studies.

External Material Inputs from the Atmosphere and Material Release to the Atmosphere

The term "precipitation" in water quality studies may be used to describe the external material deposited by wet (rainfall) and dry (dustfall) processes on water surface of a lake or reservoir. The dry dustfall is a continuous process, while wet precipitation (rain, snow, sleet, etc.) is an intermittent process. Precipitation contributes chemical substances to surface waters and should be considered (not simply ignored) in analyzing surface water quality [26]. Precipitation impurities or chemical depositions will vary from one location to another and one time period to another, for example, chloride ion concentrations are highest along coastal areas.

Acid rain has harmed lakes in Europe and the USA. Sulfate emissions from the burning of coal were identified as the cause. Another example of atmospheric input is phosphorus loading of Lake Superior, where one third of the total P input was found to be from the atmosphere [36]. On the other hand, Lake Superior cleansed itself of PCB (polychlorinated biphenyl) that had accumulated by transferring it to the atmosphere in gaseous form (volatilization) [37].

Volatilization and gas exchange between atmosphere and water are controlled by mechanisms or processes at either sides of the air–water interface. Several gas transfer theories [1], e.g., Whitman's two-film theory and surface renewal models [38], and many empirical formulas through data synthesis were developed and used to quantify gas transfer process. Air–water equilibrium concentrations of gases dissolved in water are quantified by Henry's law [1]. Resistance to gas transfer for many gases (e.g., O_2, N_2, CH_4, H_2S, and SO_2) is in the liquid phase [1]. Oxygen reaeration is an important process to surface water quality control and modeling and was studied from the early stage of water quality engineering and management research development and applications. Various formulas of oxygen reaeration in rivers, streams, through hydraulic structures, in standing waters, and estuaries were developed, e.g., summarized by Chapra [1] and Gulliver [38].

Particle Settling and Resuspension in a Lake or Reservoir

Organic or inorganic particle deposition is the most significant process that impairs sustainability of lakes and reservoirs. Particles can settle out from the water column when they are no longer kept in suspension; inorganic clay or silica particles settle relatively fast; algae, which are almost neutrally buoyant, settle in dimictic lakes after they have died in fall. The mechanics of particle settling in lakes and reservoirs are very complicated because settling is the net effect of falling and resuspension in an intermittently turbulent or laminar environment. Settling is driven by the density difference between the solid particle and the water. If the two processes are parameterized by a particle fall velocity and a vertical turbulent resuspension coefficient, their interaction can be used to describe profiles of particle distributions in lentic waterbodies [39].

Resuspension of particles in lakes and reservoirs can be an important process that affects water quality, e.g., turbidity, in lakes and reservoirs. Resuspension of sediment particles on the shoreline of a lake or reservoir can be observed when large enough waves are present. In fact, shorelines may be devoid of fine sediments because their resuspension has carried them to deeper and deeper water by turbidity currents. This process of sediment redistribution in lakes is responsible for the sorting of sediment beds such that shorelines exposed to wind and waves may be devoid of fine sediments, while very fine particulates, particularly organic particles, are settling to the deepest part of a lake [14, 40]. Because resuspension facilitates the redistribution of once deposited materials in a lake, it is an important process. The effect of wind on resuspension in shallow settling basins was investigated by Sheng and Lick [41] and modeled by Rodney and Stefan [42] among others. Erdmann et al. [43] provided data from Duluth harbor.

Because of adsorption and desorption of solutes on particle surfaces, particles also become carriers of substances. These interactions are well known, and descriptions can be found in textbooks, e.g., Thomann and Mueller [44], Clark [45], and Weber and DiGiano [46]. For example, a substantial portion of the TP load to a lake or reservoir can be adsorbed to particles. A seminal article on particle transport in lakes was written by Weilenmann et al. [47].

Flux of Solutes into or from the Pores of a Lake or Reservoir Sediment Bed

The best example of this type of flux is DO transfer from the benthic boundary layer to the surfaces of sediment bed particles, where it is used by microbial biofilms or in chemical reactions. The linkage between DO uptake rates, sometimes referred to sedimentary oxygen demand (SOD), and the flow over, through and below the sediment surface, have been established by field studies (Boynton et al. 1981) and laboratory experiments [48]. Sediment topography and surface waves have been shown to enhance physical transfer rates by "hyporheic flow" between the overlying water and the pore system significantly [49–53]. Connections to the microbial populations have been investigated experimentally, e.g., by Steinberger and Hondzo [54] and O'Connor and Hondzo [55], and modeled, e.g., by Nakamura and Stefan [56] and Higashino et al. [57]. Transfer of solutes other than DO, such as nitrate, iron, and sulfate, at the sediment–water interface have also been studied by Arnon et al. (2007) and Thumdrup et al. [58]. Transfer rates of solutes over a wide range of flow and sediment conditions have been scaled, i.e., linked to dimensionless parameters by O'Connor and Harvey [59]. Transfer of colloids and organic particles into sediment pore systems have been the subject of studies by Packman et al. (2000) and Huettel et al. (1996).

In the long-term, materials incorporated or adsorbed to particles on a lake or reservoir bed may be buried in the sediments, or they may be recycled. The recycling process of phosphate is well understood and depends on chemical reactions with iron [60]. The process has been modeled in detail, e.g., by Nakamura (1994), and as a seasonal bulk process by Lorenzen et al. [61]. The annual recycling of phosphate from the lake or reservoir sediment is referred to as "internal nutrient loading" and can sustain multiple annual algal blooms [62] or blooms over long periods of time because the fertilizer is essentially recycling. The biogeochemical and transport processes associated with the transfer and use of dissolved substances in the sediment pore system are very numerous and complex and have been described, analyzed, and modeled in books, e.g., by Boudreau and Joergensen [63], DiToro [64], Clark [45], and Zheng and Bennett (1995). An experiment on sulfate uptake in lake sediments is described by Manous et al. [65].

Material Transformations by Chemical and Biological Processes (Kinetics) Within the Water Layers of a Lake or Reservoir

Biological processes by which materials in a lake are transformed include plant growth by photosynthesis (primary productivity), grazing by zooplankton, and predation by fish and higher-order animals. Respiration, death, and microbial decomposition affect all organic materials. Chemical processes include oxidation and reduction, and adsorption and desorption round out the menu of possible interactions. Models of lake or reservoir water quality [66, 67] include algorithms that quantify the material fluxes associated by these processes. Chemical and biological kinetics used for this purpose can be simple zero or first order, or can be higher order and/or more complex, e.g., Michaelis Menten, Monod, Haldane kinetics, or processes in series. Figure 11.1 gives a very simple flowchart for the principal biological processes in a lake.

Information on the chemical kinetics of lakes can be found in the limnological literature; the literature on chemical kinetics, e.g., by Stumm and Morgan [68, 69], Gobas and McCorquodale [70], Brezonik [71], Weber and DiGiano [46], and Gulliver [38]; and the water quality modeling literature, e.g., by Thomann and Mueller [44], Chapra [1], Zison et al. [72], and Bowie et al. [73].

Studies of lakes and reservoirs have progressed to ever finer detail and are often highly specialized. Earlier studies of the physics of lakes and reservoirs dealt with readily identifiable phenomena such as the seasonal temperature stratification, wind-driven Langmuir circulation, and the thermal bar which are all described in classical textbooks [15, 16]; more recent studies rely on field measurement techniques that record temperature and turbulent velocities at high temporal and spatial resolution, as well as transparency, specific conductance, and other parameters. Similarly, the study of lake chemistry has progressed from dissolved oxygen and nutrients to subspecies of nutrients, toxic substances, and endocrine disruptors [74, 75]. Lake ecology used to concentrate on phytoplankton, zooplankton, and fish, but has evolved to include detailed measurements and analyses of macrophytes, periphyton, and microbial components (see, e.g., Scheffer [76]).

Lake and Reservoir Manipulations to Affect Fluxes of Materials

Manipulation of water quality and lakes and reservoirs has become a subject of considerable interest, and many different tools have been tried. Cooke et al. [77] give an overview of techniques used in lakes. Engineering approaches have been described by Reckhow and Chapra [78]. Examples of simulation model applications to guide lake or reservoir management decisions have been given by Henderson-Sellers [79].

Destratification by air bubble release at the lake bottom was an early tool [80] and has been modeled by Fast [81], Goossens [82], Zic and Stefan [83], and Zic et al. [84]. Instead of air, pure oxygen has been used in deeper lakes. Air bubble plume can cause phosphorus recycling from the lake sediments [85].

The US Army Corps of Engineers in Vicksburg, MS, has developed and applied various techniques to mix stratified reservoir water near dam faces with large propellers before its release through outlet structures, in order to improve the water quality in the tailwater downstream from reservoirs.

Mixing (aeration) of selective layers has been attempted and practiced, particularly for aeration of selective layers in stratified lakes [86, 87]. Hydraulic jet mixing systems are energetically attractive [88–90]. To prevent internal phosphorus loading of eutrophic lakes, various sediment treatment techniques have been developed

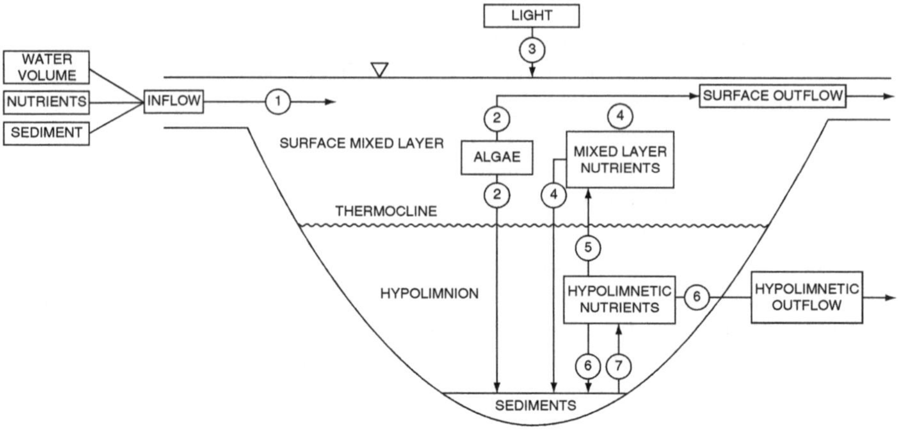

Fig. 11.3 Schematic of lake manipulation (processes 1–7) for eutrophication (water quality) control (From Henderson-Sellers [79])

(alum, RIPLOX). Herbicides have been applied to kill phytoplankton [91] or macrophytes selectively. Biomanipulation by influencing the food chain is an elegant technique that has had some success. Fish stocking and fish management in lakes are very common in some states where recreational fishing is popular.

Lake manipulation techniques (Fig. 11.3) to control the excessive growth of algae are designed to change one or more of the following:

1. The inflow of water, nutrients (mostly P), or sediments to a lake
2. The loss of algal populations by settling or grazing or outflow from a lake
3. The availability of light to an algal population of a lake
4. The nutrients in the photogenic surface layer of a lake
5. The transfer of nutrients from the hypolimnion to the surface layer of a lake
6. The concentration of nutrients in the hypolimnion
7. The release of nutrients from the sediments

Material Residence Times, Stratification, Mixing, and Currents in a Lake or Reservoir

Residence Times

Hydraulic residence time (years) is a measure of the flushing effect by flow through a lake or reservoir. It can be defined as a lake's or reservoir's volume (m^3) divided by the volumetric outflow rate (m^3/year). Some definitions use inflow instead of outflow rates. Hydraulic residence times range from a few days or weeks for

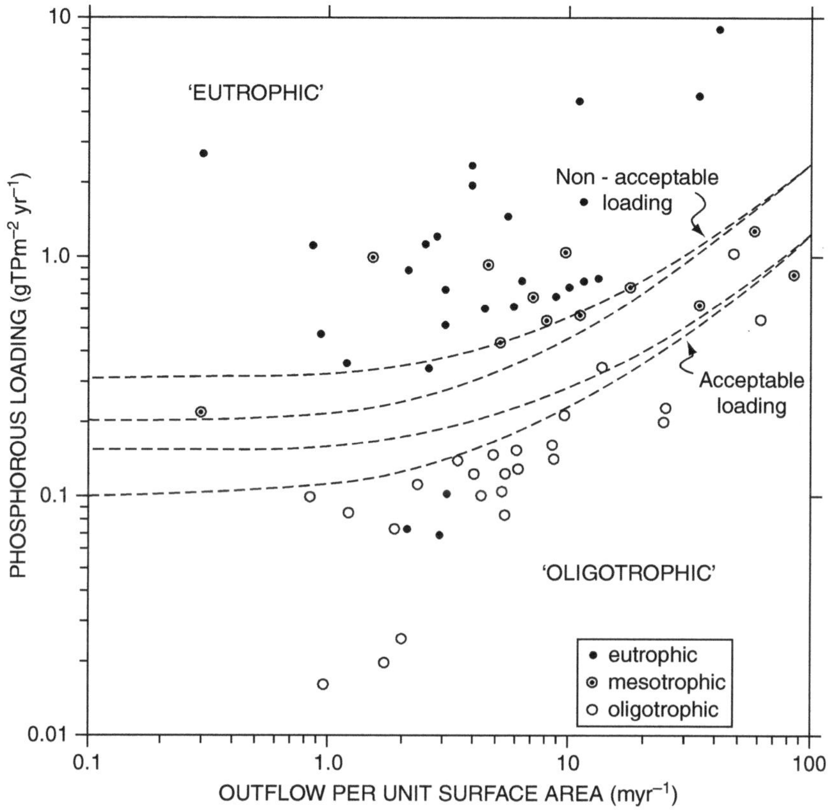

Fig. 11.4 Annual phosphorus loading tolerance of a lake as a function of annual hydraulic flushing rate (outflow per unit surface area) (From [92])

shallow river impoundments, to several years for quaternary lakes and large reservoirs, to hundreds of years of the world's largest lakes. Hydraulic residence time determines how long different chemical or biological processes can work on the removal or generation of different materials/substances. Lake and reservoir ecosystems depend on hydraulic residence times.

Hydraulic flushing or residence time determines the impact of a nutrient input to a lake. Lake or reservoir eutrophication is in many instances linked to phosphorus supply to a lake. The famous Vollenweider [92] diagram reproduced in Fig. 11.4 shows annual flushing rate (m/year) = annual outflow per lake surface area on the horizontal axis and annual volumetric P loading rate (g/m^3/year) on the vertical axis. The lines are boundaries between eutrophic, mesotrophic, and oligotrophic lakes, i.e., very high, intermediate, and low primary (plant) productivity. It can be seen that the same trophic state can be achieved with a high P loading rate when the flushing rate is higher.

The rate of removal of a material/substance/organism, e.g., by settling or by volatilization or by uptake, can be used in addition to the flushing rate to define a material residence time. If material is removed, the material residence time will be shorter than the hydraulic residence time. For first-order processes, the following relationship holds:

$$
\begin{aligned}
(\text{material residence time})^{-1} = {} & (\text{hydraulic residence time})^{-1} \\
& + (\text{settling rate coeff.}) \\
& + (\text{removal rate coeff.}) \\
& + (\text{volatilization rate coeff.})
\end{aligned}
$$

Stratification or Vertical Mixing of a Lake

For water quality in lakes or reservoirs, it makes a large difference if a lentic waterbody is density (temperature) stratified or not, in addition to the flushing effect by flow through the lake or reservoir. Temperature stratification is driven mostly by solar irradiation of the water surface and seasonal temperature variations. Stratification can occur during the day and be absent during the night, it can be intermittent for several days at a time, it can be seasonal in northern temperate climate regions with one or two full mixing events, or permanent if a saline water layer forms in a lake. Correspondingly, lakes are referred to as well-mixed, polymictic, monomictic, dimictic, or meromictic. Stratification and/or mixing depend on the physical processes that produce them.

The density of freshwater changes ever so slightly with temperature or salinity (concentration of dissolved substances). A lake or reservoir can become density stratified whenever heat or any dissolved substance is added or removed from portions of its water. Because a lake or reservoir is a lentic or standing waterbody, its water will tend to arrange itself in layers of increasing density from top to bottom. The differences are typically very slight, but significant. The stable equilibrium of a density-stratified lake is easily reinforced or upset by many physical mechanisms such as daily heating through the water surface by solar radiation, inflows of warmer or colder water, or wind action on the water surface. These processes and their effects on lake stratification have been studied extensively.

There are two major processes that produce lake mixing: convective mixing induced by density instability, i.e., denser water on top of lighter water; or forced mixing, e.g., by wind or inflows (jet mixing). Density instabilities can be caused by cooling of surface waters (cooler water is denser above 4°C) or by dissolved (e.g., road salt) or suspended (e.g., sediment) materials in inflowing waters. Wind mixing depends on the shear stress exerted by wind on the water surface. That shear causes not only clearly visible "progressive" surface waves with amplitudes dependent on shear stress but also less noticeable and relatively slow "standing" waves on the lake surface. In a stratified lake, internal waves are produced by wind and contribute to mixing at different depths when they break on the sloping beaches, invisible to the eye below the water surface.

Criteria for lake or reservoir mixing have been developed. One of the easiest to use is the "lake geometry ratio" $A_S^{0.25}/H_{max}$ [93]. It is the ratio of the fourth root of the lake surface area A_S (in m^2) divided by the maximum lake depth H_{max} (in m). In the north-central USA, lakes with a geometry ratio greater than 4 are polymictic; those with a ratio smaller than 1 are dimictic. The lake geometry ratio depends on lake bathymetry only and is climate independent. That makes it easier to use than other criteria, e.g., the Wedderburn number which requires knowledge of the wind shear stress (Imberger in Fischer et al. [23]).

Density- or Wind-Driven Currents in a Lake or Reservoir

When water temperature or dissolved or suspended material content makes a portion of the water in a lake denser than its ambient water, that water will always tend to sink lower in a stratifies lake until it reaches a layer of equal density. This sinking has a huge effect on water quality because it redistributes material at a rate that is much faster than molecular diffusion.

An organized sinking flow is a density current or turbidity current if suspended sediment makes the water heavier than its surroundings. Turbidity currents are sediment-laden underwater rivers that can occur unnoticed because they take place below a quiescent water surface. Only a plunge line visible on the water surface by the accumulation of floating wood or a color contrast between clear lake water and turbid inflow may give away the occurrence of a significant process. Reservoirs in the Yellow River Basin are famous for their turbidity currents, and the routing and the release of these turbidity currents through bottom outlets at a downstream dam site have become management tools to delay the deposition and filling of the Yellow River reservoirs by the fine loess sediments. Plunging flows and density currents in reservoirs have been investigated and modeled among others by Imberger [23], Alavian et al. [94], Akiyama and Stefan [95–98], Johnson et al. [99, 100], Farrell and Stefan [101, 102], and Fang and Stefan [103]. A density current of saline snowmelt water into an ice-covered lake was described by Ellis et al. [104].

When the inflow to a lake or reservoir is warmer than the lake, it will spread out on the surface of the lake [105–109].

Wind induces horizontal exchange flows in multibasin lakes [110–113] that can transport salinity, nutrients, or contaminants from basin to basin. Wind also controls vertical mixing of a lake by progressive waves and standing waves (seiche) both on the lake surface and internally, if a lake is density (temperature) stratified. Wind also has a determining effect on the depth of the surface mixed layer [114, 115]. In nutrient-rich (eutrophic) lakes, wind can become the determinant/controlling factor for algal growth [116]. Wind controls the wind-driven circulation, stratification, and mixing dynamics in lakes, and thereby the fate and transport of organic and inorganic materials as well as the growth of algae [116] and rooted vegetation (macrophytes) in them [117]. Wind sheltering by trees, buildings, or topography can severely hinder wind access and thereby reduce wind shear on the water surface, wave action, and vertical turbulent mixing by waves, turbulent eddies, and Langmuir circulation. The development of the atmospheric boundary layer on a lake surface and its effects on surface mixing and water quality is often ignored and only partially understood and predictable [118].

Simulation of Water Quality in a Lake or Reservoir

Water quality models are often used as tools to guide decisions on lake management and reservoir operation. Without them, decisions have to be made based on experience. Water quality in a lake or reservoir is best simulated by a deterministic model. Physical modeling attempted by the USEPA and the Army Corps of Engineers did not work out because scaling laws for the physics, chemistry, and biology in lakes or reservoirs were incompatible. The simulation models use mathematical descriptions and rely on basic principles, such as conservation of mass, heat, and mass transfer equations, and well-known relationships for chemical and biological kinetics. A review of basic principles of model formulations can be found, e.g., in textbooks by Thomann and Mueller [44], Chapra [1], and Clark [45], and in a monograph by Stefan, Ambrose, and Dortch [119] but also in some user manuals.

A review of major available hydrological and water resources simulation models was given by Wurbs [120]. One- or multidimensional models of lake or reservoir water quality, e.g., WASP (EUTRO04) [66], CE-QUAL-R1 [121], CE-QUAL-W2 [67], DYRESM [122], and MINLAKE [123], are composed of submodels that analyze and project the hydrodynamics in a lake or reservoir, the heat transfer to obtain temperature distribution and density stratification, sediment particle dynamics, chemical kinetics, biological kinetics and particle–solute interaction. Numerous boundary conditions have to be specified at the water surface or the sediment–water interface. These models require considerable data input and can therefore be tedious to use. Training on the use of some of these models is available from the custodian organizations or from consultants. Information on some of these models can be found on the internet under the name of the model.

Empirical models, e.g., to relate nutrients to lake eutrophication [27, 92, 124], are popular and widely used tools to guide nutrient management in lakes. However, the simplifications necessary in empirical models can lead to outliers and high variance in the results of these models. Empirical models are typically zero-dimensional, i.e., they do not resolve spatial differences and assume well-mixed conditions; they are usually applicable at long (monthly or annual) timescales. Sometimes they apply only to a portion, e.g., the surface mixed layer, of a lake (Fig. 11.3).

One-dimensional water quality models of stratified lakes describe water quality C(z,t) as a function of depth (z) below the water surface and of time (t). They can be quite sufficient for smaller lakes and work well even at short, e.g., daily, timescales. They do not give water quality changes in the longitudinal direction of reservoirs.

Two-dimensional reservoir water quality models have to resolve the flow through the waterbody and can show longitudinal water quality gradients. Three zones are often found by two-dimensional models [8]: a shallow inflow region where velocities are initially high, sediment settles out, and deltas form; a productive zone where nutrients in the inflow combined with sunlight lead to high primary productivity, i.e., phytoplankton growth; and finally a quiescent region where algae are consumed or settle out and water quality is increased. Two-dimensional (width-integrated) models where water quality parameters C(x, z, t) are functions of distance from the dam or inflow, depth below the water surface, and time are very appropriate for reservoirs. Such models can handle stratified flows such as density and turbidity currents, at least in their developed stages away from the plunge point. A schematic of 0-, 1-, 2-, and 3-dimensional model resolution is shown in Fig. 11.5.

Three-dimensional models are used to simulate the most demanding and complex waterbodies. A 3-dimensional model developed by HYDROQUAL has been applied to the tidal Chesapeake Bay, San Francisco Bay, and the Upper Mississippi River impoundments and Lake Pepin in the Twin Cities area of Minnesota. Over 30 water quality constituents have been simulated at daily or shorter timescales and for multiyear periods. Another 3-dimensional hydrodynamic and water quality model ELCOM-CAEDYM was developed and applied by Bruce et al. [125] to simulate the role of zooplankton on C, N, and P cycling in Lake Kinneret, Israel. Other applications of this model have been described by Gal et al. [126], Leon et al. [127], and Robson and Hamilton [128]. An introduction and CE-ROM for 3-dimensional hydrodynamic and water quality lake modeling was offered by Edinger [129].

As an example of a water quality model, the structure and use of the MINLAKE model will be described in more detail. The original MINLAKE model [123] simulates the processes shown in Fig. 11.6. A flowchart is shown in Fig. 11.7. It was applied, e.g., by West and Stefan [130–132], to several lakes in the Twin Cities, Minnesota, and elsewhere, and modified by Fang et al. [133] to MINLAKE96. This deterministic, process-oriented, dynamic and one-dimensional (vertical) year-round lake water quality model was later modified for water temperature and dissolved oxygen (DO) simulations in deep oligotrophic lakes in Minnesota, and then called MINLAKE2010 [134].

Year-Round Water Temperature and Ice-Cover Model

The MINLAKE numerical simulation model for water temperature profiles in lakes solves the one-dimensional, unsteady heat transfer equation

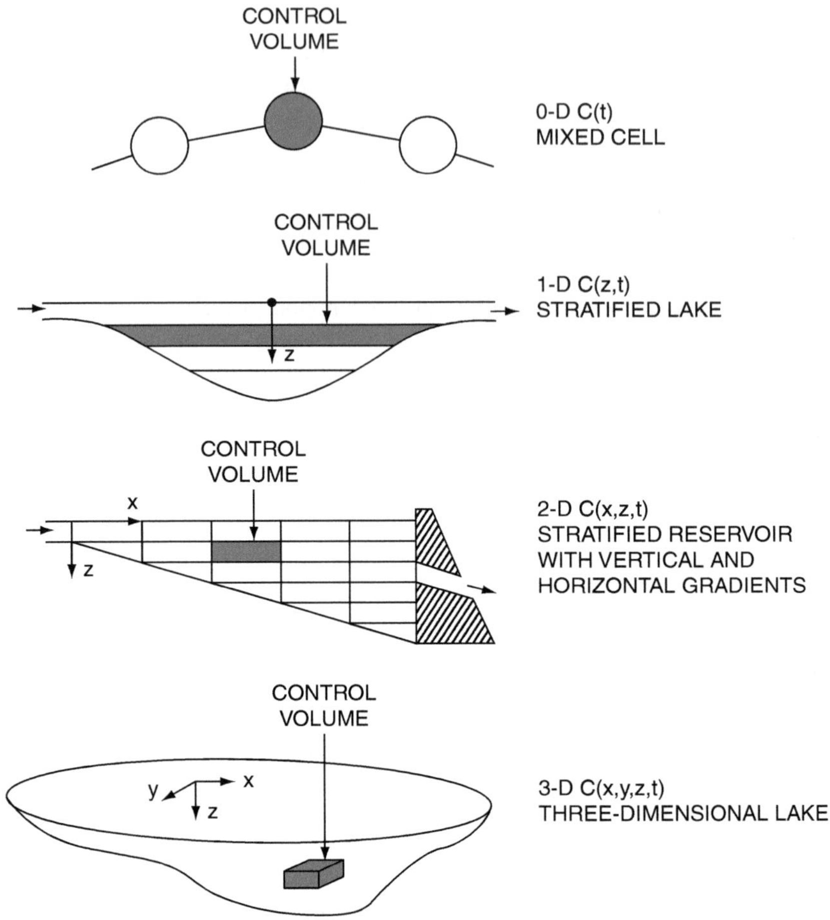

Fig. 11.5 Schematic of the spatial resolution of lake or reservoir water quality models. Control volumes are shown as shaded elements. Concentration of a water quality constituent is designated by (C) (From Stefan, Ambrose, and Dortch [119])

$$\frac{\partial T}{\partial t} = \frac{1}{A}\frac{\partial}{\partial z}\left[K_z A \frac{\partial T}{\partial z}\right] + \frac{H_w}{\rho c_p} \tag{11.1}$$

where T (°C) is the water temperature, t (day) is the time, A (m^2) is the horizontal area of a lake as a function of depth z (m), K_Z (m^2 day^{-1}) is the vertical turbulent heat diffusion coefficient, ρc_p (J m^{-3}°C^{-1}) represents heat capacity per unit volume of water and is equal to the density of water (ρ) times heat capacity of water (c_p), and H_w (J m^{-3} day^{-1}) is the heat source or sink strength per unit volume of water. Solar radiation absorption in the water column contributes to the heat source term. Heat exchange between the atmosphere and the water during the open-water season (7) is treated as a source/sink term for the topmost

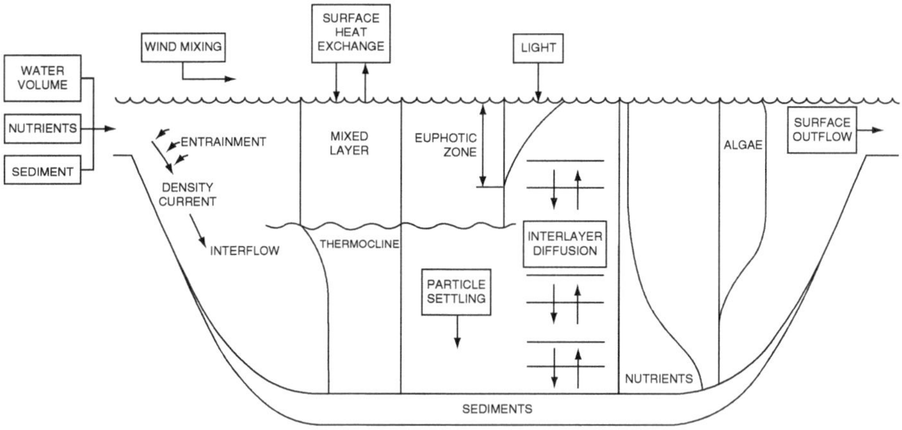

Fig. 11.6 Schematic of the variables and processes simulated by the 1-D MINLAKE model [123]

water layer in a lake. For the open-water season, the computational scheme and the determination of source and sink terms have been discussed, e.g., by Edinger et al. [135], Ford and Stefan [136], Harleman [137], Hondzo and Stefan [138, 139], among others. Equation 11.1 is solved numerically using an implicit finite difference scheme and a Gaussian elimination method, e.g., for time steps of 1 day and water layer thicknesses of 1 m. The model uses a stacked layer system to represent a lake and its environment in the open-water and winter ice-cover periods. Besides the water layers, the lake sediments, the ice cover, and the snow cover are included in the model by separate submodels (Fig. 11.8).

Climate conditions and variations over seasons are driving forces of seasonal variations of water temperature in a lake. Weather parameters needed are: daily average air temperature, dew point temperature, wind speed, sunshine percentage, total daily solar radiation, and precipitation (rainfall and snowfall). They are used as model input parameters to calculate heat fluxes across the water surface or ice and snow cover in winter. If solar radiation has not been measured, it is calculated [141].

Compared to the regional water temperature model for the open-water season [139], the year-round water temperature simulation model has been expanded significantly by simulating ice and snow covers above the water and including the heat exchange between each water layer and its adjoining sediments.

The heat flux across the sediment–water interface is treated as a contribution to the source/sink term for each water layer from the water surface to the lake bottom [142]. The direction of the heat flux between the sediment and the water reverses frequently on shorter, e.g., daily, timescales [143]. Heat can transfer into or out of the lake sediment during both the open-water season and the winter ice-cover period. The lake sediments not only provide seasonal heat storage but also add significant thermal inertia to the water column. Sediment heat fluxes are most important in shallow lakes and during winter ice-cover period.

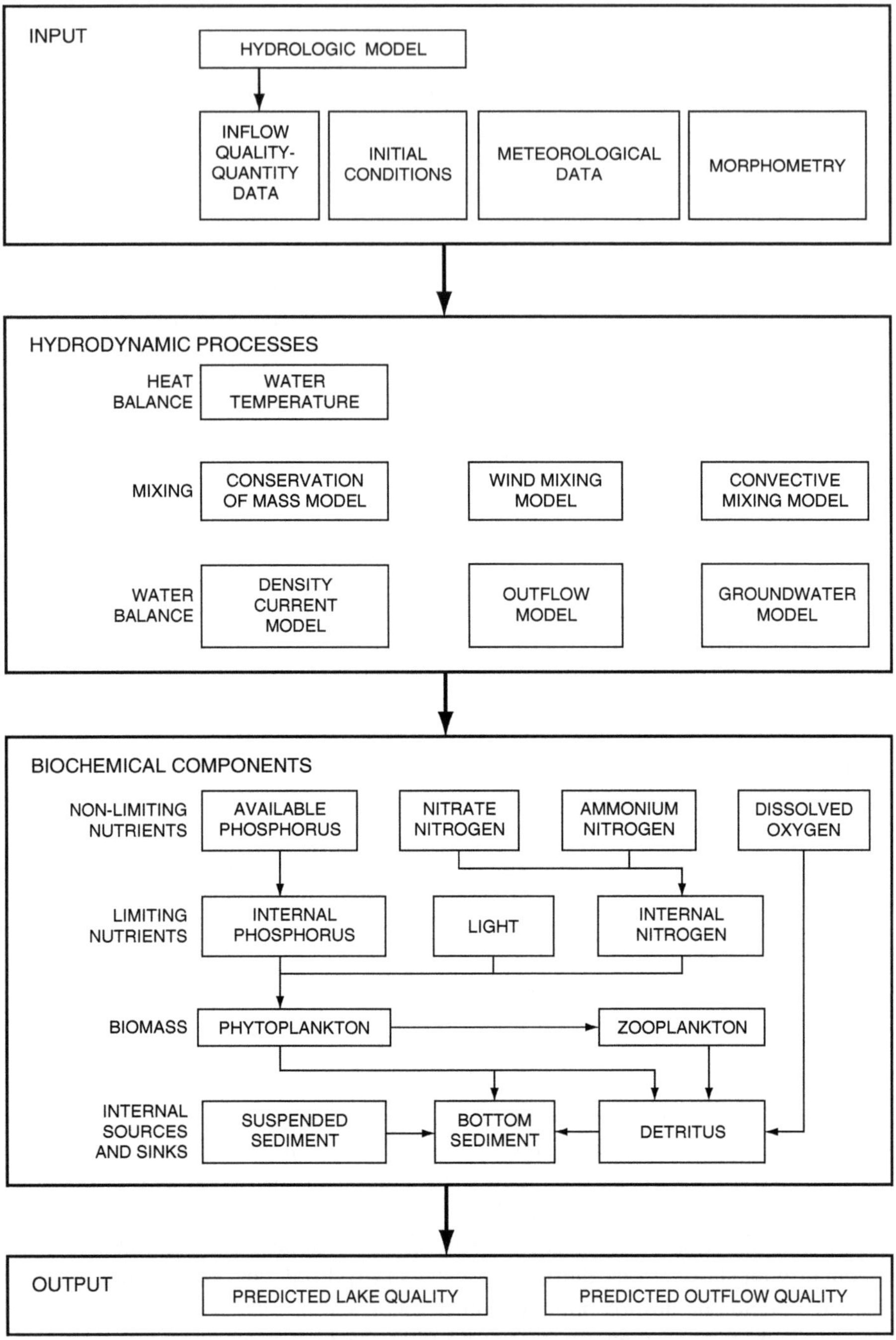

Fig. 11.7 Flowchart and major model sections and components in the original MINLAKE model [123]

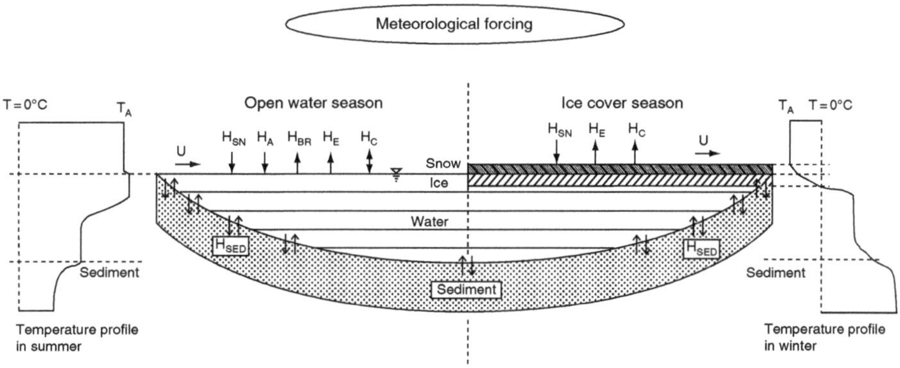

Fig. 11.8 Schematic of a stratified lake with a 10-m sediment layer showing heat transfer components and water/sediment temperature profiles in the open-water season and in the ice-cover season [140]

The snow and ice thickness submodels developed by Gu and Stefan [144] have been used with some modifications. The complete set of equations for the ice- and snow-cover simulation has been summarized by Fang and Stefan [142]. During the ice-cover period, the model simulates ice thickness and sediment temperature profiles (heat conduction equation) first, then determines the heat source/sink term H_W in (Eq. 11.1), and finally solves the heat transfer (Eq. 11.1) to obtain water temperature profiles below the ice. At the air/snow interface (or air–ice interface if snow is absent), the net heat flux from the atmosphere into or out of the snow/ice cover is calculated. Contributions of heat flux are made by solar radiation (H_{SN}), evaporation (H_E), and convection (H_C). Snow thickness is determined from snow accumulation (precipitation), followed by compaction and snow melting. The model simulates melting of snow by surface heat input (convection, rainfall, solar radiation), melting within the snow layer due to internal absorption of short wave radiation, and transformation of wetted snow to ice when cracks in the ice cover allow the water to migrate on the ice surface [133]. In the model, ice growth occurs at the ice/water interface and at the ice surface [133]; ice decay occurs and is simulated at the snow/ice interface, the ice/water interface, and within the ice layer.

Ice formation on small freshwater lakes generally occurs on a calm, cold night. Rising winds and daytime heating may subsequently break up this cover until calm and cold conditions occur again and the ice cover forms a second time [145]. A process-descriptive algorithm which replaced previous empirical and lake-size-dependent criteria to predict the date of ice formation is incorporated in the model [146]. The ice-formation algorithm uses a full heat budget equation to estimate surface cooling, quantifies the effect of forced connective (wind) mixing, and includes the latent heat removed by ice formation. The algorithm has a fine spatial resolution near the water surface where temperature gradients before freeze-over are the greatest. Inverse temperature stratification occurs in the near-surface water several hours before ice formation. The ice-formation algorithm is combined with the year-round temperature model and was tested previously against observations in Ryan Lake and eight other Minnesota lakes for multiple (9–36) years. The difference between the simulated and observed (permanent) ice-formation dates was less than 6 days for all lakes studied. Water temperature profiles are typically predicted with an RMSE on the order of 1°C.

The water temperature and ice-cover model can be run in a continuous mode over many simulation years covering sequences of both open-water seasons and ice-cover periods. The model uses a stacked layer system (Fig. 11.1); lake sediment and water layers during the summer open-water season, and additional ice-cover and snow-cover layers for winter. A lake or reservoir is divided into well-mixed horizontal layers having typical thickness of 1 m.

Year-Round Dissolved Oxygen Model

The one-dimensional, deterministic, unsteady year-round dissolved oxygen transport equation [147], which is the basis of the DO model, is

$$\begin{aligned}\frac{\partial C}{\partial t} = & \frac{1}{A}\frac{\partial}{\partial z}\left(AK_Z\frac{\partial C}{\partial z}\right) - \frac{S_b}{A}\frac{\partial A}{\partial z} \\ & + P_{MAX}\,Min[L]Chla \\ & - \frac{1}{YCHO2}\,k_r\,\theta_r{}^{T-20}\,Chla - k_b\,\theta_b{}^{T-20}\,BOD\end{aligned} \tag{11.2}$$

where C(z,t) is the dissolved oxygen concentration in mg L^{-1} as a function of depth (z) and time (t), A(z) is the horizontal area at different depths in m^2, K_z is the turbulent diffusion coefficient of DO in m^2 day^{-1}, S_b is the sedimentary oxygen demand (SOD) coefficient in mg O_2 (m^{-2} day^{-1}), P_{max} is the maximum specific oxygen production rate by photosynthesis at saturating light conditions in [mg O_2 (mg Chl-a)$^{-1}$ h^{-1}], Min[L] is the light limitation determined by the Haldane equation, Chl-a is the chlorophyll-a concentration in mg L^{-1}, YCHO2 is the yield coefficient which is the ratio of mg chlorophyll-a to mg oxygen, k_r and k_b are the first-order decay for biochemical oxygen demand (BOD) and respiration rate coefficient (day^{-1}), respectively, θ_r and θ_b are the temperature adjustment coefficient for plant respiration and BOD, BOD is the biochemical oxygen demand concentration in mg L^{-1}, and T(z, t) is the water temperatures in °C. In the model, the oxygen transfer through the water surface (reaeration) is expressed as

$$\mathrm{k_e(C_S - C)/\Delta z_S} \tag{11.3}$$

It is used as an oxygen source or sink term in the topmost water (surface) layer. Diffusive oxygen flux at the lake bottom is set equal to zero as a boundary condition. (Sedimentary oxygen demand is treated as a source/sink term.)

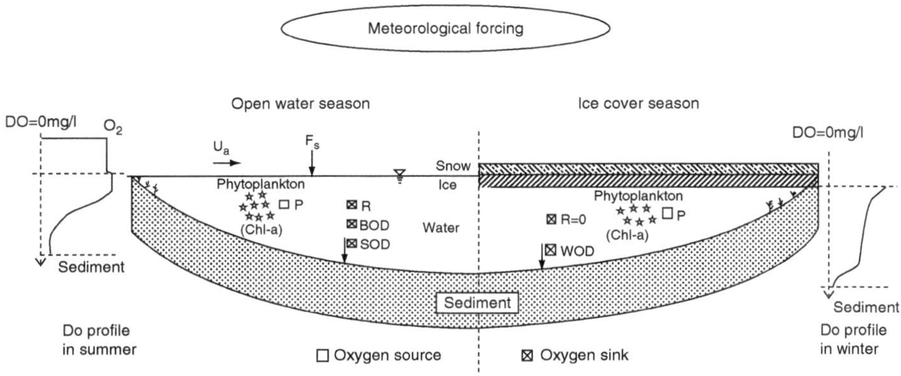

Fig. 11.9 Schematic of a stratified lake showing dissolved oxygen source/sink terms and dissolved oxygen profiles in the open-water season and in the ice-cover period [140]

For the dissolved oxygen simulations in a lake over the winter season (Fig. 11.9), some modifications had to be made in Eq. 11.2 to account for the presence of an ice cover and low temperatures. These modifications [142] include: (a) reaeration is zero (k_e is set equal to zero); (b) the respiration rate coefficient k_r is zero; (c) the water column oxygen demand, WOD, by detrital and other organic matter, is set to a constant value (0.010 g O_2 m^{-3} day^{-1}), independent of the trophic status of a lake [148]; and (d) sedimentary oxygen demand (S_b) is dependent on trophic state and set equal to 0.226, 0.152, and 0.075 (g O_2 m^{-2} day^{-1}) for eutrophic, mesotrophic, and oligotrophic lakes, respectively.

Water column oxygen demand in winter (WOD) is constant and very low, whereas the biochemical oxygen demand (BOD) in summer varies as a function of trophic status. Both WOD and BOD describe the same processes, i.e., microbial and chemical decomposition of detrital or nonliving organic material, and can therefore be expected to depend on trophic state. However, the very limited database for WOD and its small value did not justify or require an adjustment of WOD for trophic state.

Trophic state characterizes biological productivity and relates to plant density, especially phytoplankton; nutrient availability; photosynthetic oxygen production; and respiratory consumption. All of these characteristics can be used to measure trophic status. In many, but not all, lakes, trophic states are closely related to phytoplankton concentration and lake clarity. With this in mind, lakes were divided into three trophic states by the mean annual chlorophyll-a concentrations given in Table 11.1. Secchi depths (SD) of 1.2, 2.5, and 4.5 m were selected for eutrophic, mesotrophic, and oligotrophic lakes using Carlson's trophic state index [159], respectively.

The radiation attenuation in a lake is used to quantify how much of the solar energy reaching the water surface can penetrate through a water column to heat water below the surface and to support photosynthesis of phytoplankton. Total attenuation coefficient, μ (m^{-1}), is given by

$$\mu = \mu_w + \mu_{ch}\,(Chl - a) \tag{11.4}$$

where μ_w (m^{-1}) is the attenuation coefficient of the lake water excluding chlorophyll-a, μ_{ch} [m^{-1} (g m^{-3} Chl-a)$^{-1}$] is the attenuation coefficient due to chlorophyll-a [152], and *Chl-a* is chlorophyll-a concentration (g m^{-3}). An alternative is to estimate total attenuation coefficient as function of Secchi depth [139] by the Eq. 11.5. The coefficient 1.84 is empirical, and other similar values have also been proposed.

$$\mu = \frac{1.84}{SD} \tag{11.5}$$

Model Input and Output

Model parameters and coefficients are shown in Table 11.1 for the temperature submodel and in Table 11.2 for the DO submodel of MINLAKE2010.

MINLAKE2010 model input data are read through ten input files: (1) bathymetry data, (2) weather data (solar radiation optionally from Maxwell et al. 1998), (3) lake-specific model parameter, (4) fixed model parameter, (5) long-term annual and January air temperatures, (6) weather station information (elevations, longitude, and latitude), (7) dimensionless sediment temperature profile [162], (8) seasonal chlorophyll-a pattern file [147, 163], (9) path file directing the program to open model input or output files, and (10) future climate data, e.g., output data of General Circulation Models (GCM) models.

Most of the input data files are common data files for model simulations of many lakes. Only three data files are created for a specific lake: the bathymetry data file, the lake-specific model parameter input file, and the path file. Sediment temperature at 10 m below sediment–water interfaces of a water layer is a model input [162]. These 10-m sediment temperatures have been related to mean annual air temperatures [164]. Long-term air temperature in January is used to specify whether a lake will have an ice cover [162]. The station data files, including long-term air temperatures, were developed for 209 weather stations over the contiguous United States of America. Therefore, in addition to three lake-specific input files mentioned earlier, the station data file and the file with long-term air temperatures were also updated when new weather stations were used.

Because the growth of phytoplankton is not explicitly modeled in the MINLAKE2010 version of the model (it is in the original version by Riley and Stefan [123]), the seasonal pattern of standing algal crops had to be specified, as shown by the solid line in Fig. 11.10 for mesotrophic lakes. Similar patterns were developed also for eutrophic and oligotrophic lakes [134, 147].

The model output consists of daily vertical profiles of water temperature (example in Fig. 11.11) and DO in the water column. Three measured temperature profiles on

Table 11.1 Parameters and coefficient values in the hydrothermal model [147]

Coefficients and symbols		Units	Range and references	Selected value
Open-water season				
Radiation absorption for water	β_w	–	0.4[a]	0.4
Sediment specific heat	c_{psed}	kcal kg^{-1}°C^{-1}	0.2–0.3[b]	0.28
Sediment thermal conductivity	k_{sed}	kcal day^{-1}°C^{-1} m^{-1}	8.64–51.8[b]	19.25
Radiation attenuation by Chl-a	μ_{chh}	m^2 g^{-1} Chl-a	0.2–31.5[c]	20.0
Radiation attenuation by water	μ_w	m^{-1}	0.33–1.03[d]	0.51
Sediment density	ρ_{sed}	kg m^{-3}	1,650–2,300[b]	1,970
Wind sheltering	W_{str}	–	0.01–1.0[e]	Varies
Winter ice cover				
Surface reflectivity for ice	α_i	–	0.55[f]	0.55
Surface reflectivity for snow	α_{sw}	–	0.4–0.95[g]	0.80
Radiation absorption for ice	β_i	–	0.17–0.32[h]	0.17
Radiation absorption for snow	β_{sw}	–	0.17–0.34[i]	0.34
Snow compaction	c_{sw}	–	0.125–0.5[g]	0.4
Ice thermal conductivity	k_i	kcal day^{-1}°C^{-1} m^{-1}	45.8[b]	53.6
Snow thermal conductivity	k_{sw}	kcal day^{-1}°C^{-1} m^{-1}	2.16[b]	5.57
Ice density	ρ_i	kg m^{-3}	920[b]	920.0
Snow density	ρ_{sw}	kg m^{-3}	100–400[g]	300.0
Radiation attenuation by ice	μ_i	m^{-1}	1.6–7.0[j]	1.6
Radiation attenuation by snow	μ_{sw}	m^{-1}	20–40[k]	40.0
Ice latent heat of fusion	λ_i	kcal kg^{-1}	80[l]	80.0
Snow latent heat of fusion	λ_{sw}	kcal kg^{-1}	80[l]	80.0

[a]Dake and Harleman [149]
[b]Carslaw and Jaeger [150]
[c]Bannister [151]
[d]Megard et al. [152]
[e]Riley and Stefan [123]
[f]Bolsenga [153]
[g]Lock [154]
[h]Wake and Rumer [155]
[i]Scott [156]
[j]Pivovarov [157]
[k]Greene [158]
[l]Ashton [145]

July 5, 1996, in Swan Lake (Fig. 11.11) indicate that horizontal variations of water temperature at different depths are typically much smaller than vertical temperature difference (stratification) from surface to bottom; therefore, one-dimensional assumption in lake modeling is valid. An example of simulated versus observed water temperatures and DO concentrations in Minnesota lakes is given in Fig. 11.12.

Table 11.2 Parameters and coefficient values in the dissolved oxygen model [134, 147]

Coefficients and symbols		Units	Range and references	Selected value	
Independent of trophic status					
BOD decay coefficient	k_b	day^{-1}	0.02–3.4[a]	0.1	
Respiration rate coefficient	k_r	day^{-1}	0.05–0.5[a]	0.1	
BOD temperature adjustment	θ_b	–	1.047[a]	1.047	
Photosynthesis temperature adjustment	θ_p	–	1.066[b]	1.036	
Respiration temperature adjustment	θ_r	–	1.045[c], 1.047[d]	1.047	
Sediment temperature adjustment	θ_s	–	1.034–1.13[e]	1.065	
Respiration ratio	YCHO2	–	0.0083[f]	0.0083	
Water column oxygen demand during winter	WOD	$g\ m^{-3}\ day^{-1}$	0.01	0.01	
Dependent on trophic status					
Coefficients and symbols		Units	Eutrophic	Mesotrophic	Oligotrophic
Oxygen equivalent	BOD	$mg\ l^{-1}$	1.0[g]	0.5	0.2
Chlorophyll-a	Chl-a	$mg\ m^{-3}$	15[h]	6	2
SOD during open-water season	S_{b20}	$g\ m^{-2}\ day^{-1}$	1.5[g]	1.25	1.0
			H_{max} = 24 m	H_{max} = 24 m	H_{max} = 24 m
			1.95	1.75	1.2
			H_{max} = 13 m	H_{max} = 13 m	H_{max} = 13 m
			2.2	1.9	1.25
			H_{max} = 4 m	H_{max} = 4 m	H_{max} = 4 m
SOD during ice-cover period	S_b	$g\ m^{-2}\ day^{-1}$	0.226	0.152	0.075

H_{max} = 4 m = shallow lake, H_{max} = 13 m = medium depth, H_{max} = 24 m = deep lake [147]
[a]From QUAL2E [160]
[b]From [44]
[c]From EUTR04 [66]
[d]From MINLAKE [123]
[e]From [161]
[f]From [68]
[g]From Fang et al. [134]
[h]From Stefan and Fang [147]

Simulation of Sustainable Fish Habitat in Lakes Under Global Warming

Sustaining fish habitat in lakes and reservoirs is a major resource management issue. Freshwater fish habitat is constrained by physical, chemical, and biological attributes that relate to water quality, food supply, and human interference [2].

In lakes, water temperature and dissolved oxygen (DO) concentrations are two of the most significant water quality parameters affecting survival and growth of fishes [2, 3]. An increase of atmospheric greenhouse gases is projected to cause

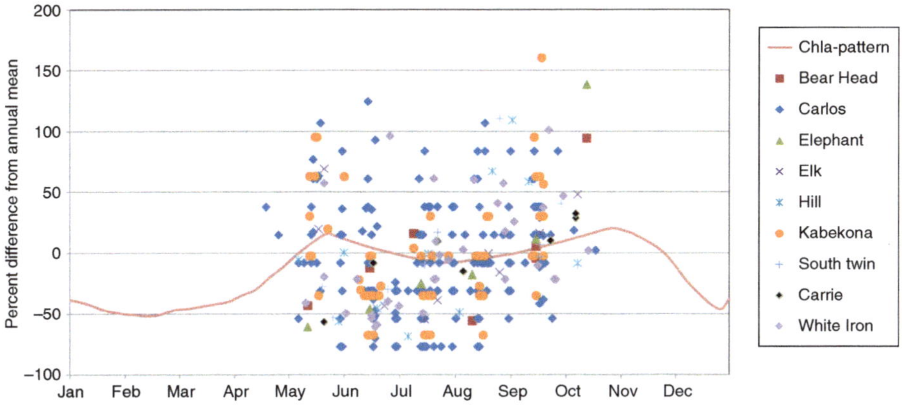

Fig. 11.10 Seasonal chlorophyll-*a* distribution specified as MINLAKE2010 model input for mesotrophic lakes. Generic distribution (*solid line*) and field data from nine study lakes

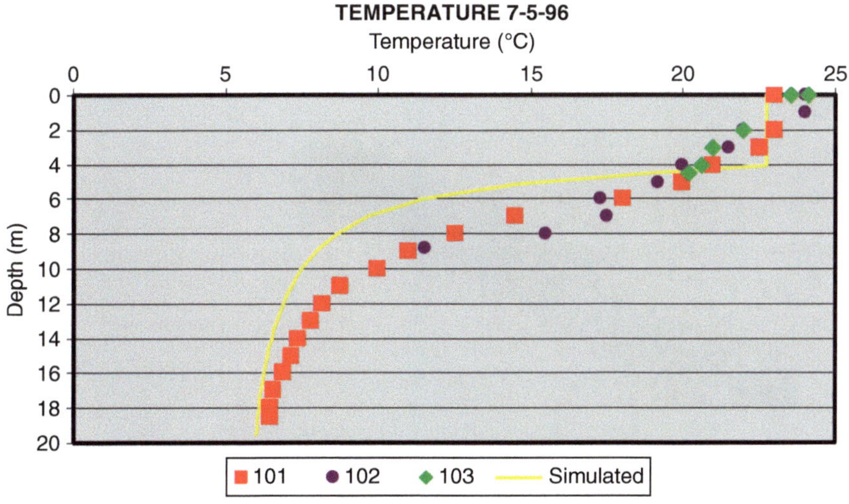

Fig. 11.11 Example of simulated (*solid line*) and three measured (sites 101,102,103) temperature profiles in Swan Lake, Minnesota

climate warming, which would alter water temperature and DO characteristics in lakes. These changes are in turn expected to have an effect on indigenous fish populations [165–169]. For example, cold-water species typically have physiological optima at temperatures near 15°C and are generally not found where summer water temperatures are higher than 20–26°C [170]. Can fish habitat, especially of cold-water species, be sustained under global warming? Cold-water fish habitat in lakes can be reduced in several ways: by direct warming of the water, by increased

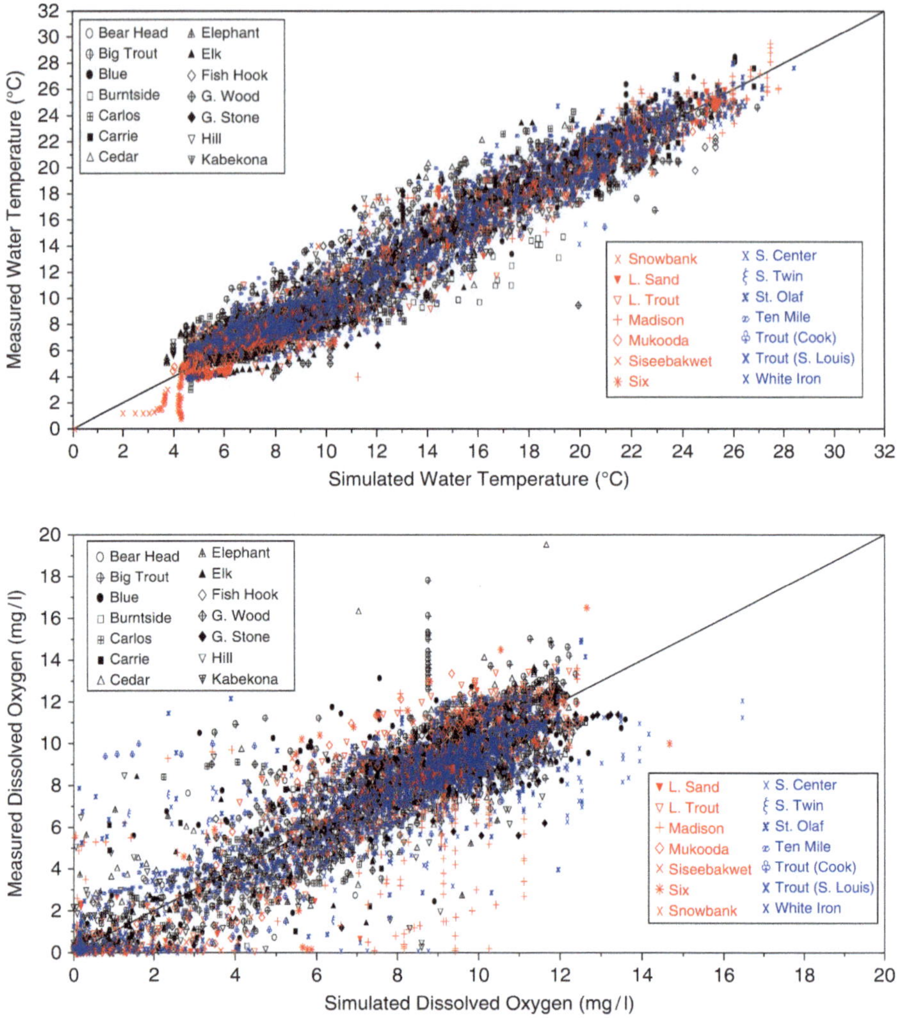

Fig. 11.12 Simulated versus measured temperature and dissolved oxygen (DO) in 28 Minnesota lakes. The solid line (1:1) would indicate perfect agreement

hypolimnetic oxygen depletion during periods of stratification, and by thermocline deepening [165, 171, 172]. Can potential refuge lakes and reservoirs be identified and protected?

Water temperatures and DO concentrations in Minnesota lakes under several projected climate scenarios were previously simulated for the open-water season and the entire year [140, 173], and the results were used to estimate potential future fish habitats [166–169, 174]. Fish habitat estimated from daily temperature and DO profiles under past climate conditions was validated against fish observations

[175, 176]. In these studies, fish species were grouped into three thermal guilds – cold-water, coolwater, and warm-water – that differ in the optimal temperature range for physiological function and ecological success.

In a follow-up study, habitat for only one fish species was investigated. Cisco, *Coregonus artedi*, is the most common cold-water stenothermal fish in Minnesota lakes. The combination of a wide geographic distribution and a requirement for cold, oxygenated water make cisco an excellent "canary in a mineshaft" species that is a sensitive indicator of climate change.

After daily water temperature and DO were simulated under past and future climate scenarios, two approaches were explored to determine if cisco habitat existed in a lake. The first approach was based on separate water temperature and DO constraints; this approach had previously been used to examine the potential impact of future climate warming on cold-water, coolwater, and warm-water fish species (guilds) in small lakes in the contiguous United States [166].

The second approach used a single oxythermal habitat variable to define suitable or unsuitable fish habitat. Jacobson et al. [177] developed a generalized oxythermal habitat variable, TDO3, called "temperature at 3.0 mg/L DO." It is determined by interpolating the water temperature at a benchmark oxygen concentration (i.e., 3.0 mg/L) from vertical temperature and DO profiles in a lake. DO = 3.0 mg/L is an oxygen concentration limit that is probably lethal or nearly so for many cold-water species [2, 178, 179]. TDO3 has a strong connection with four cold-water taxa (lake trout, cisco, whitefish, and burbot) in Minnesota lakes. Cisco were present in lakes with a broad range of maximum TDO3 values, with central borders of 4.0–16.9°C [177]. TDO3 allows to evaluate or quantify which lake is a better refuge lake for cisco under a future climate scenario.

Simulation of Cisco Habitat by the Constant Values Method (CVM)

Fish habitat for cold-water fish guilds (e.g., cisco) was estimated from simulated daily water temperature (T) and DO profiles in lakes, similar to the approach by Christie and Regier [3]. Temperature and DO criteria for survival and good growth of a fish species were applied to the simulated year-round daily water temperature and DO profiles, as shown schematically in Fig. 11.13. This schematic figure is for a lake *where cold-water fish cannot be present* during much of the open-water season. On days when the isotherm of lethal temperature (LT) and the DO limit isopleth for a fish species intersect, the entire depth of a stratified lake is uninhabitable. When the maximum water temperature is lower than the lethal temperature (LT) everywhere in a lake, the isotherm for LT will not show up. In this case, the DO survival limit becomes the only survival criterion; this occurs especially during the winter ice-cover period, as shown in Fig. 11.14.

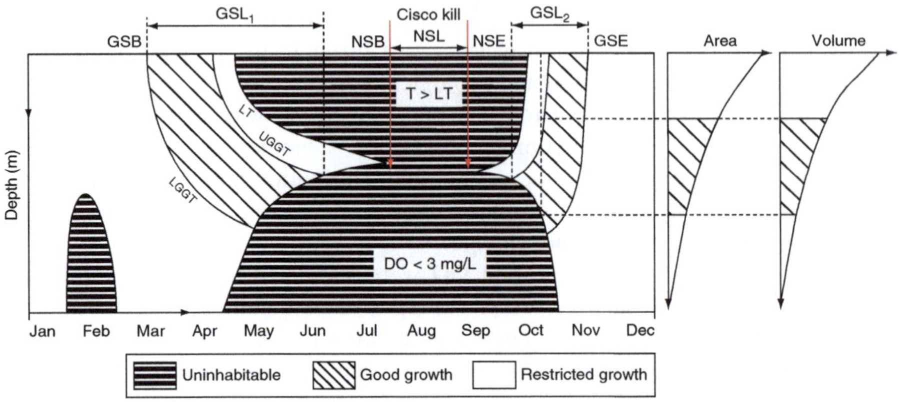

Fig. 11.13 Depth versus time plot of the critical isotherm and DO isopleth, showing projected cisco kill in a seasonally stratified lake as dark shaded areas. Habitat parameters (e.g., NSL, GSL) are discussed by Stefan et al. [166]

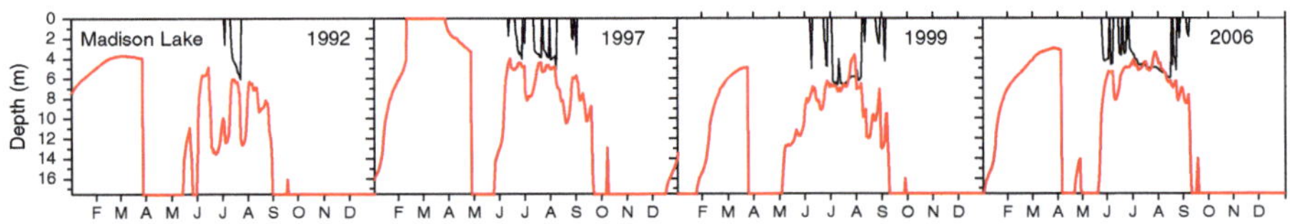

Fig. 11.14 Depth versus time contours of cisco fish habitat in Madison Lake for selected years. Thin lines are isotherms and indicate a temperature constraint for fish habitat in the upper part of the stratified lake; bolder lines are DO isopleths that indicate a dissolved oxygen constraint for fish habitat in the lower portion of the lake

In addition to the lethal temperature (LT) isotherm, Fig. 11.13 also gives the upper good-growth temperature (UGGT) and the lower good-growth temperature (LGGT) isotherms. Between these isotherms/isopleths, three fish habitats are identified:

1. *Uninhabitable space* if temperature is above or DO is below the survival limit
2. *Good-growth habitat* if temperature is between the upper and lower good-growth limits (i.e., LGGT < T < UGGT) and DO is above the survival limit
3. *Restricted growth habitat* if temperature is above the upper good-growth temperature but below the survival limit (i.e., UGGT < T < LT), or if temperature is below the lower good-growth limit of temperature (i.e., T < LGGT) and DO is above the survival limit

Projections on growth habitat and survival parameters of three fish guilds (coldwater, coolwater, and warm-water) in the U.S. were reported by Stefan et al. [166] and Fang et al. [167–169].

The LT is the water temperature to which fish cannot be acclimated without causing death. The LT = 23.4°C [170] used for the cold-water fish guild in previous studies [166] was the mean value of LT values for ten cold-water fish species (pink salmon, sockeye salmon, chinook salmon, chum salmon, coho salmon, brown trout, rainbow trout, brook trout, lake trout, mountain whitefish).

The DO = 3.0 mg/L requirement for the cold-water fish guild, below which mortality is more likely to occur or growth is impaired [179], was developed from an available US EPA database [180]. The DO = 3.0 mg/L was for the open-water season. In ice-covered shallow, eutrophic or mesotrophic lakes DO can drop below 2.0 mg/L near the end of an extended ice-cover period due to bottom (sediment) and water column oxygen demands. For ice-covered lakes, DO criteria for fish survival could be set at lower values than 3.0 mg/L, but data from previous studies are still inadequate to establish specific tolerance limits of DO for fish winterkill [181]. The use of a single year-round DO criterion had no adverse effect on projections of winterkill under a 2 × CO_2 climate scenario; no winterkill was projected despite using higher DO limits [166, 182].

Figure 11.14 is an example of contour plots in Madison Lake, Minnesota, which has a maximum depth of 18.0 m, surface area of 4.5 km^2, and mean summer Secchi depth of 0.88 m (eutrophic lake). Isotherms at the lethal temperature of 23.4°C are drawn as thin black lines, and isopleths at the DO limit of 3.0 mg/L are drawn as thicker red lines. The CVM projected fish kill in 42 out of 47 simulation years (1961 to 2008), summer kill in 1999 and 2006, and winterkill in 1997 in Fig. 11.14 are samples because 2006 was the warmest year and 1997 was the coldest year in the simulation period. No fish kill was projected in 1992, as shown in Fig. 11.14. The maximum continuous kill days in Madison Lake were projected to last 34 days and occurred in 2006. Madison Lake is classified as a non-cisco lake by Minnesota Department of Natural Resources. The maximum continuous kill days in 15 lakes studied were projected to occur either in 2006 for summer kill (with a few exceptions) or in 1996 or 1997 for winterkill (the coldest 2 years of the simulation period).

Simulation of Cisco Habitat by the Equation Method (EM)

Jacobson et al. [177] developed an equation for the lethal-niche boundary of adult cisco by remapping the measured DO concentrations and temperatures from the profiles in 16 Minnesota lakes that experienced cisco mortality in midsummer of 2006. The equation is given as

$$DO_{lethal} = 0.40 + 0.000006e^{0.59T_{lethal}} \tag{11.6}$$

where DO_{lethal} and T_{lethal} are the dissolved oxygen concentration and the water temperature that define the lethal-niche boundary. DO_{lethal} is the required or needed

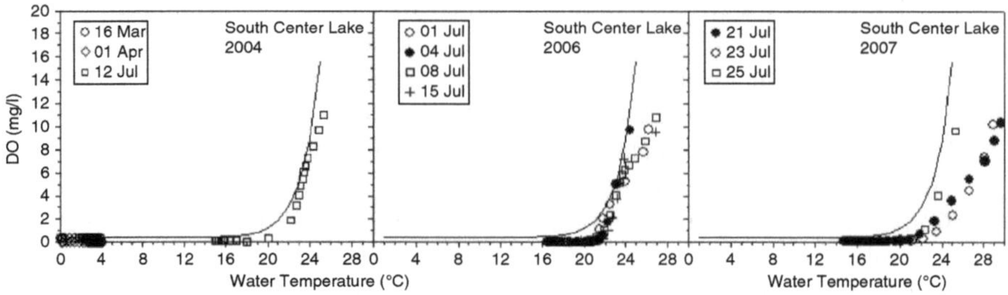

Fig. 11.15 Simulated DO concentration versus simulated temperature in South Center Lake on selected summer days in 2006 with potential cisco mortality against the lethal-niche-boundary curve developed by Jacobson et al. [177]

minimum DO concentration at a given temperature T_{lethal} for cisco to survive. Equation 11.6 gives a DO survival limit for cisco that is not constant but instead depends on water temperature. The Equation Method (EM), therefore, uses survival limits for temperature and DO that are variable and related to each other.

The study by Jacobson et al. [177] leading to the EM was done specifically for cisco, whereas the study by Eaton et al. [170] that lead to the CVM was done on the entire cold-water fish guilds and not specifically on cisco (though cisco is a cold-water fish).

Equation 11.6 was implemented in a fish habitat simulation program to determine potential fish mortality in 15 study lakes. For each simulation day, the required DO concentrations, DO_{lethal}, were computed from simulated water temperatures in all water layers using Eq. 11.6 and compared with DO concentrations simulated by MINLAKE2010 in the same layers. Fish kill was assumed to occur if the simulated DO was less than the DO_{lethal} value at all water depths on that day. If simulated DO was larger than DO_{lethal} in some of the water layers, fish mortality was not assumed to occur because cisco could swim to the water layers with suitable DO and temperature conditions.

Sample plots of simulated DO concentrations versus simulated temperatures at the same location in four study lakes on selected summer days when cisco mortality (kill) potential was high were developed. The lethal-niche-boundary curve (Eq. 11.6) developed by Jacobson et al. [177] was added for reference. Examples of results for South Center Lake are shown in Fig. 11.15. All simulated data points in Fig. 11.15 are located to the right of the curve, indicating that adult cisco kill could indeed occur in all water layers (depths) on the days shown [183]. Fish kill of adult cisco was simulated to occur also on some days in four other lakes (Carrie Lake, Hill Lake, White Iron Lake, and Madison Lake) (Fang et al. [183]), and this result agreed with field observations because no cisco had been found in those five lakes. This means that these lakes are not sustainable to cisco or potential other cold-water fish species.

Projection of Sustainable Cold-Water Fish Habitat in Lakes Under Future Climate Scenarios

Fish habitat projections under future climate scenarios were made for cisco in Minnesota lakes using the two methods described above and three future climate scenarios based on the output of three GCMs of the global atmosphere: CCCma GCM 2.0, CCCma CGCM 3.1 (A1B scenario), and MIROC 3.2 (A1B scenario) from the Special Report on Emission Scenarios of the IPCC (2001). Descriptions of the evolving stages of the GCMs and examples of applications were given by Boer et al. [184], McFarlane et al. [185], and Kim et al. [186, 187]. MIROC was described by Hasumi and Emori [188]. One output of the cisco habitat simulations was the projected number of annual cisco kill days; these were summarized and interpreted for the different climate scenarios.

Results obtained by both the CVM and the EM showed a considerable increase in the number of annual cisco kill days under future climate scenarios. Projections of the number of annual kill days under all three future climate scenarios were consistent. The GCM 2.0 scenario projected higher numbers of cisco kill days because the future atmospheric CO_2 concentration was higher for this earlier climate scenario model. Future climate scenarios based on the two more recent CGCMs (CCCma CGCM 3.1 and MIROC 3.2) gave almost the same numbers of projected cisco kill days [183].

As a preliminary conclusion, 19 out of 21 simulated cisco lakes were projected to be viable cisco refuge lakes under a warmer climate. In lakes not supporting cisco habitat, there is a shift in cisco kill days from winter to summer.

Figure 11.16 shows the distribution of 620 current cisco lakes in Minnesota and 21 selected cisco study lakes in a coordinate system of lake geometry ratio (GR) versus Secchi depth (SD). Under the CCCma CGCM 3.1 future climate scenario, 19 of the 21 simulated cisco lakes were projected to have no cisco kill under future climate scenarios. There is a region on the (SD) versus (GR) plot in Fig. 11.15 with "refuge" lakes that can support cisco habitat under future climate scenarios. Such refuge lakes should have an SD greater than 2.5 m (mesotrophic or oligotrophic lakes) and a GR less than 3.0 $m^{-0.5}$ (typically seasonally stratified, dimictic lakes). Once refuge lakes are identified, watershed protection efforts can be initiated at refuge lakes to prevent deterioration of water quality by anthropogenic activities.

The findings in Fig. 11.16 were confirmed by projections using the oxythermal habitat parameter TDO3. Jacobson et al. [177] proposed a single variable to quantify oxythermal habitat that allows for comparison across several cold-water fish species (lake trout, cisco, lake whitefish, and burbot) that have different requirements for cold, oxygenated water. The single generalized oxythermal habitat variable is defined as the water temperature at 3 mg/L of DO and is called TDO3. The TDO3 can be determined by interpolating the temperature of water at the DO concentration of 3 mg/L from measured or simulated vertical temperature and DO profiles. When

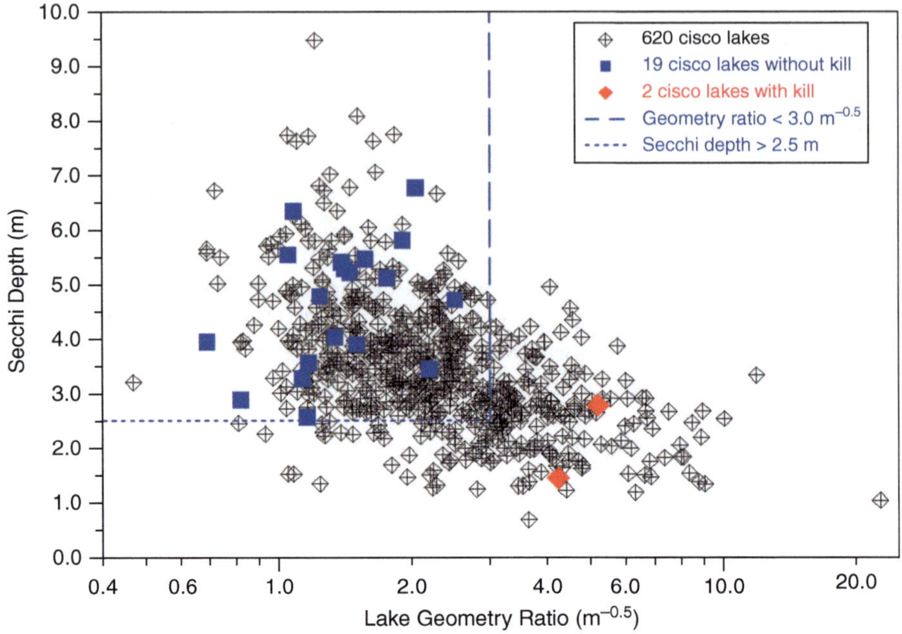

Fig. 11.16 Distribution of 21 study lakes among 578 current cisco lakes in Minnesota projecting no future cisco kill in 19 and potential cisco kill in two study lakes under the CCCma CGCM 3.1 future climate scenario

nonmonotonic profiles generate low oxygen concentrations with more than one TDO3 value, the coldest TDO3 is used [177].

The oxythermal parameter TDO3 measures environmental conditions and stress in a lake. Low values of summer TDO3 indicate excellent oxythermal habitat for cold-water fish, i.e., fish have a wide range of temperatures available in the hypolimnion with sufficient oxygen concentrations. High values of TDO3 indicate poor oxythermal habitat for cold-water fish, with little or no cold water with sufficient oxygen. Very high values of TDO3 indicate hypolimnia that are anoxic or are found in unstratified lakes. The conclusion of the fish habitat study [183] was that under fairly stringent selection criteria (DO $>$ 3 mg/L and TDO3 $\leq$ 17°C), at least one fourth to one third of the lakes that currently have cisco populations are projected to maintain cisco habitat under projected future climate scenarios. Many of the "refuge" lakes are located in northeastern and central Minnesota.

Refuge lakes with TDO3 < 17°C under the future climate scenario MIROC 3.2 have Secchi depths greater than $\approx$ 2.5 m, lake geometry ratios less than $\approx$ 2.5 m, maximum depths greater than $\approx$ 15 m, and surface areas less than $\approx$ 30 km^2 [183].

Cisco refuge lakes were also ranked in narrow (2°C) bands of TDO3 values (Fig. 11.17). AvgATDO3$_{VB}$ in Fig. 11.17 is the average of mean daily TDO3 values over the fixed 31-day benchmark period (Julian day 209–239) in the simulation period (one value for 1962–2008 period). Names and locations of lakes in each

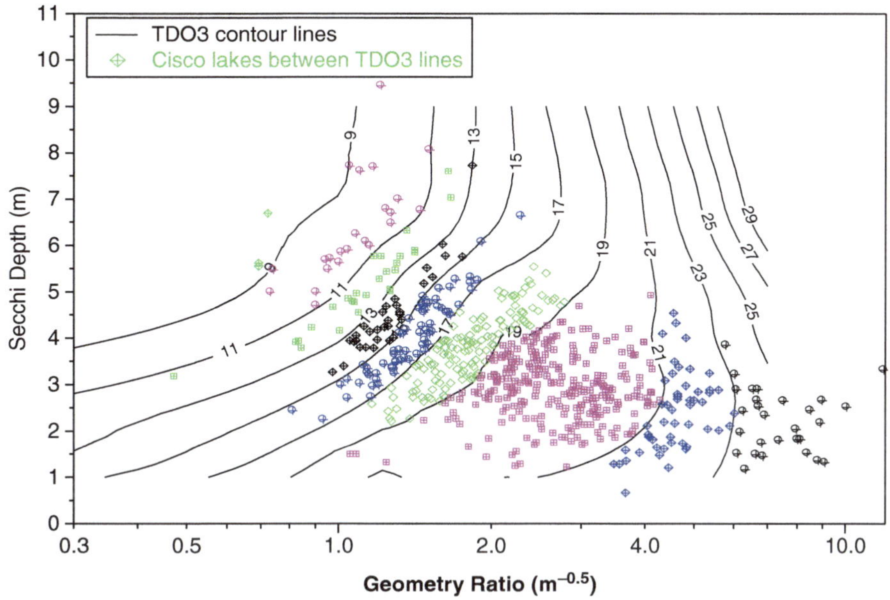

Fig. 11.17 Distribution of 620 cisco lakes in Minnesota between *contour lines* of AvgATDO3$_{VB}$ for variable benchmark periods simulated for the MIROC 3.2 future climate scenario using Duluth weather data

band were given to the Minnesota Department of Natural Resources for management decisions. Lakes with the lowest TDO3 values were the best candidates for cisco refuge lakes because a low TDO3 value is presumed to reflect the least stress on adult cisco.

Future Directions

There are many unanswered questions regarding transport and transformation processes that control water quality and habitat in lakes or reservoirs because these processes are numerous and often interacting. This lack of knowledge and information hinders meaningful decision making on lake and reservoir management and can lead to waste of effort and resources. Examples where future research efforts can pay sustained benefits are:

1. Continuous automated data collection in lakes and reservoirs. Existing and future sensors of physical, chemical, and biological lake parameters will enable the collection and recording of time series of information at high temporal and spatial resolution. The installation of measuring stations can be expensive, but the return will be information that can really decipher the timescales and vertical scales of processes.

2. Development of simulation models for critical lakes as a permanent tool for decision making. Existing models can be tailored to specific, valuable lakes and can be used first to simulate the outcome of contemplated actions. In the long-term, model development, modification and validation can benefit greatly when paired with automated data collection advocated above (1).
3. Research on the intermediate-scale coupling between the atmosphere and the hydrosphere of lakes and reservoirs. A missing link in our understanding of lakes is their coupling to the atmospheric boundary layer. Atmospheric boundary layers over lakes have been all but ignored roughness transitions from the land to water and the transfer of momentum from wind, and the transfer rates of gases at lake and reservoir surfaces have been measured and analyzed as if the atmospheric boundary layer over a lake surface did not have to respond to wakes of canopies, buildings, and hills, and adjust to the roughness transition from land to water.
4. Research on the large-scale coupling between the atmosphere and the hydrosphere of lakes and reservoirs. "Large-scale" here means responses to climate (warming). Climate/weather effects on heat transfer at a water surface and the consequences for stratification and mixing can be estimated. However, the intensification of the hydrologic cycle by global warming has effects on lake and reservoir water budgets, on material import from watersheds, and material budgets of lakes and reservoirs; it also affects internal material transport and transformation processes, and thereby habitat in ecosystems.
5. Interactions and exchange processes at the sediment–water interface are highly diverse and difficult to model in a comprehensive way. Flow of water in the boundary layer can be highly intermittent, possibly density stratified, and possibly affected by sediment settling and/or sediment resuspension. The porous sediment surface allows for some exchange of mass and momentum but also much damping. The pore scale is far shorter than the internal lake modeling scale, and the chemistry of the waters in the sediment pore system can be complex; sediment particle interactions with solutes can be significant, and there are also microbial biofilms. At present, the fluxes of materials across the sediment interface that are needed as boundary conditions in lake water quality models are mostly estimated empirically. Processes in the flowing water boundary layer above the sediment layer and in the sediment pore system need to be better coupled to in lake models.
6. Microbial processes and solute–particle interactions in lakes may have far more impact on water quality and material fluxes than is acknowledged in current lake models. The microstructure of lake and reservoir stratification needs to be better understood because it controls some of these processes.
7. Criteria used to assess fish habitat in lakes are currently fairly crude. What are the proper timescales? Criteria need to be established for different life stages of fish, not only adult fish. More quantitative information on the response of fish to environmental parameters has a bearing on fish survival projections under environmental stress such as global warming.

8. Lake and reservoir management, including goal-setting, data collection and storage, planning, decision making, resource allocation, and implementation, are complicated not only because lakes and reservoirs are complicated but also the multitude of agencies and organizations that are in charge gives an administrative and organizational structure that is not optimal. A single, multidisciplinary entity in charge of a lake or reservoir may do a better job than the compromises developed under current administrative structures. Simplification of the organizational structure and the resulting decision-making process are appropriate.
9. One of the many management issues has to do with internal nutrient loading. Many TMDL studies of external inputs to lakes and reservoirs accomplish little when lake ecosystems function on the basis of internal nutrient recycling.
10. Biomanipulation is practiced in a very empirical way and needs a better basis.
11. Lake and reservoir improvement technologies need to be reviewed and can probably be improved. Guidance manuals need to be developed for these techniques.

Bibliography

Primary Literature

1. Chapra SC (1997) Surface water-quality modeling. McGraw-Hill, New York
2. Fry EFJ (1971) The effect of environmental factors on the physiology of fish. Academic, New York
3. Christie CG, Regier HA (1988) Measurements of optimal habitat and their relationship to yields for four commercial fish species. Can J Fish Aquat Sci 45:301–314
4. Zipparro VJ, Hasen H (eds) (1993) Handbook of applied hydraulics. McGraw-Hill, New York
5. USBR (U. S. Bureau of Reclamation) (1987) Design of small dams. U. S. Govmt. Printing Office, Denver
6. Novák P, Nalluri C (1990) Hydraulic structures. Unwin Hyman, London
7. Press H (1958) Talsperren (in German), 2nd edn. Ernst & Sohn, Berlin
8. Thornton KW, Kimmel BL, Payne FE (1990) Reservoir limnology. Wiley, New York
9. Ackerman WC, White GF, Worthington EB (eds) (1973) Man-made lakes: their problems and environmental effects. American Geophysical Union, Washington, DC
10. Stefan HG (ed) (1980) Surface water impoundments. ASCE, New York
11. NRC (national Research Council) (1987) Glen canyon environmental studies review committee. National Academy, Washington, DC
12. Hutchinson GE (1957) A treatise on limnology. Wiley, New York
13. Frey DG (1966) Limnology in North America. The University of Wisconsin Press, Madison
14. Lerman A, Baccini P (1978) Lakes – chemistry, geology, physics. Springer, New York
15. Wetzel RG (1983) Limnology, 2nd edn. WB Saunders, Philadelphia
16. Horne A, Goldman C (1994) Limnology. McGraw-Hill, New York
17. Lampert W, Sommer U (1997) Limnoecology: the ecology of lakes and streams. Oxford University Press, New York
18. Henderson-Sellers B (1984) Engineering limnology. Pitman Advanced Publication Program, Boston

19. Straskraba M, Gnauck AH (1985) Freshwater ecosystems. Modeling and simulation. Elsevier, Amsterdam
20. Rouse H, Ince S (1957) History of hydraulics. Iowa Institute of Hydraulic Research, Iowa City
21. Chien N, Wang Z (1998) Mechanics of sediment transport. ASCE Press, Reston
22. Dortch MS (1992) Modeling water quality of reservoir tailwaters, Technical Report W-92-1. US Army Corps of Engineers, Waterways Experiment Station, Vicksburg
23. Fischer HB (1979) Mixing in inland and coastal waters. Academic, New York
24. Gray WG (1986) Physics-based modeling of lakes, reservoirs, and impoundments. ASCE Press, Reston
25. Herb WR, Janke B, Mohseni O, Stefan HG (2009) Runoff temperature model for paved surfaces. J Hydrol Eng 14:1146–1155
26. Wanielista MP, Yousef YA (1993) Stormwater management. Wiley, New York
27. Hino M (1994) Water quality and its control. International Association for Hydraulic Research, AA Balkema, Rotterdam
28. Baker LA (1984) A model for phosphorus cycling in lakes. International Memorandum No. 100, St Anthony Falls Laboratory, University of Minnesota, Minneapolis
29. Stefan HG, Dhamotharan S, Schiebe FR, Fu AY, Cardoni JJ (1989) Dynamic simulation of turbidity and its correction in Lake Chico, Arkansas. In: Henderson-Sellers B (ed) Water quality modeling, vol IV. CRC Press, Baco Raton
30. Novotny EV, Stefan HG (2010) Projections of chloride concentrations inurban lakes receiving road de-icing salt. Water Air Soil Pollut 211:261–271
31. Gassman PW (2007) The soil and water assessment tool: historical development, applications, and future research direction. Trans ASABE 50(4):1211–1250
32. James W, Huber W, Rossman L et al (2005) Water systems models: user's guide to SWMM. CHI, Guelph
33. Rossman L (2008) Storm Water Management Model. User's manual Version 5.0. US Environmental Protection Agency, Cincinnati
34. USGS (2008) SNTEMP stream temperature model, http://smig.usgs.gov/cgi-bin/SMIC/model_home_pages/model_home?selection=sntemp
35. Herb WR, Mohseni O, Stefan HG (2009) Simulation of temperature mitigation by a stormwater detention pond. J Am Water Resour Assoc 45:1164–1178
36. PLUARG (1978) Environmental management strategy for the great lakes system. Final Report of the Pollution from Land Use Activities Reference Group (PLUARGE) to the International Joint Commission. Windsor, Ontario
37. Hornbuckle KC, Jeremiason JD, Sweet CW, Eisenreich SJ (1994) Seasonal variations in air-water exchange of polychlorinated biphenyls in Lake Superior. Environ Sci Technol 28:1491–1501
38. Gulliver JS (2007) Introduction to chemical transport in the environment. Cambridge University Press, Cambridge
39. Dhamotharan S, Gulliver JS, Stefan HG (1981) Unsteady one-dimensional settling of suspended sediment. Water Resour Res 17:1125–1132
40. Johnson TC (1980) Sediment redistribution by waves in lakes, reservoirs and embayments. In: Stefan HG (ed) Proceeding of international symposium on surface water impoundements. ASCE, Reston, Minneapolis, pp 1307–1317
41. Sheng YP, Lick W (1979) The transport and resuspension of sediments in a shallow lake. J Geophys Res 84:1809–1826
42. Rodney MW, Stefan HG (1987) Conceptual model for wind-generated sediment resuspension in shallow ponds. In: Proceedings of the national symposium on mining, hydrology, sedimentology and reclamation. University of Kentucky, Lexington, pp 263–269
43. Erdmann JB, Stefan HG, Brezonik PL (1994) Analysis of wind and ship induced sediment resuspension in Duluth-Superior harbor. J Am Water Resour Assoc 30:1043–1053

44. Thomann RV, Mueller JA (1987) Principles of surface water quality modeling and control. Harper & Row, New York
45. Clark MM (1996) Transport modeling for environmental engineers and scientists. Wiley, New York
46. Weber W Jr, DiGiano FA (1995) Process dynamics in environmental systems. Wiley, New York
47. Weilenmann U, O'Melia CR, Stumm W (1989) Particle transport in lakes: models and measurements. Limnol Oceanogr 34:1–18
48. Mackenthun AA, Stefan HG (1998) Effect of flow velocity on sediment oxygen demand: experiments. J Environ Eng 124:222–230
49. Qian Q, Clark JJ, Voller VR, Stefan HG (2009) Modeling of vertical dispersion of a solute in a stream or lake bed enhanced by wave induced interstitial flow. J Am Water Resour Assoc 45:343–354
50. Qian Q, Clark JJ, Voller VR, Stefan HG (2009) Depth-dependent dispersion coefficient for modeling of vertical solute exchange in a lake bed under surface waves. J Hydraul Eng ASCE 135:187–197
51. Elliott AH, Brooks NH (1997) Transfer of non-sorbing solutes to a streambed with bed forms: theory. Water Resour Res 33:123–136
52. Packman AI, Salehin M, Zaramella M (2004) Hyporheic exchange with gravel beds: basic hydrodynamic interactions and bedform-induced advective flows. J Hydraul Eng 130:647–656
53. Roy H, Huttel M, Jorgensen BB (2002) The role of small-scale sediment topography for oxygen flux across the diffusive boundary layer. Limnol Oceangr 47:837–847
54. Steinberger N, Hondzo M (1999) Diffusional mass transfer at sediment-water interface. J Environ Eng 125:192–200
55. O'Connor BL, Hondzo M (2008) Dissolved oxygen transfer to sediments by sweep and eject motions in aquatic environments. Limnol Oceanogr 53:566–578
56. Nakamura Y, Stefan HG (1994) Effect of flow velocity on sediment oxygen demand: theory. J Environ Eng 120:996–1016
57. Higashino M, Clark JJ, Stefan HG (2009) Pore water flow due to near-bed turbulence and associated solute transfer in a stream or lake sediment bed. Water Resour Res 45
58. Thamdrup B, Fossing H, Jorgensen BB (1994) Manganese, iron and sulfur cycling in a coastal marine sediment, Aarhus Bay, Denmark. Geochimica et Cosmochimica Acta 58:5115–5129
59. O'Connor BL, Harvey JW (2008) Scaling hyporheic exchange and its influence on biogeochemical reactions in aquatic ecosystems. Water Resour Res 44:W12423
60. Mortimer CH (1971) Chemical exchanges between sediments and water in the Great Lakes-speculations on probable regulatory mechanisms. Limnol Oceanogr 16:387–404
61. Lorenzen CJ, Welschmeyer NA, Copping AE, Vernet M (1983) Sinking rates of organic particles. Limnol Oceanogr 28:766–769
62. Stefan HG, Hanson MJ (1981) Phosphorus recycling in five shallow lakes. J Environ Eng Div 107:713–730
63. Boudreau BP, Jørgensen BB (2001) The benthic boundary layer: transport processes and biogeochemistry. Oxford University Press, Oxford
64. DiToro DM (2001) Sediment flux modeling. Wiley, New York
65. Manous JD, Gantzer CJ, Stefan HG (2007) Spatial variation of sediment sulfate reduction rates in a saline lake. J Environ Eng-ASCE 133:1106–1116
66. Ambrose RB, Wool TA, Connolly JP, Schanz RW (1988) WASP (EUTRO4), A hydrodynamic and water quality model – model theory, user's manual, and programmer's guide. US Environmental Protection Agency, Athens, Georgia
67. Cole TM, Wells SA (2004) CE-QUAL-W2: a two-dimensional, laterally averaged, hydrodynamic and water quality model, version 3.2, user's manual. Environmental Laboratory, US Army Engineer Research and Development Center, Vicksburg

68. Stumm W, Morgan JJ (1981) Aquatic chemistry. Wiley-Intersciences, New York
69. Stumm W, Morgan JJ (1996) Aquatic chemistry: chemical equilibria and rates in natural waters, 3rd edn. Wiley, New York
70. Gobas FAPC, McCorquodale JA (1992) Chemical dynamics in fresh water ecosystems. Lewis, Chelsea
71. Brezonik PL (1993) Chemical kinetics and process dynamics in aquatic systems. Lewis, Boca Raton
72. Zison SW, Mills WB, Diemer D, Chen C (1978) Rates, constants, and kinetics formulations in surface water quality modeling. Tetra Tech Incorporation, SEPA, ORD, Athens, Georgia
73. Bowie GL, Mills WB, Porcella DB et al (1985) Rates, constants, and kinetics formulations in surface water quality modeling (2nd edn), EPA/600/3-85/040. Environmental Research Laboratory, Office of Research and Development, US Environmental Protection Agency
74. Novak P, Arnold W, Blazer V et al (2011) On the need for a national (U.S) research program to Elucidate the potential risks to human health and the environment posed by contaminants of emerging concern. Environ Sci Technol 45:3829–3830
75. Calabrese EJ (1985) Age and susceptibility to toxic substances. Wiley, New York
76. Scheffer M (1998) Ecology of shallow lakes. Chapman and Hall, London
77. Cooke GD (1993) Restoration and management of lakes and reservoirs. Lewis, Boca Raton
78. Reckhow KH, Chapra SC (1983) Engineering approaches for lake management, vol 1, Data analysis and empirical modeling. Ann Arbor Science, Boston
79. Henderson-Sellers B (1991) Water quality modeling, vol IV, Decision support techniques for lakes and reservoirs. CRC Press, Boca Raton
80. Burns F, Powling I (1981) Destratification of lakes and reservoirs to improve water quality. Australian Government, Canberra
81. Fast AW (1971) The effects of artificial aeration on lake ecology. US Environmental Protection Agency, Washington, DC
82. Goossens L (1979) Reservoir destratification with bubble columns. Delft University Press, Delft
83. Zic K, Stefan H (1988) Lake Aerator effect on temperature stratification analyzed by "MINLAKE" model. Lake Reserv Manag 4:85–90
84. Zic K, Stefan H, Ellis C (1992) Laboratory study of water destratification by a bubble plume. J Hydraul Res 30:7–27
85. Gulliver JS, Stefan HG (1982) Lake phytoplankton model with destratification. J Environ Eng Div-ASCE 108:864–882
86. Ellis CR, Stefan HG (1990) Hydraulic design of a winter lake aeration system. J Environ Engin-ASCE 116:376–393
87. Stefan HG, Bender MD, Shapiro J, Wright DI (1987) Hydrodynamic-design of a metalimnetic lake aerator. J Environ Eng-ASCE 113:1249–1264
88. Stefan HG, Gu R (1992) Efficiency of jet-mixing of temperature-stratified water. J Environ Eng ASCE 118:363–379
89. Stefan HG, Gu R (1990) Jet mixing in lake or reservoir stratification simulations. Lake Reserv Manag 6:165–174
90. Stefan HG, Gu RC (1991) Conceptual design procedure for hydraulic destratification systems in small ponds, lakes, or reservoirs for water-quality improvement. Water Resour Bull 27:967–978
91. Hanson MJ, Stefan HG (1984) Side effects of 58 years of copper sulfate treatment of the fairmont lakes, Minnesota. Water Resour Bull 20:889–900
92. Vollenweider R (1968) Scientific fundamentals of the eutrophication of lakes and flowing waters, with particular reference to phosphorus and nitrogen as factors in eutrophication. Organisation for Economic Cooperation and Development Technical Report OAS/CSI/68.27, Paris

93. Gorham E, Boyce FM (1989) Influence of lake surface area and depth upon thermal stratification and the depth of the summer thermocline. J Great Lakes Res 15:233–245
94. Alavian V, Jirka GH, Denton RA, Johnson MC, Stefan HG (1992) Density currents entering lakes and reservoirs. J Hydraul Eng 118:1464–1489
95. Akiyama J, Stefan HG (1984) Plunging flow into a reservoir – theory. J Hydraulic Eng ASCE 110:484–499
96. Akiyama J, Stefan H (1985) Turbidity current with erosion and deposition. J Hydraul Eng 111:1473–1496
97. Akiyama J, Stefan HG (1987) Onset of underflow in slightly diverging channels. J Hydraul Eng ASCE 113:825–844
98. Akiyama J, Stefan HG (1988) Turbidity-current simulation in a diverging channel. Water Resour Res 24:579–587
99. Johnson TR, Ellis CR, Farrell GJ, Stefan HG (1987) Negatively buoyant flow in a diverging channel. Part 2: flow field description. J Hydraul Eng ASCE 113:731–742
100. Johnson TR, Farrell GJ, Ellis CR, Stefan HG (1987) Negatively buoyant flow in a diverging channel. Part1: flow Regimes. J Hydraul Eng ASCE 113:716–730
101. Farrell GJ, Stefan HG (1988) Mathematical-modeling of plunging reservoir flows. J Hydraul Res 26:525–537
102. Farrell GJ, Stefan HG (1989) Two-layer analysis of a plunging density-current in a diverging horizontal channel. J Hydraul Res 27:35–47
103. Fang X, Stefan HG (2000) Dependence of dilution of a plunging discharge over a sloping bottom on inflow conditions and bottom friction. J Hydraul Res 38:15–25
104. Ellis CR, Champlin J, Stefan HG (1997) Density current intrusions in an ice-covered urban lake. J Am Water Resour Assoc 33:1363–1374
105. Stefan HG (1970) Modeling spread of heated water over lake. J Power Div 96:469–482
106. Stefan H, Schiebe FR (1970) Heated discharge from flume into tank. J Sanit Eng Div 96:1415–1433
107. Stefan HG (1970) Stratification of flow from channel into deep lake. J Hydraul Div 96:1417–1434
108. Stefan HG, Chu CS, Wing H (1972) Impact of cooling water on lake temperatures. J Power Div 98:253–272
109. Stefan H, Vaidyaraman P (1972) Jet type model for the three-dimensional thermal plume in a crosscurrent and under wind. Water Resour Res 8:998–1014
110. Demetracopoulos AC, Stefan HG (1983) Model of Mississippi river pool: dissolved oxygen. J Environ Eng ASCE 109:1020–1034
111. Stefan HG, Anderson KJ (1980) Wind-driven flow in Mississippi river impoundment. J Hydraul Div 106:1503–1520
112. Demetracopoulos AC, Stefan HG (1983) Model of Mississippi river pool: mass transport. J Environ Eng 109:1006–1019
113. Manous JD, Stefan HG (2004) Sulfate distribution in a multi-basin, saline lake. Hydrobiologia 529:169–185
114. Stefan H, Ford DE (1975) Temperature dynamics in dimictic lakes. J Hydraul Div ASCE 101:97–114
115. Ford DE, Stefan HG (1980) Stratification variability in three morphometrically different lakes under identical meteorological forcing. J Am Water ResourAssoc 16:243–247
116. Stefan HG, Megard RO, Skoglund T (1976) Wind control of algae growth in eutrophic lakes. J Environ Eng Div 102:1201–1213
117. Herb WR, Stefan HG (2005) Model for wind-driven vertical mixing in a shallow lake with submersed macrophytes. J Hydraul Eng 131:488–496
118. Markfort CD, Perez ALS, Thill JW et al (2010) Wind sheltering of a lake by a tree canopy or bluff topography. Water Resour Res 46:W03530
119. Stefan HG, Ambrose RB, Dortch MS (1990) Surface water quality models:modeler's perspective. In: Proceedings international symposium on water quality modeling of agricultural

non-point sources, US Department of Agriculture, American Society of Agricultural Engineers. St. Joseph
120. Wurbs RA (1998) Dissemination of generalized water resources models in the United States. Water Int 23:190–198
121. Anon (1986) A numerical one-dimensional model of reservoir water qulity. user's manual. Instruction Report E-82-1, US Army Corps of Engineers,Vickburg
122. Imberger J, Patterson JC (1981) A dynamic reservoir simulation model – DYRESM. In: Fischer HB (ed) Transport models for inland and coastal waters. Academic, New York, Chapter 9
123. Riley MJ, Stefan HG (1988) Minlake: a dynamic lake water quality simulation model. Ecol Model 43:155–182
124. Middlebrooks EJ, Falkenbory DH, Maloney TE (1976) Modeling the eutrophication process. Ann Arbor Science, Boston
125. Bruce LC, Hamilton D, Imberger J et al (2006) A numerical simulation of the role of zooplankton in C, N and P cycling in Lake Kinneret, Israel. Ecol Model 193:412–436
126. Gal G, Hipsey M, Parparov A et al (2009) Implementation of ecological modeling as an effective management and investigation tool: Lake Kinneret as a case study. Ecol Model 220:1697–1718
127. Leon LF, Lam D, Schertzer W, Swayne D (2005) Lake and climate models linkage: a 3D hydrodynamic contribution. Adv Geosci 4:57–62
128. Robson B, Hamilton D (2004) Three-dimenstional modelling of a microcystis bloom event in the Swan River estuary, Western Australia. Ecol Model 174:203–222
129. Edinger JE (2001) Waterbody hydrodynamic and water quality modeling, workbook and CD-ROM. ASCE Press, Reston
130. West D, Stefan HG (1998) Simulation of lake water quality using a one-dimensional model with watershed input. Model description and application to lake riley and lake elmo, Project Report No.430. St. Anthony Falls Laboratory, University of Minnesota, Minnesota
131. West D, Stefan HG (1999) Modeling of watershed input and potential climate change effects on water quality in agricultural reservoirs in the little Washita River watershed, Project Report No.437. St. Anthony Falls Laboratory, University of Minnesota, Minnesota
132. West D, Stefan HG (2000) Simulation of water quality and primary productivity control strategies for lake McCarrons, Project Report No.426. St. Anthony Falls Laboratory, University of Minnesota, Minnesota
133. Fang X, Stefan HG (1996) Development and validation of the water quality model MINLAKE96 with winter data. Anthony Falls Laboratory, University of Minnesota, Minneapolis
134. Fang X, Alam SR, Jacobson P, Pereira D, Stefan HG (2010) Simulations of water quality in Cisco Lakes in Minnesota. Anthony Falls laboratory, University of Minnesota, Minneapolis
135. Edinger JE, Duttweiler DW, Geyer JC (1968) The response of water temperatures to meteorological conditions. Water Resour Res 4:1137–1143
136. Ford DE, Stefan HG (1980) Thermal prediction using internal energy model. J Hydraul Div ASCE 106(HY1):39–55
137. Harleman DRF (1982) Hydrothermal analysis of lake and reservoir. J Hydraul Div ASCE 108(HY3):39–49
138. Hondzo M, Stefan HG (1993) Lake water temperature simulation model. J Hydraul Eng 119:1251–1273
139. Hondzo M, Stefan HG (1992) Water temperature characteristics of lakes subjected to climate change. St Anthony Falls Hydraulic Laboratory, University of Minnesota, Minneapolis
140. Fang X, Stefan HG (2009) Simulations of climate effects on water temperature, dissolved oxygen, and ice and snow covers in lakes of the contiguous United States under past and future climate scenarios. Limnol Oceangr 54:2359–2370
141. Maxwell EL (1998) METSTAT – the solar radiation model used in the production of the national solar radiation database (NSRDB). Solar Energy 62:263–279

142. Fang X, Stefan HG (1994) Temperature and dissolved oxygen simulations for a lake with ice cover. Anthony Falls Hydraulic Laboratory, University of Minnesota, Minneapolis
143. Fang X, Stefan HG (1996) Dynamics of heat exchange between sediment and water in a lake. Water Resour Res 32:1719–1727
144. Gu R, Stefan HG (1990) Year-round temperature simulation of cold climate lakes. Cold Reg Sci Technol 18:147–160
145. Ashton GD (1986) River and lake ice engineering. Water Resources, Highlands Ranch
146. Fang X, Ellis CR, Stefan HG (1996) Simulation and observation of ice formation (freeze-over) in a lake. Cold Reg Sci Technol 24:129–145
147. Stefan HG, Fang X (1994) Dissolved oxygen model for regional lake analysis. Ecol Model 71:37–68
148. Mathias JA, Barica J (1980) Factors controlling oxygen depletion in ice-covered lakes. Can J Fish Aquat Sci 37:185–194
149. Dake JMK, Harleman DRF (1969) Thermal stratification in lakes: analytical and laboratory studies. Water Resour Res 5:484–496
150. Carslaw HS, Jaeger JC (1959) Conduction of heat in solids. Oxford University Press, New York
151. Bannister TT (1974) Prediction equations in terms of chlorophyll concentration, quantum yield, and upper limit of production. Limnol Oceangr 19:1–12
152. Megard RO, Combs WS Jr, Smith PD, Knoll AS (1979) Attenuation of light and daily integral rates of photosynthesis attained by planktonic algae. Limnol Oceanogr 24:1038–1050
153. Bolsenga SJ (1977) Preliminary observations of the daily variation of Ice Albedo. J Glaciol 18:517–521
154. Lock GSH (1990) The growth and decay of ice. Cambridge University Press, Cambridge
155. Wake A, Rumer RR Jr (1979) Modeling ice regime of Lake Erie. J Hydraul Div 105:827–844
156. Scott JT (1964) A comparison of the heat balance of lakes in winter. Department of Meteorology, University of Wisconsin, Madison
157. Pivovarov AA (1972) Thermal Condition in Freezing Lakes and Rivers. Wiley, New York
158. Greene GM (1981) Simulation of ice-cover growth and decay in one dimension on the upper St. Lawrence River. NoAA Technical Memorandum Erl GLERL-36
159. Carlson RE (1977) A trophic state index for lakes. Limnol Oceanogr 22:361–369
160. Brown LC, Barnwell TO (1987) The enhanced stream water quality models QUAL2E and QUAL2W-UNCAS: documentation and user manual. U.S. Environmental Protection Agency, Athens
161. Zison SW, Mills WB, Diemer D, Chen CW (1978) Rates, constants and kinetic formulations in surface water quality modeling. Tetra Tech Incorporation, USEPA, ORD, Athens, Georgia
162. Fang X, Stefan HG (1998) Temperature variability in the lake sediments. Water Resour Res 34:717–729
163. Marshall CT, Peters RH (1989) General patterns in the seasonal development of chlorophyll a for temperate lakes. Limnol Oceanogr 34:856–867
164. Todd DK (1980) Groundwater Hydrology, 2nd edn. John Wiley, New York, NY
165. Magnuson JJ, Webster KE, Assel RA et al (1997) Potential effects of climate changes on aquatic systems: Laurentian great lakes and precambrian shield region. Hydrol Process 11:825–871
166. Stefan HG, Fang X, Eaton JG (2001) Simulated potential thermal/DO fish habitat changes in lakes in response to projected climate warming. Trans Am Fish Soc 130:459–477
167. Fang X, Stefan HG, Eaton JG, McCormick JH, Alam SR (2004) Simulation of thermal/dissolved oxygen habitat for fishes in lakes under different climate scenarios: Part 1. Cool-water fish in the contiguous US. Ecol Model 172:13–37
168. Fang X, Stefan HG, Eaton JG, McCormick JH, Alam SR (2004) Simulation of thermal/dissolved oxygen habitat for fishes in lakes under different climate scenarios: Part 2. Cold-water fish in the contiguous US. Ecol Model 172:39–54

169. Fang X, Stefan HG, Eaton JG, McCormick JH, Alam SR (2004) Simulation of thermal/dissolved oxygen habitat for fishes in lakes under different climate scenarios: Part 3. Warm-water fish in the contiguous US. Ecol Model 172:55–68
170. Eaton JG, McCormick JH, Goodno BE et al (1995) A field information based system for estimating fish temperature requirements. Fisheries 20:10–18
171. Schindler DW, Bayley SE, Parker BR et al (1996) The effects of climate warming on the properties of boreal lakes and streams at the experimental lakes area, Northwestern Ontario. Limnol Oceanogr 41:1004–1017
172. Stefan HG, Hondzo M, Fang X, Eaton JG, McCormick JH (1996) Simulated long-term temperature and dissolved oxygen characteristics of lakes in the north-central United States and associated fish habitat limits. Limnol Oceanogr 41:1124–1135
173. Stefan HG, Fang X (1995) A methodology to estimate year-round effects of climate change on water temperature, ice and dissolved oxygen characteristics of temperate zone lakes with application to Minnesota. Anthony Falls Laboratory, University of Minnesota, Minneapolis, MN, St, 55414
174. Stefan HG, Hondzo M, Eaton JG, McCormick JH (1995) Predicted effects of global climate change on fishes in Minnesota lakes. Fish Aquat Sci 121:57–72 (Special Publication)
175. Stefan HG, Hondzo M, Eaton JG, McCormick JH (1995) Validation of a fish habitat model for lakes. Ecol Model 82:211–224
176. Fang X, Stefan HG, Alam SR (1999) Simulation and validation of fish thermal DO habitat in north-central US lakes under different climate scenarios. Ecol Model 118:167–191
177. Jacobson PC, Stefan HG, Pereira DL (2010) Coldwater fish oxythermal habitat in Minnesota lakes: influence of total phosphorus, July air temperature, and relative depth. Can J Fish Aquat Sci 67:2003–2013
178. Evans D (2007) Effects of hypoxia on scope-for-activity and power capacity of lake trout (Salvelinus namaycush). Can J Fish Aquat Sci 64:345–361
179. US EPA (1976) Quality criteria for water. United States Environmental Protection Agency (US EPA), Washington, DC
180. Chapman G (1986) Ambient aquatic life criteria for dissolved oxygen. US Environmental Protection Agency, Washington, DC
181. Barica J, Mathias JA (1979) Oxygen depletion and winterkill risk in small Prairie lakes under extended ice cover. J Fish Res Board Can 36:980–986
182. Fang X, Stefan HG (2000) Projected climate change effects on winterkill in shallow lakes in the northern U.S. Environ Manag 25:291–304
183. Fang X, Alam SR, Jiang LP et al (2010) Simulations of Cisco Fish Habitat in Minnesota Lakes under Future Climate Scenarios. Anthony Falls laboratory, University of Minnesota, Minneapolis, St
184. Boer GJ, McFarlane NA, Lazare M (1992) Greenhouse gas-induced climate change simulated with the CCC second-generation general circulation model. J Clim 5:1045–1077
185. McFarlane G, Boer J, Blanchet JP, Lazare M (1992) The Canadian climate centre second-generation general circulation model and its equilibrium climate. J Clim 5:1013–1044
186. Kim S-J, Flato GM, Boer GJ, McFarlane NA (2002) A coupled climate model simulation of the Last Glacial maximum, Part 1: transient multi-decadal response. Clim Dynam 19:515–537
187. Kim S-J, Flato GM, Boer GJ (2003) A coupled climate model simulation of the Last Glacial maximum, Part 2: approach to equilibrium. Clim Dynam 20:635–661
188. Hasumi H, Emori S (2004) K-1 coupled model (MIROC) description. University of Tokyo, Center for Climate System Research

Books and Reviews

Eaton JG, McCormick JH, Stefan HG, Hondzo M (1995) Extreme-value analysis of a fish temperature-field database. Ecol Eng 4(4):289–305

Ellis CR, Stefan HG (1989) Oxygen demand in ice covered lakes as it pertains to winter aeration. J Am Water Resour Assoc 25(6):1169–1176

Ellis CR, Stefan HG (1991) Field testing of an ice-preserving winter lake aeration system. J Am Water Resour Assoc 27(6):903–914

Ellis CR, Stefan HG, Gu R (1991) Water temperature dynamics and heat-transfer beneath the ice cover of a lake. Limnol Oceanogr 36(2):324–335

Fang X, Stefan HG (1995) Interaction between oxygen transfer mechanisms in lake models. J Environ Eng 121(6):447–454

Fang X, Stefan HG (1996) Long-term lake water temperature and ice cover simulations/ measurements. Cold Reg Sci Technol 24(3):289–304

Fang X, Stefan HG (1997) Simulated climate change effects on dissolved oxygen characteristics in ice-covered lakes. Ecol Model 103(2–3):209–229

Fang X, Stefan HG (1998) Potential climate warming effects on ice covers of small lakes in the contiguous U.S. Cold Reg Sci Technol 27(2):119–140

Fang X, Stefan HG (1999) Projections of climate change effects on water temperature characteristics of small lakes in the contiguous US. Clim Change 42(2):377–412

Gao S, Stefan HG (1999) Multiple linear regression for lake ice and lake temperature characteristics. J Cold Reg Eng 13(2):59–77

Gao S, Stefan HG (2004) Potential climate change effects on ice covers of five freshwater lakes. J Hydrol Eng 9(3):226–234

Gu R, Stefan HG (1993) Validation of cold climate lake temperature simulation. Cold Reg Sci Technol 22(1):99–104

Gu R, Luck FN, Stefan HG (1996) Water quality stratification in shallow wastewater stabilization ponds. J Am Water Resour Assoc 32(4):831–844

Henneman HE, Stefan HG (1998) Snow and ice albedo measured with two types of pyranometers. J Am Water Resour Assoc 34(6):1487–1494

Henneman HE, Stefan HG (1999) Albedo models for snow and ice on a freshwater lake. Cold Reg Sci Technol 29(1):31–48

Herb WR, Stefan HG (2004) Temperature stratification and mixing dynamics in a shallow lake with submersed macrophytes. Lake Reserv Manag 20(4):296–308

Herb WR, Stefan HG (2005) Dynamics of vertical mixing in a shallow lake with submersed macrophytes. Water Resour Res 41(2):W02023, doi:10.1029/2003wr002613

Herb WR, Stefan HG (2006) Seasonal growth of submersed macrophytes in lakes: the effects of biomass density and light competition. Ecol Model 193(3–4):560–574

Hondzo M, Stefan H (1991) Three case studies of lake temperature and stratification response to warmer climate. Water Resour Res 27(8):1837–1846

Hondzo M, Stefan HG (1993) Regional water temperature characteristics of lakes subjected to climate change. Clim Change 24:187–211

Hondzo M, Stefan HG (1996) Long-term lake water quality predictors. Water Res 30(12):2835–2852

Hondzo M, Stefan HG (1996) Dependence of water quality and fish habitat on lake morphometry and meteorology. J Water Resour Plan Manag 122(5):364–373

Hondzo M, Ellis CR, Stefan HG (1991) Vertical diffusion in small stratified lake: data and error analysis. J Hydraul Eng 117(10):1352–1369

Horsch GM, Stefan HG (1988) Convective circulation in littoral water due to surface cooling. Limnol Oceanogr 33(5):1068–1083

Johnson SL, Stefan HG (2006) Indicators of climate warming in Minnesota: Lake ICE covers and snowmelt runoff. Clim Change 75(4):421–453. doi:10.1007/s10584-006-0356-0

Manous JJD, Stefan HG (2003) Projected sulfate redistribution as impacted by lake level stabilization scenarios: devils lake, North Dakota. J Water Resour Plan Manag 129(5):399–408
Rasmussen AH, Hondzo M, Stefan HG (1995) A test of several evaporation equations for water temperature simulations in lakes. J Am Water Resour Assoc 31(6):1023–1028
Stefan HG, Fang X (1993) Model simulations of dissolved oxygen characteristics of Minnesota lakes: past and future. Environ Manag 18(1):73–92
Stefan HG, Fang X (1997) Simulated climate change effects on ice and snow covers on lakes in a temperate region. Cold Reg Sci Technol 25(2):137–152
Stefan HG, Fang X, Hondzo M (1998) Simulated climate changes effects on year-round water temperatures in temperate zone lakes. Clim Change 40:547–576
Stefan HG, Fang X, David W, Eaton JG, McCormick JH (1995) Simulation of dissolved oxygen profiles in a transparent, dimictic lake. Limnol Oceanogr 40(1):105–118
Stefan HG, Hondzo M, Fang X (1993) Lake water quality modeling for projected future climate scenarios. J Environ Qual 22(3):417–431
Stefan HG, Horsch GM, Barko JW (1989) A model for the estimation of convective exchange in the littoral region of a shallow lake during cooling. Hydrobiologia 174(3):225–234
Stefan H, Cardoni J, Schiebe F, Cooper C (1983) Model of light penetration in a turbid lake. Water Resour Res 19(1):109–120. doi:10.1029/WR019i001p00109
Stefanovic DL, Stefan HG (2002) Two-dimensional temperature and dissolved oxygen dynamics in the littoral region of an ice-covered lake. Cold Reg Sci Technol 34(3):159–178
Williams G, Layman KL, Stefan HG (2004) Dependence of lake ice covers on climatic, geographic and bathymetric variables. Cold Reg Sci Technol 40(3):145–164
Williams SG, Stefan HG (2006) Modeling of lake ice characteristics in North America using climate, geography, and lake bathymetry. J Cold Reg Eng 20(4):140–167

Chapter 12
Oceanic Fate and Transport of Chemicals

Robert P. Mason

Glossary

Anthropogenic	Produced by or derived from human-related activities.
Biogeochemical cycling	The overall transport of chemicals through the ocean waters as modified by chemical, physical, and biological processes.
Chemical inputs	The external sources of chemicals to the ocean from the atmosphere or from deep ocean environments or from rivers and other terrestrial sources.
Chemical sinks	All elements and compounds can be removed from the ocean by various processes. The relative ratio of their input to their removal provides an indication of how they are distributed in the ocean, and whether human activity has increased their ocean concentration.
Major ions and nutrients	Those chemicals present in the ocean at high concentrations and the major nutrients (nitrogen, phosphorous, and silica).
Metalloid	An element in the periodic table that acts both as a metal and a nonmetal, depending on the chemical environment.
Micronutrients and trace elements	Those chemicals present in the ocean at low concentrations but which still have an important impact of ocean biological productivity, either because they are

This chapter was originally published as part of the Encyclopedia of Sustainability Science and Technology edited by Robert A. Meyers. DOI:10.1007/978-1-4419-0851-3

R.P. Mason (✉)
Department of Marine Sciences & Chemistry, University of Connecticut,
Groton, CT 06340, USA
e-mail: robert.mason@uconn.edu

	essential nutrients (e.g., iron and zinc) or toxic compounds (e.g., lead and mercury).
Organic chemicals	In the context of this chapter these are chemicals manufactured purposely (e.g., PCBs, pesticides) or inadvertently (e.g., dioxins) by humans, and also released to the environment by human activity (petroleum compounds).
Persistent organic pollutants (POPs)	Organic compounds that are stable in the environmental and not rapidly degraded and which are also bioaccumulative through the food chain and potentially toxic to organisms.

Definition of the Subject and Its Importance

or their consumers, are nonessential elements or compounds, or are major components with essentially uniform concentration across the ocean system. Given that there are over 100 elements on Earth, and a finite but very large number of organic chemicals, derived either from an original biochemical source or manufactured by humans, it would be impossible to deal with all these compounds and elements in any detail. Therefore, while this chapter contains a description of the cycling of compounds through the marine environment and identifies the main pathways of movement and the major sources and sinks for compounds, in both the solid and dissolved phases, the focus of the discussion is a smaller list of/or their consumers, are nonessential elements or compounds, or are major components with essentially uniform concentration across the ocean system. Given that there are over 100 elements on Earth, and a finite but very large number of organic chemicals, derived either from an original biochemical source or manufactured by humans, it would be impossible to deal with all these compounds and elements in any detail. Therefore, while this chapter contains a description of the cycling of compounds through the marine environment and identifies the main pathways of movement and the major sources and sinks for compounds, in both the solid and dissolved phases, the focus of the discussion is a smaller list of compounds whose inputs to the ocean and whose fate and transport have important consequences for oceanic life and the environmental health of the ocean and the biosphere in general. The four main groups of elements/compounds and their fate and transport, and how these have changed in the recent and geological past, and the consequences of these changes on the ocean are described.

The four main classes of compounds are: (1) nutrients, including both the major nutrients (nitrogen (N), phosphorous (P), silica (Si)), and the minor nutrients, which are constituents of important enzymes and biochemicals (e.g., iron (Fe), zinc (Zn) and other transition metals); (2) toxic metals and metalloids, such as mercury (Hg), lead (Pb), arsenic (As) and silver (Ag); (3) organic compounds, primarily those manufactured or extracted from the Earth by humans (petroleum-derived compounds, pesticides and fungicides, pharmaceutical products and related chemicals); and (4) radioactive compounds and other environmental pollutants not covered in the other classes. While the addition of carbon dioxide (CO_2), released by fossil fuel burning and other human activities, to the ocean has and will have a large and currently uncertain impact on the basic ocean chemistry through its exacerbation of ocean acidification, and will impact, for example, the ability of calcifying organisms to precipitate their external shells, this will not be covered in this chapter.

Much of the discussion focuses on chemicals whose concentration in the ocean has been exacerbated by human activities. It concentrates on their dominant sources to the ocean, the factors that control their overall distribution and fate, with a brief discussion of the mechanisms for their accumulation into the marine food chain. Human health and environmental health impacts of chemicals assimilated into marine biota will not be directly discussed but this is clearly the most important concern of the increased ocean concentrations of many compounds. Historically, it was thought that the "solution to pollution was dilution" and that the ocean was a large and limitless reservoir that could not be severely impacted by the input of human wastes and other chemicals into the environment. This notion was prevalent in the early part of the twentieth Century, and before, but it has become increasing clear that this idea is a fallacy and that human activity has markedly impacted the ocean, as it has the other reservoirs of the biosphere.

Therefore, there has been a resultant change in attitude that has resulted in numerous conventions and international activities to mitigate past anthropogenic impacts and to ensure a sustainable future for marine resources. The debate however continues about the degree to which the ocean can be used as a "waste disposal site." Some scientists and ecosystem managers focus on the "precautionary principle" while many others think that the inputs can be effectively managed and assessed through the quantification of loading and an understanding of the level of the chemical in a particular environment that can be tolerated. These two approaches require different levels of understanding and knowledge. Therefore, there is the need to further examine and ascertain the degree to which the ocean can buffer ongoing inputs of chemicals. However, many such debates lack an appreciation for the complexity of the ocean system and how the inputs of chemicals in one location can impact the ocean more generally, and how chemicals are transported from one location to the other, and how long they actually reside in the ocean water column before being removed. This chapter attempts to provide the background and insights necessary to make such informed decisions.

Introduction

All the elements are found in the ocean, but their concentrations differ by many orders of magnitude in total concentration. The concentration of many elements and compounds has not changed substantially for thousands of years while those chemicals that have a strong anthropogenic component to their emissions to the atmosphere and to coastal waters have changed substantially in the last 100 years. The overall processes important in the cycling of chemicals in the ocean and across the important interfaces between the ocean and the rest of the biosphere are shown in Fig. 12.1. Exchange of chemicals occurs across the air–sea interface due to deposition of gases and particulate, as both wet and dry deposition. For gases, the direction of exchange – from the water to air or vice versa – depends on the relative concentrations of the compounds in each phase, and the relative solubility or volatility of the compound. Inputs of chemicals from the terrestrial environment are also important via riverine discharge as well as subsurface inputs from groundwater. Removal of chemicals from the surface ocean is due to mixing of ocean currents and internal cycling as well as the transport of chemicals through the sinking of particulate material to the deep ocean. For most chemicals that reach the deep ocean, the ultimate sink is removal to the sediment, and the relative rate of sedimentation to input from the surface ocean determines the net difference in surface and deep ocean concentrations. This is discussed further in the remainder of the chapter.

The major ions are sodium (Na) and chloride (Cl), both which are present at around 0.5 M concentration, with the other major ions (Calcium (Ca), magnesium (Mg), potassium (K), sulfate, and bicarbonate) being present at concentrations of 5–20 times less (Table 12.1) [77, 100]. In contrast, the major nutrients are present around 10^{-6} M or lower (often more than a thousand times less), the major transition metals are present at 10^{-8}–10^{-10} M and some of the more rare metal elements are present at $<10^{-12}$ M (a trillion times less) (Table 12.1). Total organic matter ranges from 10^{-3} to 10^{-5} M (on a carbon basis). Concentrations of many trace species are highest in coastal waters compared to the deep ocean and for elements and chemicals that interact and bind strongly with suspended particulate matter in the water, their total concentration is determined to a large degree by the suspended solids (TSS) levels, which vary from >100 mg/L for some rivers and estuaries to <1 mg/L in open ocean waters [100].

In terms of organic compounds, the focus of the discussion is contaminant and toxic compounds rather than the large suite of biochemicals and their degradation products found in seawater. The most important from an environmental health point of view are the so-called "persistent organic pollutants" (POPs), as defined by the Stockholm Convention. The POP designation includes a series of pesticides and other compounds initially termed the "dirty dozen" in earlier classifications of toxic and bioaccumulative organic compounds (aldrin, chlordane, dieldrin, endrin, heptachlor, DDT, and its breakdown products (designated here as ∑DDT), hexachlorobenzene, mirex, polychlorinated biphenyls (PCBs), polychlorinated dibenzo-*p*-dioxins (referred to as dioxins in this chapter), polychlorinated dibenzofurans, and toxaphene). Since then, polycyclic aromatic hydrocarbons (PAHs) and certain brominated flame-retardants (e.g., polychlorinated diphenyl ethers (PBDEs)), as well as some organometallic compounds such as tributyltin (TBT) and related products (referred to here as alkylated tin compounds) have been added [57, 64].

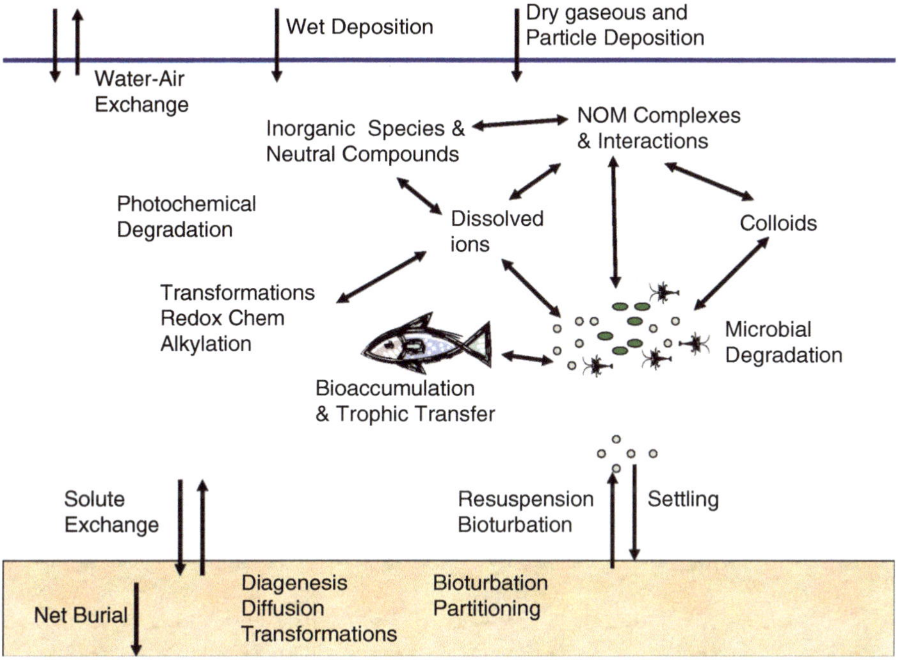

Fig. 12.1 Diagram showing the major sources and sinks for chemicals in the ocean and the major processes that are involved in determining their horizontal and vertical distributions and concentration

While not specifically on the POP list, other organic chemicals that are of concern will be mentioned throughout the chapter. Concentrations of the classes of organic compounds are given in Table 12.1 and are typically low (pg/L range) [58, 64]. Given the range in compounds in each class of organic chemicals in the Table, and discussed in the chapter, concentrations are given on a mass basis.

As noted above, this chapter cannot cover all aspects of chemicals in the ocean. Important topics of marine environmental pollution that will not be dealt with in detail include the transport of sediments themselves to coastal waters due to enhanced erosion and human activities within the watershed, such as deforestation, agriculture, or due to activities in coastal waters, such as dredging and bottom trawling, and the impact of the resultant excessive sediment on the ecosystem health of coastal waters [12, 45]. Excessive turbidity and sedimentation in such waters will impact the growth and survival of bottom-dwelling (benthic) plants (sea grasses and other plants) and animals (e.g., bivalves), either as a result of light limitation for plants, or due to smothering of benthic organisms and habitat degradation. Excessive sediment input is an important environmental concern. Additionally, sediment transport is an important vehicle for contaminants and other chemicals [100], and this aspect will be considered in detail.

Table 12.1 Estimated range and average concentrations for open ocean waters for major ions, trace elements and organic chemicals, and for the elements, the estimated residence time of each metal in the global ocean, and the major dissolved inorganic species in solution. The data for the organic chemicals are for the early 1990s. Taken from the literature ([17, 18, 21, 64, 111])

Element	Ocean Conc. Range (nM)	Average Conc. (nM, except where indicated)	Residence time ($\times 10^3$ year)	Major dissolved inorganic species for elements
Major ions	Na^+, Cl^-	0.49, 0,55 M	$>10^5$	Free ions
	K^+, Ca^{2+}, Mg^{2+}	9.7, 8.3, 4.4 mM		
	SO_4^{2-}, HCO_3^-	13, 1.8 mM		
Nitrate	<100–100,000	–	–	Free ions, acid–base pairs
Phosphate	<100–5,000	–		
Silicate	1,000–150,000	–		
Al	0.3–40	20	0.6	$Al(OH)_x^{3-x}$, x = 3, 4
Cr	3–5	4	8.2	CrO_4^{2-}
Mn	0.08–5	0.3	1.3	Mn^{2+}
Fe	0.01–2	0.5	0.05	$Fe(OH)_x^{3-x}$, x = 2, 3
Co	<0.01–0.3	0.02	0.34	Co^{2+}
Ni	2–12	8	8.2	Ni^{2+}
Cu	0.5–4.5	4	0.97	$CuCO_3^0$
Zn	0.05–9	5	0.51	Zn^{2+}
As	20–25	23	39	$HAsO_4^{2-}$
Mo	–	105	820	MoO_4^{2-}
Ag	<0.01–0.04	0.02	0.35	$AgCl_x^{1-x}$, x = 1–4
Cd	<0.01–1	0.5	–	$CdCl_x^{2-x}$, x = 1–4
Hg	<0.01	0.002	0.56	$HgCl_x^{2-x}$, x = 1–4
Pb	<0.01–0.15	0.1	0.81	$PbCO_3^0$
Se	0.5–2.3	1.7	26	SeO_4^{2-}
Total PCBs	1–50 pg/L	15 pg/L	NA	–
∑DDT	0.2–10 pg/L	2 pg/L	NA	–
∑Chlordanes	2–20 pg/L	5 pg/L	NA	–
∑HCH	0.05–2.1 ng/L	0.2 ng/L	NA	–

The chapter will also not directly consider the complex mixture termed "natural" organic matter (OM), either particulate or dissolved (POM or DOM) that is primarily derived from natural sources and biochemical processes, either those occurring in situ in the ocean or being derived from terrestrial inputs. Some of the OM comes from human activities such as sewage releases and dumping, and enhanced inputs to coastal waters have helped to drive water column oxygen depletion, although the major culprits are excessive amounts of nutrients. So, for this reason, OM will not be discussed directly although it has an impact on the cycling of the other chemicals to be discussed, and this will be included in the discussion.

As the chapter is focused on chemicals, there will be no discussion of the fate and transport of bacteria and related organisms, and their potential impact on human health [12, 45]. The most common and well-known contamination of this kind is from *Escherichia coli* bacteria, a mammalian intestinal bacteria, and the number of such bacteria in a water sample is often used as a pollution indicator as they do not survive for long periods in seawater. Shellfish can also accumulate these and other toxic bacteria and their chemical products, and therefore cause toxicity onto their consumers. Additionally, there is the potential for the harmful algal blooms (HAB) in marine waters as a result of the rapid growth of dinoflagellates and other organisms, which again can result in human health impacts and disease. Overall, infections as a result of contaminated seafood can be viral, parasitic, or bacterial.

Finally, marine litter and debris is an ever-increasing ocean problem, especially in terms of small plastic materials and other smaller products that are more easily ingested by marine organisms. Many ships dump garbage and wastes while at sea, which is not biodegradable and accumulates on the ocean surface. Additionally, gear is lost from ships while at sea, and beach litter and other material are washed to sea during storms. Sewage inputs and storm-water overflow into coastal waters can be a source of household, medical, and other wastes. Plastic particles can be large or small ($<$1 mm in size) [110]. Given ocean circulation patterns, there are now regions of the ocean where there is an ever-growing accumulation of marine debris that is extensive in size, with densities of $>$10,000 particles per square kilometer. The main impact of large plastics is on marine animals, such as turtles, and birds. Recent action has lead to regulations banning dumping plastics and other non-degradable products at sea. Additionally, it is now being recognized that the smaller plastic particles can accumulate hydrophobic organic compounds [44] and therefore could have a role in the bioaccumulation of these compounds through their ingestion by benthic organisms and filter feeders. Further discussion on this topic can be found in a number of publications ([110]; etc.) and the important problem of ocean debris will not be further considered in this chapter.

The Transport of Chemicals to the Open Ocean

The three important sources of chemicals to the ocean are: (1) inputs from the atmosphere of chemicals in the gas phase and attached to particles (termed dry deposition) and inputs of chemicals with precipitation (wet deposition); (2) inputs from the terrestrial environment (watershed inputs, surface runoff, and point and nonpoint discharges and groundwater inputs); and (3) inputs from the deep ocean via release from sediments and/or from hydrothermal vents and other natural point source inputs (Fig. 12.1). Additionally, some chemicals present in seawater as a dissolved gas can build up in ocean surface waters to supersaturated concentrations and thereafter can be lost to the atmosphere via gas exchange processes; this gas exchange can be thought of as negative dry deposition. The relative importance of each source for the elements and organic chemicals is collated in Table 12.2.

Table 12.2 Estimated inputs of some elements and chemicals to the ocean from the atmosphere, coastal/riverine inputs and from hydrothermal vents. The atmospheric inputs include both wet and dry (gaseous and particulate) net deposition. The relative importance of each source is indicated for each metal. Data taken from the literature. Fluxes are in Gmol year^{-1} for the elements and in Mg year^{-1} for the organic compounds. Taken from the literature as noted in the text

Element	Fluvial flux	Pluvial flux	Hydro-therm.	%Pluvial[a]
N	1.5–5	3.5–10	–	~70
Al	5,445	2,156	0.4	28
Mn	35	11	25	16
Fe	1345	469	75	25
Co	0.65	0.23	0.070	24
Ni	3.0	1.5	–	34
Cu	3.4	0.65	0.75	14
Zn	6.7	1.3	2.0	13
As	5.0	0.08	–	2
Ag	0.60	0.050	0.001	8
Cd	0.042	0.04	0.007	45
Hg	0.003	0.017	0.001	81
Pb	0.78	1.3	0.001	62
∑PCB	~60	~230	NA	~80
∑DDT	~4	~170	NA	~98
∑HCH	~60	~4700	NA	~99
HCB	~4	~80	NA	~95
Dieldrin	~4	~40	NA	~91
Chlordane	~4	~22	NA	~85

NA not applicable
[a]Estimates of percentage are the fraction of the total flux

Atmospheric Inputs

The information collected on the input of elements and chemicals to the ocean surface from the atmosphere (Table 12.2) is difficult to summarize for two reasons. One, sampling has occurred over a span of nearly 30 years and during these times there have been large changes in their deposition rates due to changes in anthropogenic inputs and loadings, and changes in human activity globally. For example, data from Bermuda suggest that the concentration of Pb in deposition has decreased by about 90% between 1982/93 and 1996/97 (Fig. 12.2c) [15, 106] and a similar decrease in ocean surface water concentrations has been found over the same period (Fig. 12.2a, b; ~60% decrease). Similarly, cadmium (Cd) has decreased by 80%, Zn by about 55%, copper (Cu) and nickel (Ni) by about 60%, and the decreases for the coastal east coast USA is of the same order [60].

This is also true for the POPs. The data for organic chemicals given below are from papers published in the early 1990s [33, 47, 111] and as many of these chemicals are now banned, these values have changed substantially as well. For example, emissions of α-HCH (lindane) has decreased by more than 90% since the early 1980s; emissions of PCBs more than 90% since their peak emissions in the

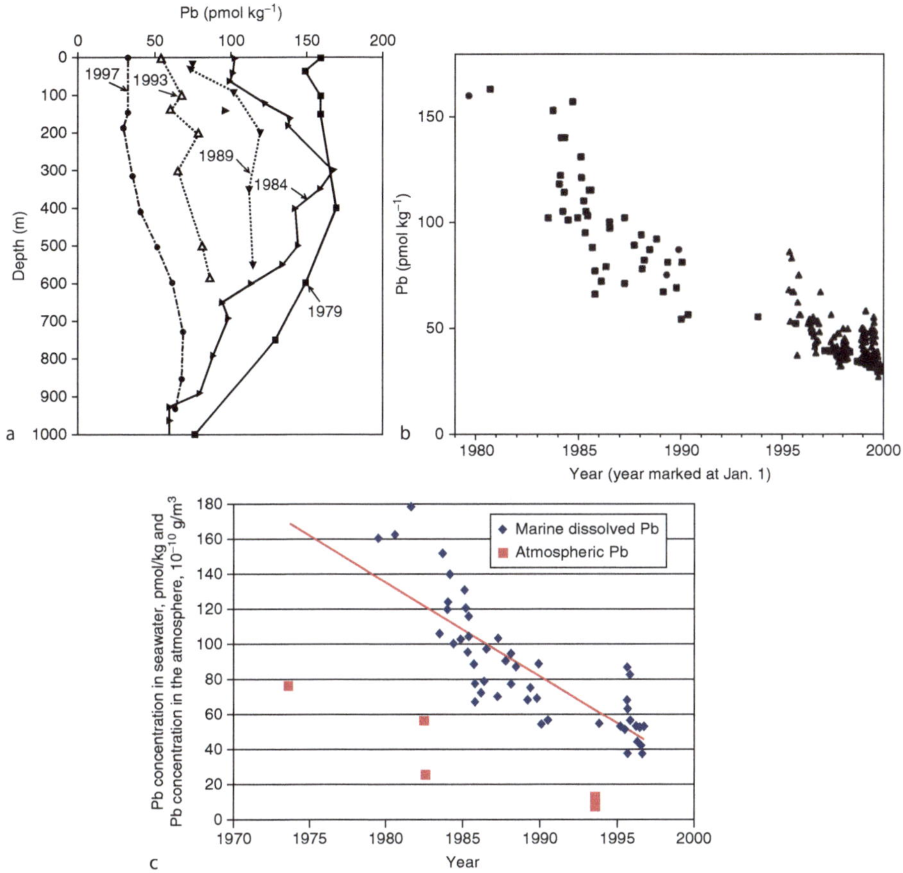

Fig. 12.2 (**a**) Vertical concentration profiles for lead in the North Atlantic at a location close to Bermuda over time; (**b**) A plot of the data from surface samples collections at the same location over time; and (**c**) A compilation of data for both the surface ocean and the atmosphere (atmospheric particle concentrations) from the same location. The first two figures are reprinted from "Anthropogenic Trace metals in the Ocean," a chapter by E.A. Boyle, pp. 273–280, with permission from Elsevier, and the third figure from "Atmospheric Input of Pollutants," a chapter by R.A. Duce, pp. 281–290, with permission from Elsevie, both chapters were published in *Marine Chemistry and Geochemistry*, Steele, J.H., Thorpe, S.A. and Turekian, K.K. (Eds), a derivative of the *Encyclopedia of Ocean Sciences*, 2nd Ed., Elsevier, Amsterdam, copyright (2010). Data are taken from various publications as indicated in the original figures

1970s; and DDT emissions have decreased by 50% in the last decade [57]. Overall, the data also suggest that the rate and timing of the changes are not all similar, reflecting the different anthropogenic sources to the atmosphere, changes in usage at different times and the differences in locations of emissions over time (e.g., for many chemicals, currently decreasing for North America and Europe; increasing for Asia, or in Africa).

Table 12.3 Range in the flux of metals from the atmosphere to the ocean surface. All values are given in μmol m^{-2} $year^{-1}$. Data taken from the literature and converted as necessary ([3–6, 21, 34, 60 74, 87, 95, 121, 123])

	Open ocean			Coastal/seas
Element	Wet	Dry	Total	Total[a]
Al	40–555	7–110	100–2,000	590–10^4
Cr	1.2		1.7–3.3	1.7–40
Mn	0.55–1.6	0.11	0.73–10.4	10–70
Fe	5.0–100	3.0	132–600	220–4570
Co	0.003–0.017	0.001	0.042–0.46	0.60–6.6
Ni			1.4–5.1	6.3–66
Cu	0.32–1.1	0.55	0.63–16	5.0–205
Zn	0.31–10	0.30–0.15	0.37–21	50–1,370
Cd	0.036		0.21–0.80	0.20–38
Pb	0.97–4.1	0.005	0.068–7.0	1.0–130
Hg	0.001–0.048	<0.001	0.001–0.048	0.05–0.15
Ag	0.28	0.009	0.18–0.32	0.28–0.83
As	0.13–0.4	0.09	0.36–0.4	1.3–38
Se	0.051–0.51	0.13	0.13–1.8	2.8–6.1

Secondly, compiling data is difficult as it is often reported in different units and not all the same metals have been the focus of the various investigations. Thus, for some metals, there is a relatively large number of datasets to work from, while for other metals, data limitation is a problem [33, 34]. Also, for the crustal elements, the concentrations vary depending on the relative importance of dust input and therefore there are large ranges in the reported concentrations. For example, studies on Bermuda have shown that there are large seasonal differences in the concentration of metals in rain and in aerosols [60] and this reflects the dominance of inputs from the east, with a relatively high dust component in summer, and inputs from the west, off the North American continent, being more dominant in fall-winter. Additionally, seasonal differences in rainfall amounts are obviously also important.

In addition to the total inputs, differences in the extent of input are also present given the dominance of anthropogenic sources near cities, and the locations of many large urban environments in coastal locations. This is illustrated by the compilation of data for the metals and the metalloids (Table 12.3) [21, 34]. An examination of the data indicates firstly that for the open ocean, wet deposition is the more important source, especially for the "anthropogenic metals" (Pb, Zn, As, Hg) and this reflects the typically low aerosol concentrations found over the open ocean. Thus, the dry deposition component is relatively small, and scavenging of the particles from the atmosphere by precipitation is the main mechanism of their removal. For the crustal metals (Al, Fe, Mn), the wet deposition is of somewhat lesser importance but still the dominant atmospheric flux, while for the other metals, the wet is typically greater than 70% of the total deposition. Total deposition fluxes vary most for the main crustal elements, while the variability appears to be lower for metals such as Cr, V, Ni, and Co, which have both a crustal source and an anthropogenic source. Metals such as Pb, Cu, and Cd show a large range in their values and this reflects the changes in their inputs over time, as noted above. For the crustal metals, changes on a decadal basis have been relatively small in the last 30 years.

There is a strong contrast between the fluxes for the open ocean and those for the coastal areas, and marginal seas (Table 12.3) (see references in Table legend). This is true both for the crustal metals and for the anthropogenically derived metals, and is likely also true for the organic chemicals. The increased flux for the crustal metals is primarily due to the higher aerosol concentrations in the coastal environments, while the higher fluxes for the other metals are mostly a consequence of their proximity to anthropogenic sources. Dust fluxes decrease exponentially offshore and most of the open ocean has inputs that are two orders of magnitude lower that the surrounding coastal environments, except for regions of high dust input (to the North Pacific from Asia; to the tropical Atlantic from Saharan Africa). Additionally, for those chemicals with a dominant anthropogenic signal (Pb, Hg, organic chemicals), inputs are highest in the Northern Hemisphere and have historically been higher into the North Atlantic, although this is changing depending on the specific chemical or element. Overall, the remote Atlantic, being a smaller ocean, has higher inputs, and there is clearly a strong gradient between the Northern and Southern Hemispheres.

Additionally, many cities are located in coastal areas or close to the coast, and therefore contribute substantially to the coastal fluxes, which are highest within a local radius of 50–100 km. For the anthropogenically dominated elements and organic chemicals, it has been shown that the fluxes measured in the urban environment are substantially greater than those representing the regional average value. In addition, continents with high urbanization and which typically have offshore air mass flow for all or part of the year (Asia and North America in the Northern Hemisphere, in particular) result in a substantial input of contaminants to the coastal waters. The much higher fluxes for the coastal regions of Asia is a result of the current rapid urbanization in the region, as well as the substantial input of dust, particularly during the spring "dust period" (March–May) when a large fraction of the annual dust input occurs. In the case of Europe, it has been shown that due to meteorological conditions, there are periods where there is substantial transport from the continent to the Arctic region, and also clear evidence of contamination of the North Atlantic coastal waters and the Mediterranean, due to anthropogenic inputs [1, 33, 34, 48, 90]. Overall, there is much uncertainty in the flux estimates as a result of the sporadic nature of the measurements and the lack of consistency in the metals that are quantified in specific studies, and due to the relatively short duration of many studies.

For organic chemicals, their interaction between the atmosphere and the surface ocean is complex and this is particularly so for the chemicals where the fluxes are dominated by gas exchange as these chemicals are volatile enough to be reemitted to the atmosphere after deposition. Scientists therefore discuss the environmental fate of organic chemicals in terms of their overall mode of long-range transport. Based on their chemical properties, organic chemicals are classified as "fliers," "multi- hoppers," "single hoppers," and "swimmers" ([57, 64, 105]) and this designation was developed mostly to distinguish the ability of a chemical to be transported and deposited in the Arctic region. Most lower molecular weight organics that are volatile are "multi-hoppers" as after they are deposited to the ocean and land surfaces they can be efficiently reemitted on short timescales and this enhances their overall lifetime in the biosphere and the extent of their long-range transport. Such multiple exchanges are mostly driven by seasonal and spatial temperature differences. Less volatile chemicals are "single hoppers" as they tend to associate with particles and are not reemitted for the most part after deposition, except if the particles to which they are attached are resuspended. "Swimmers" are those chemicals that are relatively soluble and therefore are transported spatially mostly through movement of ocean currents while "fliers" are those compounds that do not have substantial air-sea exchange and their transport and fate is mostly within the atmosphere.

Deposition of N to the ocean is related to the distribution of its sources (urban areas and agriculture) as the primary sources of N are from release of nitrogen compounds (collectively termed NO_x) from coal and other fuel combustion, and from the use of N-containing fertilizers, as well as other sources related to human activity [34, 48]. It is estimated that 30–50% of the N input to the ocean is from anthropogenic sources [33]. Nitrogen fixation in the ocean provides an input that is about 50% of the external sources. Currently, inputs of N are changing from being dominated by inputs from North America and Europe to having a larger Asian signal, as is found for many other compounds which have an anthropogenic component. Given the importance of agriculture as a source, inputs from Africa are also expected to rise in the future [33].

Inputs from the Terrestrial Environment

The export of material from rivers to the coastal zone is not evenly distributed globally because of the dominance of export from large rivers such as the Amazon, and the large Asian rivers, and also as a result of spatial differences in terrain [42, 75]. In addition, material export is relatively higher for rivers draining steep topography compared to those rivers draining regions with a large coastal plain, especially for chemicals associated with particulate material. The export of particulate material from rivers to the ocean has been estimated by various groups. The three regions of high particulate discharge are: (1) the Amazon River region (~14% of the total input); (2) southern Asia (34%), and (3) eastern Asia (36%). These regions all have large rivers and periods of high rainfall that drive the high particulate discharge. Much of the particulate material is however removed in estuaries and at the coastal boundaries due to the mixing of fresh and saline waters, changes in water flow rate and coagulation processes [21, 75]. The difference in ionic strength, pH, and the concentrations of dissolved solids, causes a general coagulation and removal of organic material in the mixing zone. The continents with the widest coastal plain therefore tend to be locations where particulate removal in the estuary is enhanced. The estimates in Table 12.2 are for the net transport of elements and chemicals from the coastal zone to the open ocean.

While it is most likely that many particle-reactive chemicals will be removed from the water column during estuarine mixing, it is possible for them to pass through without their concentration changing dramatically (so-called conservative elements) or they may even be added to the water column as a result of estuarine processing. If there is a linear change in concentration of a particular element or chemical relative to salinity through the estuary, then it is assumed that the concentration is the result of dilution of the riverine signal with seawater. In estuaries, salinity changes are not spatially linear with most of the salinity change occurring in a small mixing zone where the freshwater and saline waters meet. This region is also a region of high turbidity and particle settling.

For some elements and chemicals, the estuarine mixing of freshwater and seawater and the resultant chemical interactions that occur result in the elements being added or removed from the water column by precipitation, adsorption to suspended matter and resultant sinking to the sediment, or from release from particles with increasing salinity. Many metals and hydrophobic organic chemicals are highly particle reactive and are strongly retained in estuaries as most ($\sim$95%) of the suspended matter entering from rivers is not exported out of the coastal zone [21]. The role of other factors, such as biological productivity in surface waters and microbial degradation at depth also needs to be considered. While many of the redox processes that occur in estuaries can occur abiotically, it is now known that many of these processes are biotically mediated. Phytoplankton productivity plays an important role in estuarine processing of elements and chemicals. Phytoplankton productivity is driven by the supply of nutrients, which are typically in higher concentration in river waters than in coastal waters, and thus the input of freshwater and the estuarine mixing regime determines to a large degree the location where maximum productivity occurs, but it is often below the region of maximum turbidity, where light levels are low. Many elements and chemicals are actively accumulated by algae in this region, or bind strongly to the plankton, detrital material, and other particles. All these processes lead to the removal of chemicals in the high turbidity zone and in the maximum region of primary production, therefore potentially decreasing the dissolved concentration in surface waters. Their ultimate fate is sinking into the deeper waters and accumulation at the sediment–water interface. Thus, there are multiple mechanisms for the removal of dissolved constituents in the low salinity mixing regime of estuaries.

Chemical Inputs from Hydrothermal Vents, Ocean Seeps, and Sediments

The discovery of the presence of hydrothermal vents on the deep ocean floor, near the mid-ocean ridges and spreading centers in the late 1970s, and demonstration of the input of high temperature fluids from these systems changed our understanding of the sources and sinks of major and minor elements to the ocean [43, 104]. The percolation of seawater through the ocean crust leads to its heating followed by the addition or subtraction of chemicals due to reactions of the heated water with the crustal material. Temperatures can be as high as 350–400°C. The water becomes less dense and rises back to the surface where it is emitted and mixed with the surrounding cold seawater. Such processes are very important in the overall oceanic chemistry and fate of the major ions. Detailed chapters on hydrothermal vents and their importance to ocean geochemistry have been recently published and these provide much more information than can be detailed here [43, 54].

Magnesium and, to a lesser degree, Na and K, are removed during hydrothermal activity, while Ca, Fe, Mn (both in reduced forms), and other metals are added to the fluids. Sulfate is converted to sulfide in the low oxygen environment and bicarbonate (HCO_3^-) is converted into methane. The low pH (<4) leads to the conversion of dissolved carbonate species (primarily HCO_3^-) into CO_2 [43, 100]. The enriched elements in hot hydrothermal fluids are subsequently precipitated as the fluids cool or as these sulfidic, hot solutions are emitted into the oxygenated bottom ocean waters. Sulfides of Fe and other metals are precipitated, forming precipitate chimneys through which the fluids vent. Any remaining Fe and Mn will subsequently be oxidized and precipitate as their (hydr)oxides. Given the low solubility of many metal sulfides and/or their propensity to bind to and be scavenged by Fe-sulfide minerals and metal oxides precipitates, it is likely that most trace metals in hydrothermal solutions are removed to the solid phase close to the emission source.

It has been difficult to accurately estimate the importance of these sources for trace metals to ocean waters (Table 12.2). The situation is further complicated by the fact that there is still limited measurement of the concentrations of trace metals in such plumes, and the measurements that have been made have found a large variation in concentration, reflecting the heterogeneous nature of this overall process. Average values for the concentration of trace metals in hydrothermal fluids are given in Table 12.2. Overall, it is apparent that the concentrations of some metals, such as Fe, Mn, Al can be enriched by many orders of magnitude over their concentration in seawater [43]. Most transition metals, and heavy metals such as Ag, Tl, Pb, and Hg, and the metalloids (As, Se, and Sb), are also enriched in vent fluids. Overall, however, it is concluded that few of the first row transitions metals and the heavy metals (Ag, Cd, Sn) have a strong net hydrothermal source (Table 12.2).

Inputs to the deep ocean from the sediment through changes in chemical form for elements due to diagenesis within the sediment are thought to be small compared to the other sources discussed. As discussed above, the mobility of some elements is a function of their oxidation state and many can be released into solution in low oxygen environments. However, as discussed above for hydrothermal systems, these elements are rapidly removed from solution to the particulate phase on encountering oxygenated waters above the sediment and therefore, even if released from the sediment, are not more than a local source. However, in contrast to metals and most elements, there are important natural sources of organic petroleum chemicals in some ocean regions, specifically natural petroleum hydrocarbon seeps. These are found in the Gulf of Mexico and off the coast of California, for example. The amount of oil that is released from these seeps (~15% of total inputs; [46]) is difficult to estimate accurately but is not trivial compared to other sources releasing petroleum products into the environment.

Comparison of Sources

Generally, for the ocean, based on the estimates in Table 12.2, an overall estimation of the importance of atmospheric versus riverine inputs can be made. For the human-derived chemicals, hydrothermal inputs are not considered. In making these estimates, different values in the literature needed to be reconciled (e.g., [7, 9, 10, 13, 20, 21, 33, 34, 36, 37, 41, 43, 55, 57, 58, 70, 74, 82, 86, 95, 100]). Additionally, in making these calculations, it was assumed that 90% of the riverine suspended material was trapped in estuaries and did not get transported to the open ocean, and it was assumed that if there was no data available for hydrothermal inputs for a particular element, it was not significant. Clearly, this assumption may not be valid. Also, there is little information on the potential importance of groundwater inputs or the release of metals from coastal sediments and their transport offshore, which has recently been suggested to be an important source of open ocean Fe. Overall, the estimated values and relative contributions give an idea of the important sources to the ocean and allow a comparison of the relative sources and the biogeochemical reasons for these differences.

It is evident from Table 12.2 that riverine inputs are the dominant source for most trace metals to the ocean. Based on these values, Hg has the highest relative atmospheric input from the atmosphere at 85% of the total flux. The input from riverine sources is higher than the hydrothermal input, but this flux is not well known. For Pb, the atmosphere is also the dominant source, with the estimate being 63% of the total (Table 12.2). Again, the global cycle of Pb has been substantially perturbed by human emissions of Pb to the atmosphere and this is likely the reason for the importance of the atmosphere as a source [36]. In addition, Pb is highly particle reactive and is therefore removed to a high degree during transport through the estuary. Cadmium is another metal with a relatively high atmospheric signal, which is also a relatively volatile metal with important anthropogenic sources. Hydrothermal sources are also relatively important for Cd. Additionally, hydrothermal sources are somewhat important sources for Co, Cu, and Zn, but they are the dominant input for Mn, accounting for about 40% of the total inputs, almost as much as is entering the ocean from riverine sources. The atmospheric source is not important for Mn, or any of the other crustal metals, especially Al, Fe, and Co.

For the organic chemicals, the data in Table 12.2, even though they are from an early compilation of data by Duce et al. [34], confirm that the atmosphere is the dominant source for these chemicals to the ocean. Therefore, these compounds are determined to a substantial degree by long-range transport from their source regions to the locations of their deposition. The extent of their global versus regional deposition is a function of their residence time in the atmosphere. For petroleum-derived products, the inputs to the ocean come both from long-range transport (atmospheric and terrestrial inputs) and from local and regional inputs that are significant, and often highly variable [46]. Total inputs per year are in the region of 2–3 million tons. These include the inputs from marine accidents, discharges associated with offshore drilling, marine storage facilities and operations, and discharges from tankers and other ships.

For most metals and organic chemicals, removal to deep ocean sediments is the major and dominant sink for any of the compound transported into the deep ocean from the surface [21, 57, 64, 105]. However, this observation is only really true for those chemicals that are present in the dissolved phase as ionic species, or those that do not have a significant vapor pressure. This is true for most elements except those that have compounds that are volatile (e.g., Hg, Se) and the organic compounds. For Hg, which can exist in surface waters as a dissolved gas (elemental Hg), and many organic compounds loss to the atmosphere via gas exchange is the dominant loss term for the ocean [74]. For the majority of the organic chemicals, this is also true and the surface ocean concentration of these chemicals is mostly driven by the partitioning between the atmosphere and the surface waters. As discussed further below, the ocean can therefore be a source or sink of these compounds via gas exchange depending on the relative concentrations in each phase. As many of these chemicals have changed in atmospheric concentration rapidly compared to the mixing time of the upper ocean (years to decades, depending on location), many oceans are now a source of organic chemicals to the atmosphere where previously they may have been a net sink.

Ocean Cycling of Chemicals

The concentration and distribution of an element or chemical in the ocean is not directly related to its crustal abundance or its concentration in the surface soils, or its rate of release into the biosphere, but is rather determined by the complex interaction between its solubility, and its volatility, its ability to bind to or be incorporated into other phases, such particulate and dissolved organic matter (POM & DOM) or TSS (POM is a fraction of the TSS), and its potential to be present as colloidal material [17]. Scientists mostly measure the "dissolved" fraction of a chemical in seawater by filtration through membranes with openings of 0.2–0.8 μm, and therefore the filtered fraction can contain some colloidal material. Much of the DOM is likely in colloidal suspension and many insoluble metals such as Fe are potentially present as colloidal material as well in ocean waters.

A chemical's rate of biotic and abiotic transformation is also important and there are many microbial processes that can degrade organic chemicals or transform them into a different, and perhaps more toxic or mobile state. Microbial processes can also transform the toxicity of metals (e.g., methylation of Hg into methylmercury (CH_3Hg) produces a more toxic and bioaccumulative compound while methylation of As appears to reduce its toxicity). Intracellular transformations of organic compounds by Cytochrome P450 and other cellular degradation pathways can lead to more or less toxic products. The fate of some elements is also determined by the system oxidation state as their solubility is a strong function of this (Fe is more soluble when present as reduced Fe^{II}, chromium (Cr) is more soluble when oxidized as hexavalent Cr).

Besides the impact of biogeochemistry, particles, and organisms on the chemical distributions, there is also the impact of ocean circulation [17]. Ocean waters mix in a complex manner and this has a large impact of the fate and transport of chemicals. Therefore, a brief description of ocean circulation is required to understand the ocean distribution of chemicals. Vertically, the ocean is divided into: (1) the mixed layer (a few hundred meters deep), which is efficiently mixed by winds and currents, and where most of ocean photosynthesis occurs, and where most organisms reside; (2) the thermocline region, where there is a rapid change in temperature from the warmer upper waters to the colder deeper waters, and where there is strong density stratification and limited vertical mixing; and (3) the deep ocean. Below the thermocline, around 1,000 m in many oceans, is a region of oxygen depleted waters where much of the remineralization of organic particles transported vertically from the surface ocean occurs. In the deep ocean, waters are cold and relatively uniform in characteristics. The complexities of the ocean circulation can be dived up into two classes – "wind-driven" and "thermohaline" circulation [85]. Surface currents are mostly driven by heating and wind-related forcing, as the passage of the wind over the ocean surface will move the upper water with it. For many ocean regions, winds are consistently in one direction (e.g., the trade winds) and this has set up an overall surface ocean large-scale circulation, which also results from the fact that the water is being moved on a the Earth spinning on its axis, and therefore the water movement is also influenced by the "Coriolis effect." This results in the water moving to the right of the wind direction in the Northern Hemisphere, and vice versa. In addition to this direct effect, when there is movement of water away from a coast or other barrier, conservation of mass will require the transport of water from depth to the surface to replace this water. Such "upwelling" of deeper water is an important part of the upper ocean circulation. Additionally, upwelling can occur at the equator as water masses are moving away from the equator due to the Coriolis force (there is a water mass divergence at the equator). Similarly, at higher latitudes, convergence and sinking upper air masses results in divergent air flow over the ocean surface, which forces convergence of ocean waters, and sinking of water masses which feed the intermediate waters of the main ocean basins. These sinking intermediate waters travel laterally toward the equatorial regions.

Such wind-driven circulation is enhanced and complemented by the thermohaline circulation that is driven by density differences in water masses, which is primarily a function of temperature but also related to salinity. As surface waters move away from the Equator they are heated and net evaporation increases salinity.

When these water masses reach the mid-latitudes in winter, their cooling increases their density so that they become more dense that the waters below and sink. These sinking waters then travel in the main thermocline region (500–100 m) back toward the equator forming a cyclonic circulation that feeds upwelling at the equator. In the high North Atlantic and around the Antarctic Peninsula, surface waters become dense enough to sink to the deep ocean as formation of polar ice additionally increases the salinity of the surface waters, and cooling is greater. These sinking waters reach the deep ocean (>2,000 m) and supply and drive the deep water circulation of the ocean, and transport chemicals on a journey that takes 1,000 years on average. Overall, sinking water in the North Atlantic is transported south into the South Atlantic, skirting the Southern Ocean and passing through the Indian Ocean, and finally moving northward from the South to the North Pacific. To complete the water mass cycle, waters are generally transported upward (net upwelling) in the North Pacific Ocean to the surface, completing the water cycle.

The distribution of the major important characteristics of ocean waters that results from this circulation is illustrated by the data in Fig. 12.3 [93]. The salinity distribution along this Atlantic Ocean transect shows the presence of the intermediate water that has sunk from the surface in the South Atlantic and is moving north toward the equator. A similar process occurs in the North Atlantic and is also shown by the difference in salinity although the distribution is more diffuse. Sinking water in the Southern Ocean near Antarctica can be seen as deep water in the Southern Hemisphere. Similarly, water sinks in the North Atlantic around Iceland but this signal is not as evident in the salinity, but is shown in the phosphate distribution. The phosphate distribution also shows very clearly that its concentration has been depleted in the surface waters and is higher at depth. It also shows the two sinking water masses from the southern ocean regions which have higher phosphate concentrations than their surrounding waters. The oxygen distribution clearly shows the very low concentrations in the intermediate waters around the equator – due to oxygen consumption during organic matter remineralization. The south flowing deeper waters, that forms through water sinking in the North Atlantic Ocean, show a decreasing oxygen concentration as they progress south, further illustrating the consumption of oxygen in these deep waters due to organic matter degradation.

The movement of the water masses in the deep ocean can be tracked using the major chemical and nutrient distributions, as discussed in reference to Fig. 12.3, but there are also many other ways to track and age these water masses. The so-called "tracers" are generally of two types: (1) radioactive chemicals whose concentration will change over time due to radioactive decay [61, 78, 112] and (2) chemicals whose input to the ocean is well known and transient [38]. The radioactive elements can be used as "clocks" as their rate of decay is first order and well known and therefore the ratio of the source ("parent") and product ("daughter") can give information on the time since the element was added to the ocean. Also, as many of the parent-daughter components have different reactivity or solubility, their relative concentrations can tell something about ocean processes. One example is the decay of uranium (U) that is very soluble to thorium (Th), which is very particle

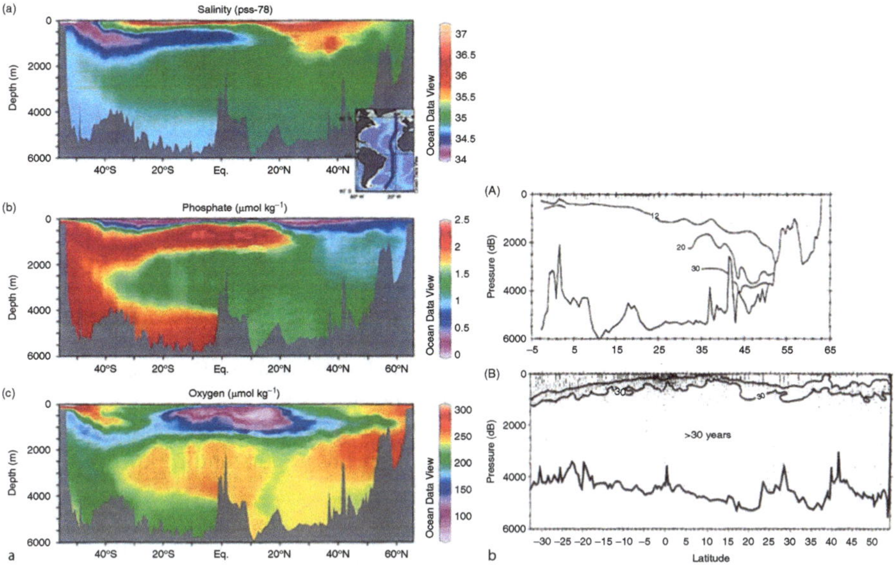

Fig. 12.3 (**a**) Meridional sections along the WOCE section A16 in the Atlantic Ocean for salinity, phosphate, and oxygen, with the cruise taken shown in the insert, indicating the vertical and horizontal distributions in the ocean, which result from ocean circulation, biological activity, and particulate remineralization; (**b**) Sections in the eastern Atlantic Ocean of the ratio of two chlorofluorocarbons (CFC-11 to CFC-12) along 20°W from 65°N to 5°S in summer 1998 and along 135°W in the eastern Pacific ocean from 54°N to 32°S in summer 1991. The first figure set is reprinted from "Inverse Modeling of Tracers and Nutrients," a chapter by R. Schlitzer, pp. 188–199, and the second figure set from "CFCs in the Ocean," a chapter by R.A. fine, pp.155–163, both published in *Marine Chemistry and Geochemistry*, Steele, J.H., Thorpe, S.A. and Turekian, K.K. (Eds), a derivative of the *Encyclopedia of Ocean Sciences*, 2nd Ed., Elsevier, Amsterdam, copyright (2010). Data are taken from various publications as indicated in the original figures

reactive and removed from the ocean by sinking particles [2, 19]. The resultant deficit in the ratio compared to that expected based on their decay rates can be used as a measure of the rate of particle removal from the ocean [22].

Some human-produced chemicals are also useful tracers. An example of this is the chlorofluorocarbons (CFSs), which were widely used for a number of industrial applications but which are now banned. Their extent of use and release to the environment is well known and so these compounds can be used to trace ocean water mass movement. An example is shown in Fig. 12.3, which shows the distribution of the ratio of CFC-11 and CFC-12, and the estimated age of the water in years based on this ratio [38]. This shows that the waters of the deep North Atlantic are relatively "young" as they are formed by sinking of surface ocean waters within the last 50 years. In contrast, in the Pacific, only the upper water masses show the presence of these chemicals and the deep waters show little

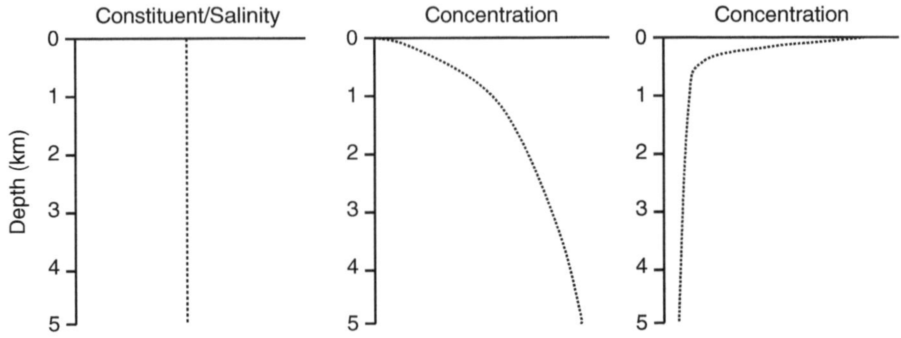

Fig. 12.4 Diagrammatic depiction of the distribution of a chemical that has (**a**) a conservative; (**b**) a nutrient-like; and (**c**) a scavenged profile in the ocean

evidence. Again, these distributions concur with the understanding of deep ocean circulation.

Other important tracers of ocean circulation are the radioactive elements released due to above ground atomic weapon testing (^{137}Cs, ^{14}C, ^{3}H) and the radioisotopes derived from the decay of U and Th (their specific isotopes and those of radium (Ra) (multiple radioisotopes), radon (^{222}Rn), ^{210}Pb and others) [112], and cosmogenic radioisotopes (^{14}C, ^{3}H, beryllium (Be) isotopes and many others) [113, 114].

Factors Controlling Ocean Chemical Distributions

The distribution of an element or chemical in the ocean has a characteristic distribution that relates to its solubility, degree of reaction, its requirement as a nutrient, and other factors (Fig. 12.4). Additionally, as discussed further below, many elements and chemicals are converted between different forms due to photochemical processes in the surface ocean, due to redox changes in the low oxygen intermediate waters, or due to their association with DOM. For many metals, complexation to dissolved ligands plays an important role in defining their concentration and distributions. The importance of mixing and other physical transport processes must also be considered, as detailed above. The overall range and average concentrations for a number of important elements or compounds are given in Table 12.1. For the elements, their major form and state in seawater is also given. A compilation of distributions for all the elements in the periodic table for the North Pacific has been compiled by Nozaki [81] (Fig. 12.5).

The major ions are found at a constant concentration across all the major ocean basins, and their distribution in the ocean has been termed a "conservative" distribution (Fig. 12.4) [17, 21]. Except for rare and specific circumstances in specific locations, all the major ions have this type of distribution. To exhibit such

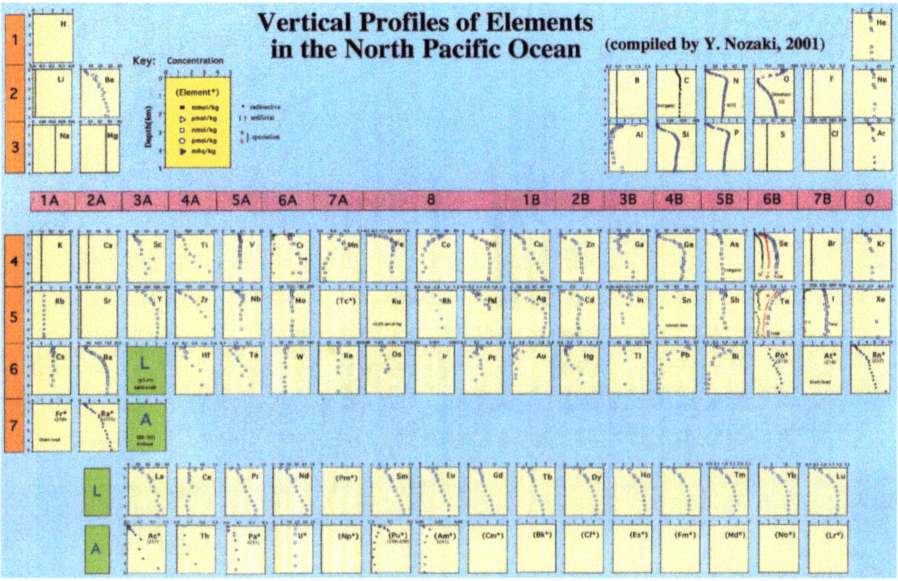

Fig. 12.5 A depiction of the vertical distributions of the elements in the North Pacific Ocean. Originally published in EOS supplementary material in 1997, for an article by Y. Nozaki "A fresh look at element distributions in the North Pacific", copyright 1997, American Geophysical Union. Reprinted with permission from the American Geophysical Union

a distribution, their residence time of the element or compounds must be much longer than the mixing time of the ocean (which is around 1,000 years) so that they can become well mixed throughout the ocean basins, and most conservative elements have residence times of $>10^6$ years [17]. Most conservative elements are found at relatively high concentrations compared to their crustal abundance and this is because of their slow rate of removal from the ocean.

Elements or chemicals with a "nutrient"-type distribution are those whose concentration is depleted in the surface ocean due to their uptake into biomass, especially into photosynthetic microorganisms, and their removal from the upper waters by the sinking of particulate material (dead cells and organisms, fecal pellets). Decomposition of this material in deeper waters leads to a rapid increase in concentrations at mid-depth in the ocean and higher and relative uniform concentrations in the deep ocean, and this decomposition depletes the oxygen in these mid-depth regions. Furthermore, if these compounds and elements persist in the deeper waters and are not effectively removed by deep ocean particle scavenging, their concentration in the deep Pacific Ocean will be greater than that of the deep North Atlantic given the increased input of these chemicals over time from above as the water masses move through the ocean basins. Both nitrate and phosphate show such a distribution, as do other micronutrients [21].

A correlation between the concentrations of Zn and Si, and cadmium (Cd) and P, in the ocean was noted in the early 1980s (Fig. 12.6), suggesting that similar

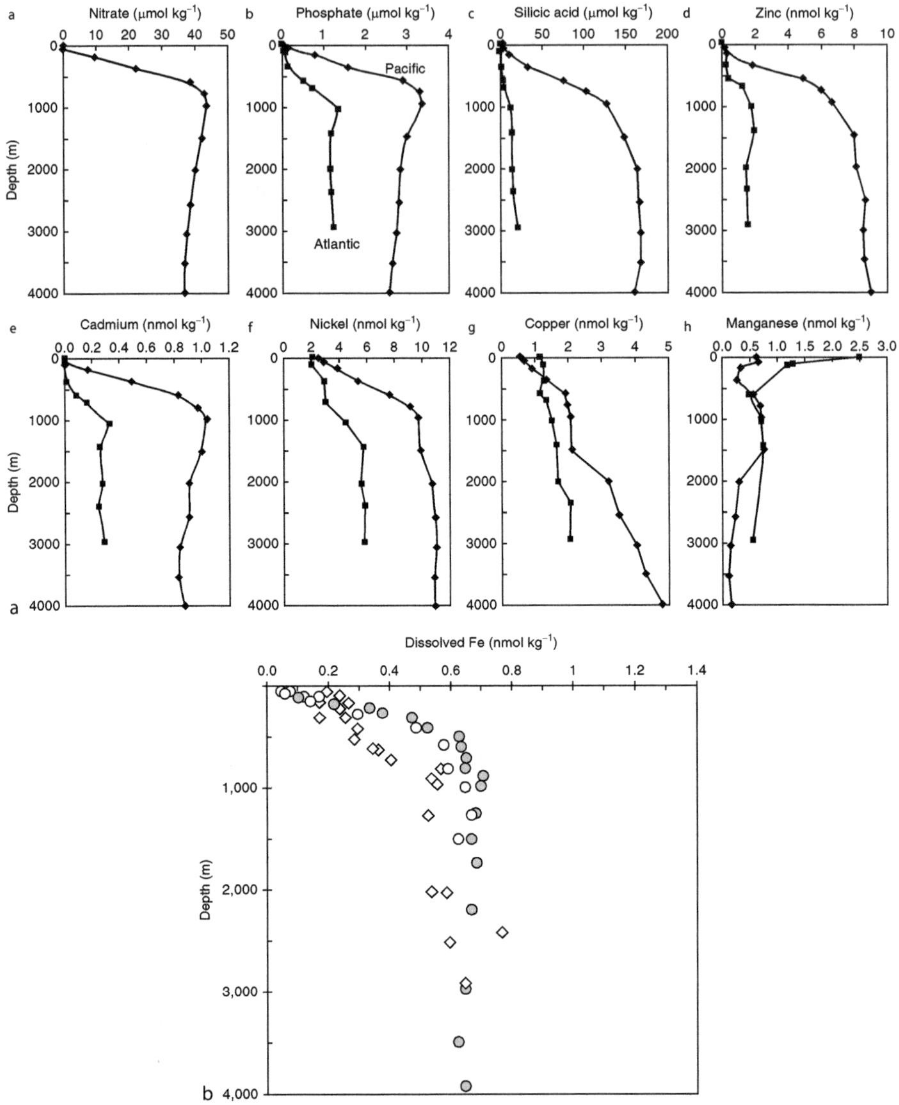

Fig. 12.6 (**a**) The vertical distributions of nutrients (N. P, Si) and metals (Zn, Cd, Cu, Ni, Mn) in the North Pacific and North Atlantic Oceans; (**b**) A similar plot for Fe for both oceans. The first figure set is reprinted from "Trace Metal Nutrients," a chapter by W.G. Sunda, pp. 17–28, published in *Marine Chemistry and Geochemistry*, Steele, J.H., Thorpe, S.A. and Turekian, K.K. (Eds), a derivative of the *Encyclopedia of Ocean Sciences*, 2nd Ed., Elsevier, Amsterdam, copyright (2010), with permission from Elsevier. The iron figure is from K. Bruland and M.C. Lohan "Controls of trace metals in seawater", pp. 23–47, a chapter in *The Oceans and Marine Geochemistry*, H. Elderfield (Ed), volume 6 in Holland, HD and Turekian, KK (Exec Eds), *Treatise on Geochemistry*, Amsterdam, Elsevier. Copyright (2004). Data are taken from various publications as indicated in the original figures

mechanisms control their ocean distribution and this has since been attributed to their "nutrient-like" behavior [21, 32]. Zinc has multiple roles in cellular metabolism and, most importantly, it is part of an important enzyme, carbonic anhydrase in marine phytoplankton, which is required to interconvert CO_2 and HCO_3^- within cells [76]. Thus, Zn is actively accumulated into plankton in surface waters and is released back into the water at depths when the organic matter decomposes, and has a similar profile with depth to the nutrient, Si. Deep water concentrations of both Si and Zn are higher in the Pacific Ocean compared to the Atlantic Ocean (Fig. 12.6) and this reflects the continued buildup of these elements during the "deep conveyor belt" circulation [17]. The correlation of Cd and P profiles was initially puzzling as Cd was thought of as a toxic metal but it has been subsequently demonstrated that Cd can actually substitute for Zn and other transition metals in carbonic anhydrase and therefore has a biological role in the ocean environment [76]. Given this fact, the correlation with P is explained.

Iron and Copper (Cu) are other micronutrient elements required for enzymes and other biochemicals but they have a more complex vertical profile as their relatively high particle reactivity and the insolubility of Fe results in their scavenging from deep ocean waters by particles. This results in a lack of increase in concentration in the North Pacific relative to the North Atlantic [18, 76]. Additionally, for Fe, its redox chemistry can result in changes in redox state in the low oxygen waters that can further complicate its ocean distribution. Finally, Fe, Cu, and other metals are released into the deep ocean by hydrothermal activity and this can also influence the ocean distribution of these metals. Finally, in a similar manner to Cd, Cu is both an essential element and a potential toxic element and therefore there is a fine balance between its concentration needed to support growth and that which can hinder organisms. Other elements in this category are Se, and to a lesser degree As.

Another oceanic distribution is characterized by elements or chemicals that are particle reactive and/or relatively insoluble, and especially those metals or organic compounds that also have the majority of their input from the atmosphere (Fig. 12.4). These are referred to as "scavenged elements" and they have deep water concentrations that tend to decrease along deep water flow paths between the Atlantic and Pacific. Hence, their distributional features are often dominated by their point of entry and removal in the ocean. These elements and compounds have short residence times ($<10^3$ years). For example, surface water concentrations of Hg and Pb are typically elevated compared to deep waters [14, 40, 92], reflecting their dominant atmospheric input pathway to the ocean. Such metals are scavenged and removed from the surface ocean via particle transport, released back into the water column during the dissolution of the biological particles, but are often re-scavenged from deep ocean waters and therefore do not have a continuing increasing concentration with depth, and their concentration is depleted in deep ocean waters, or does not increase with depth, or with time.

Organic chemicals are added to the ocean via a variety of processes. Human-made chemicals such as pesticides, polychlorinated biphenyls (PCBs) and other industrial chemicals are typically added to the ocean from both direct runoff and atmospheric deposition. Some compounds, such as petroleum-derived compounds, are naturally formed in the deep Earth but are pollutants when added to the ocean in high concentrations. Most of these compounds are highly insoluble in water and have a high tendency to associate with particles (POM) and are taken up into microorganisms, and therefore are effectively removed from the surface ocean by particle sinking. Such chemicals are also likely strongly removed from solution in estuaries and coastal waters and therefore their input to the ocean is dominated by atmospheric inputs. These compounds likely have a scavenged distribution in the ocean water column. Other pharmaceuticals and organic chemicals are manufactured to be highly soluble in water but typically have a relatively short half-life before they are degraded by photochemical processes and microbes, and as many of these compounds have only been recently (on a ocean mixing timescale) added to the ocean, their distribution and fate are poorly known. Many of these compounds are accidentally added to the ocean or are released from coastal sources, such as treatment plant effluents and stormwater outflows.

In terms of the nutrients, it is often their excessive addition to the ocean as a result of anthropogenic inputs and human activity that have been the focus of study and environmental management. For example, excessive amounts of nitrogen, in particular, in coastal waters has lead to an increase in extent, duration, and severity of oxygen depletion in deep waters, and the persistence of hypoxic (oxygen content $<$4 mg/L) and anoxic waters (undetectable oxygen/presence of sulfide), which have been termed "dead zones" in many instances in the media and elsewhere. Oxygen depletion in coastal waters has important consequences and while the formation of such zones did occur in the historic past, human activity has definitely increased the problem [115]. However, it is also important to understand the role of nutrient limitation in ocean chemistry and environmental health. Nutrient limitation is an important concern for open ocean ecosystems, and this is mainly due to the limitation of Fe input to the ocean. Such limitation has been little impacted by human activities in the last century and has its origins in the geological evolution of the Earth in the very distant past. In contrast to Fe, Si limitation is also important in some ocean locations and this can be linked to human activity as one important source of Si to the ocean is riverine inputs. While not all microorganisms require silica, it is an important nutrient for one class of phytoplankton (diatoms) and for other organisms that construct silicate shells, and therefore its limitation can have an important impact on coastal and ocean primary productivity. Construction of reservoirs on many rivers around the world has decreased the Si input to the ocean and therefore could be an important factor in determining the extent of this limitation [11, 23].

Detailed Description of the Cycling of Important Elements and Chemicals

Iron, Manganese, and Aluminum Cycling in Open Ocean Waters

Atmospheric and riverine inputs are the main sources of Fe, Mn, and Al (all crustally abundant) to the oceans (Table 12.1) and much of this input is in the particulate or colloidal fraction. While Fe is primarily present as Fe^{III} complexes and solids in oxic seawater, Mn can be found in its more reduced +2 oxidation state, and the reason for the persistency of the reduced Mn forms is its relatively slow oxidation rate compared to Fe [63, 65, 66]. Aluminum (Al) ranges widely in concentration and is elevated in regions of enhanced atmospheric input, such as the equatorial zones where rainfall is high, and is very low in polar surface waters. It is present as Al^{III} ($Al(OH)_4^-$, $Al(OH)_3$) [84]. Concentrations are higher in the North Atlantic (8–30 nM) than the North Pacific (<2 nM) due to its depletion via particle scavenging. It has high concentrations in the surface Mediterranean (>100 nM) due to the enhanced atmospheric (dust) inputs in that region. Aluminum is the element with the largest ocean variation.

While riverine inputs are important, most of it is removed in estuarine and coastal waters as all these metals are highly particle reactive. Given their increased insolubility with pH, these elements have low ocean concentrations even though they are some of the most abundant elements on the surface of earth. The relative insolubility of both Fe^{III} and Mn^{IV} leads to the rapid removal of these metals from the surface ocean. While both Mn and Fe have a role in the biochemistry of marine organisms [76], the concentration of Mn is relatively high in ocean waters compared to its nutrient requirement, and depletion of its surface concentration due to biological activity is not often found. In contrast, as noted above, dissolved Fe can show surface ocean depletion in regions of high primary productivity.

Representative profile for Fe in the major ocean basins are shown in Fig. 12.6c. In seawater, the concentration of dissolved Fe is higher than would be predicted based on its primary solubility in terms of Fe (hyd)oxides due to the presence of Fe-binding ligands in solution [18, 76]. In the absence of these ligands, the predicted concentration in equilibrium with $Fe(OH)_3$ (s) is around 0.1 pM, even considering that it is strongly hydrolyzed in seawater with the dominant inorganic complexes being $Fe(OH)_4^{2-}$, $Fe(OH)_3$ and $Fe(OH)_2$. The average measured concentration is 0.5 nM, three orders of magnitude higher (Fig. 12.6c). This is consistent with the electrochemical measurements which estimate that 99% or more of the Fe in surface ocean waters is complexed to organic ligands. Additionally, given that the measured values are for filtered waters, it is likely that this value reflects the presence of both organic and inorganic Fe-containing colloidal phases. In contrast, Mn forms only weak associations with both inorganic and organic ligands, and its dissolved concentration is much lower than the concentration expected from precipitation of Mn^{IV} hydroxide or carbonate phases.

The distributions of Fe and Mn in the water column are modified in the regions of low oxygen or anoxic waters. In some ocean locations, oxygen concentrations are sufficiently low that both Fe and Mn can be reduced, either through biologically mediated pathways or abiotically, and this has been observed, for example, in the oxygen minimum zone of the equatorial Pacific Ocean [21]. In addition to these sub-oxic environments, the distributions of Fe and Mn are highly modified in anoxic environments with elevated levels of the reduced forms of the metals being present. However, the distributions are complex for Fe because, while in low oxygen environments Fe(II) is formed with an increase in solubility as sulfide levels increase, Fe(II) is precipitated as Fe-sulfide phases (FeS and FeS_2) and this leads to a maximum in dissolved Fe in low oxygen, but low sulfide environments.

In permanently or seasonally stratified systems, the redox cycling of metals can lead to a large gradient in their concentration, as well as for metals and other constituents that are strongly associated with the oxic particulate phases of Fe and Mn. The concentration of particulate Fe and Mn, and associated metals, are enriched above the interface due to the diffusion of dissolved, reduced Fe and Mn from below, and their subsequent oxidation and precipitation in the higher oxygen waters. Sinking of these particulate materials and their dissolution below the interface leads to a peak in dissolved species below the interface. This cycling across the interface has been termed the "ferrous wheel" in analogy to the amusement park ride.

Another location where increased concentrations of dissolved Mn and Fe can be found is in association with hydrothermal vents, especially for Mn given the slower rate of Mn^{II} oxidation. Elevated concentrations have been found to persist in the vicinity of hydrothermal vents, in plumes that are enriched in other tracers of such activity (^3He) [21, 43]. While hydrothermal vents are sources of many metals as discussed above, the removal of metals with the precipitation of Fe and Mn oxides after emission and oxidation of fluids from hydrothermal systems results in the local scavenging and removal, and decreases the importance of hydrothermal sources as input to the global ocean. These processes are important in the formation of oxide deposits and also in the formation of metal-rich nodules.

The Cycling of the Other First Row Transition Metals, Zinc, and Cadmium

The distribution of Zn in open ocean waters appears to be strongly controlled by biogeochemical processes, and its distribution is that of a "classic" nutrient metal. The release of Zn from decomposing organic material in the sub-thermocline waters results in higher concentrations in deeper waters and the relationship between the concentration of Zn and that of Si [116]. As the main source of Zn to the ocean is riverine and terrestrial input (Table 12.1), and the atmospheric input is a relatively small component, depletion of surface waters is more likely. Zn is relatively soluble in seawater in the presence of inorganic ligands and has an average concentration of 5 nM (Table 12.1), greater than that of Fe and Mn, which are much more abundant elements in the terrestrial environment and in riverine inputs. Additionally, there is no important redox chemistry for Zn in ocean waters, and Zn forms weak complexes with inorganic ligands. Complexation of Zn by dissolved organic ligands likely also stabilizes its dissolved concentration and it is suspected that these ligands are directly or indirectly produced by microbial organisms, even if they may not be produced specifically for Zn complexation. Even given its direct biological role, Zn has a relatively long residence time compared to the other first row transition metals that exist in solution predominantly as cations.

As Zn is present in the absence of organic ligands mostly as the free ion in seawater (Table 12.1), it is relatively easily acquired by organisms and therefore complexation with specific organic ligands to enhance uptake is likely not needed. Therefore, complexation of Zn by organic matter does not appear to provide a unique advantage to its uptake in contrast to Fe, which is a demonstrated limiting nutrient and therefore is acquired actively through organic complexation by some microbes, and Cu, which can be toxic to some organisms at higher ocean concentrations, and complexation can result in a reduction in toxicity.

In a similar fashion to Zn, the distribution of Cd, which is from the same group in the Periodic Table, in open ocean waters is strikingly similar to that of phosphate (Fig. 12.6b). As with some other elements, Cd can be both a toxic metal at high concentration and have a beneficial role at low concentration because of its ability to be incorporated into and act as an active metal center in enzymes. Therefore, Cd is considered a nutrient metal and there is evidence of its depletion from surface waters in many ocean regions. Similarl to Zn, Cd inputs to the ocean are not dominated by the atmosphere, but in contrast to Zn, the inorganic chemistry of Cd in the ocean is dominated by chloride complexation and the relative fraction of Cd as the free metal ion is a few percent. It is also apparent that Cd is complexed to organic ligands in the ocean but the relative degree of complexation is lower than that of Zn [21] Overall, the major difference in speciation in seawater is likely not due to differences in the relative binding capacities to organic matter, but the differences in the binding to Cl. The concentration of Cd in the ocean is relatively low compared to other transition metals. Its residence time in the ocean is not well characterized as there is little information on its distribution through all the ocean basins. Given its sources and biogeochemical cycling, it is probable that it has a similar residence time to Zn and other relatively soluble transition metals.

Cobalt has received recent attention due to the potential importance of Co as a cofactor in cobalamin and other enzymes. The uptake of Co by microbes, especially in surface waters, could lead to differences in the distributions of Co and Ni, another transition metal with similar chemistry and sources. Copper is also a required nutrient metal and shows a distribution that is a mixture of that expected for a nutrient element, modified because of the tendency of Cu to complex with organic matter which enhances its solubility and residence time in deep waters.

Many studies have demonstrated the importance of organic complexation for Cu [21, 31]. It has been speculated that these ligands are biologically derived by certain microorganisms, such as cyanobacteria which are susceptible to Cu toxicity. This has been demonstrated in a few instances (e.g., [35]). Inorganically complexed Cu is a very small fraction of the total in deeper waters. As noted in Table 12.1, inorganic Cu is primarily present as the neutral $CuCO_3$ complex in seawater.

In contrast, both Ni and Co form much weaker complexes with inorganic ligands and are present predominantly as free metal ions in seawater. These differences in speciation likely impact the mechanisms of accumulation of these metals into microorganisms. Most transition metals have been shown to have a biochemical role, with Cu being incorporated into many enzymes, while the involvement of Ni is relatively small. The concentration of Ni ($\sim$8 nM) is higher than that of Cu, while the concentration of Co is sub-nM. While both Ni and Co are found organically bound, the relative fraction is smaller than for the other transition metals discussed, being 30–50%, or greater, in various open ocean locations.

The presence of cobalamin in surface ocean waters has been recently demonstrated although it is present at low pM concentrations [83]. It is possible that most of the organically complexed Co in seawater has a biological origin. The identification of specific biochemicals containing metals raises the possibility that some of the metal, identified as complexed to organic ligands, has been released into the environment as a result of cell leakage, purposeful cellular export (e.g., metals bound to metallothioniens and phytochelatins), cell death, from the release due to grazing or even from fecal material. Alternatively, these ligands could be purposely released into solution to aid in metal assimilation, as occurs for Fe (e.g., release of siderophores), or to reduce metal toxicity, as discussed for Cu above. Another example is the presence and assimilation of heme (Fe-containing) compounds as it has been shown that microbes can assimilate heme in laboratory cultures [56].

Anthropogenic Metals: Lead, Silver, and Mercury

The three most important heavy metals in the marine environment are Pb, Hg, and Ag. Studies of Pb reflect its known input to the atmosphere from Pb compounds in leaded gasoline, as well as its presence in other anthropogenic sources. The examination of Hg is driven by human and wildlife health concerns from the toxicity and bioaccumulation of CH_3Hg in marine food chains. Silver is also known to be toxic to organisms but its levels in the open ocean are generally low (<1 nM) [88]. Historically, elevated Ag in the coastal environment was considered an indicator of sewage and anthropogenic-related inputs [91], because of its extensive use in the photographic industry. This is now less the case due to changes in photographic technology (i.e., the emergence of digital cameras).

All these metals bind strongly to particles and are emitted to the atmosphere from anthropogenic sources although both Pb and Ag will be removed to a degree from stacks by particulate emission control devices. Of these elements, Hg has the highest relative natural source as it can be emitted from various terrestrial environments in its gaseous elemental form. For the open ocean, atmospheric inputs are important but additionally coastal inputs also contribute both Hg and Pb from terrestrial runoff and point source inputs. Mercury is different as is can be present as a dissolved gas (Hg^0) in seawater and therefore air-sea exchange involves both deposition and gas evasion.

Concentrations of Ag in the ocean vary from low pM values typically found in the surface ocean to 100 pM or more in deep waters, and concentrations vary substantially between ocean basins [79, 89, 108, 109] (Fig. 12.7). Silver typically has a nutrient-type profile and it has been shown that its vertical distribution mimics that of Si, suggesting its incorporation into the more recalcitrant tissues of microbes and other organisms in the surface ocean and its relatively slow release from sinking particulate material [89, 109]. In the deep ocean, there is continual net increase as Ag is being continuously released from settling particulate matter but is not being scavenged in deep waters in a similar fashion to other metals, making the distribution of Ag similar to that of Al. Additionally, both elements are supplied to the ocean mainly from river and coastal sources, with atmospheric deposition being a secondary source [49, 88]. Concentrations are increasing over time in the oceans. A comparison of intermediate waters in the North Pacific Ocean suggests that Ag concentrations have increased in the last 20 years, reflecting increased anthropogenic inputs [89]. In contrast, a similar comparison for Hg concluded that there had not been a substantial increase in concentration over that time period [63], although more recent evidence suggests that concentrations have increased more recently [101].

A number of ongoing and detailed studies have documented the overall contamination of the ocean by anthropogenic Pb and have demonstrated the resultant decrease in ocean Pb as a result of the phasing out of Pb additives for gasoline [107] (Figs. 12.2 and 12.8). However, there is still Pb input to the atmosphere from other anthropogenic sources and current studies are evaluating the extent of this input. One unique aspect of examining Pb geochemistry is the fact that it has numerous isotopes, and these isotopes are daughters of the uranium-thorium (U–Th) decay series: ^{206}Pb is the stable product of the ^{238}U decay series, ^{207}Pb the stable product of the ^{235}U decay series, and ^{208}Pb is the stable product of the ^{232}Th decay series. Thus, the isotopic ratio of Pb in the environment is altered due to the presence of either U or Th in the medium. It has been demonstrated that Pb from different locations has ratios which are different enough to track the sources of the Pb in the ocean and other environments as the different isotope ratios allow for more resolution of the source signals (Fig. 12.8b). Data for the South and equatorial Atlantic Ocean collected in 1996 is plotted against the various potential source signals and demonstrates a range of potential sources, including inputs from North America, large rivers, and coastal sources. In contrast, the data for the deep waters of the North Pacific, which represent the "oldest" marine water masses, shows a Pb

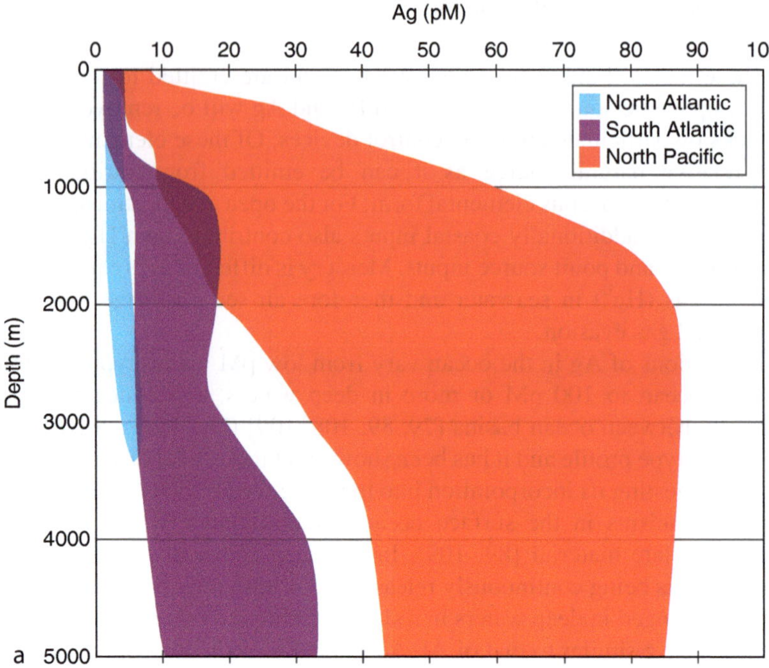

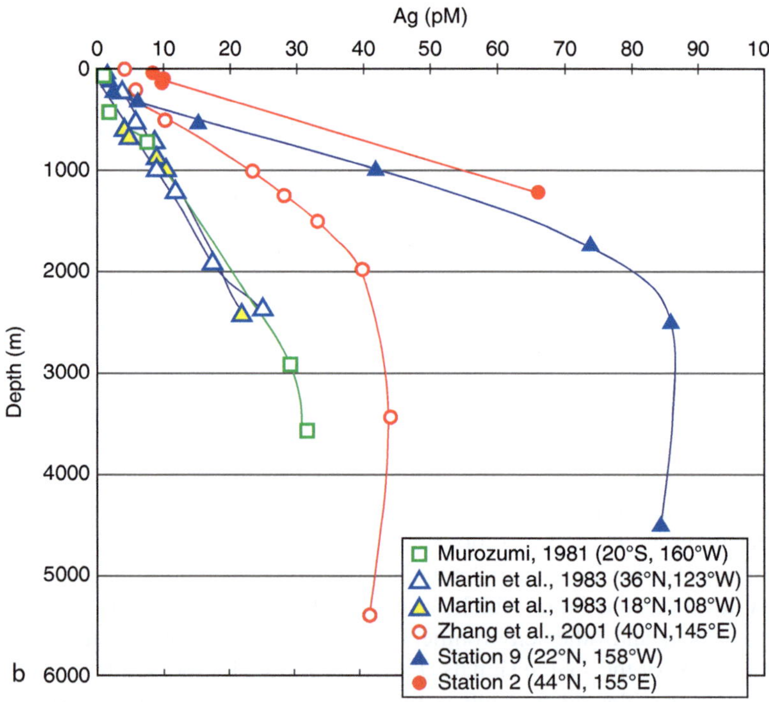

Fig. 12.7 (**a**) Distributions of silver in the various ocean basins and (**b**) Changes in the concentrations within the water column in the North Pacific over time. Taken from M.A. Ranville and A. R. Flegal "Silver in the North Pacific Ocean", published in *Geochemistry Geophysics Geosystems* volume 6, Art. No. Q03M01. Copyright (2005), American Geophysical Union. Reprinted with permission as stated by American Geophysical Union

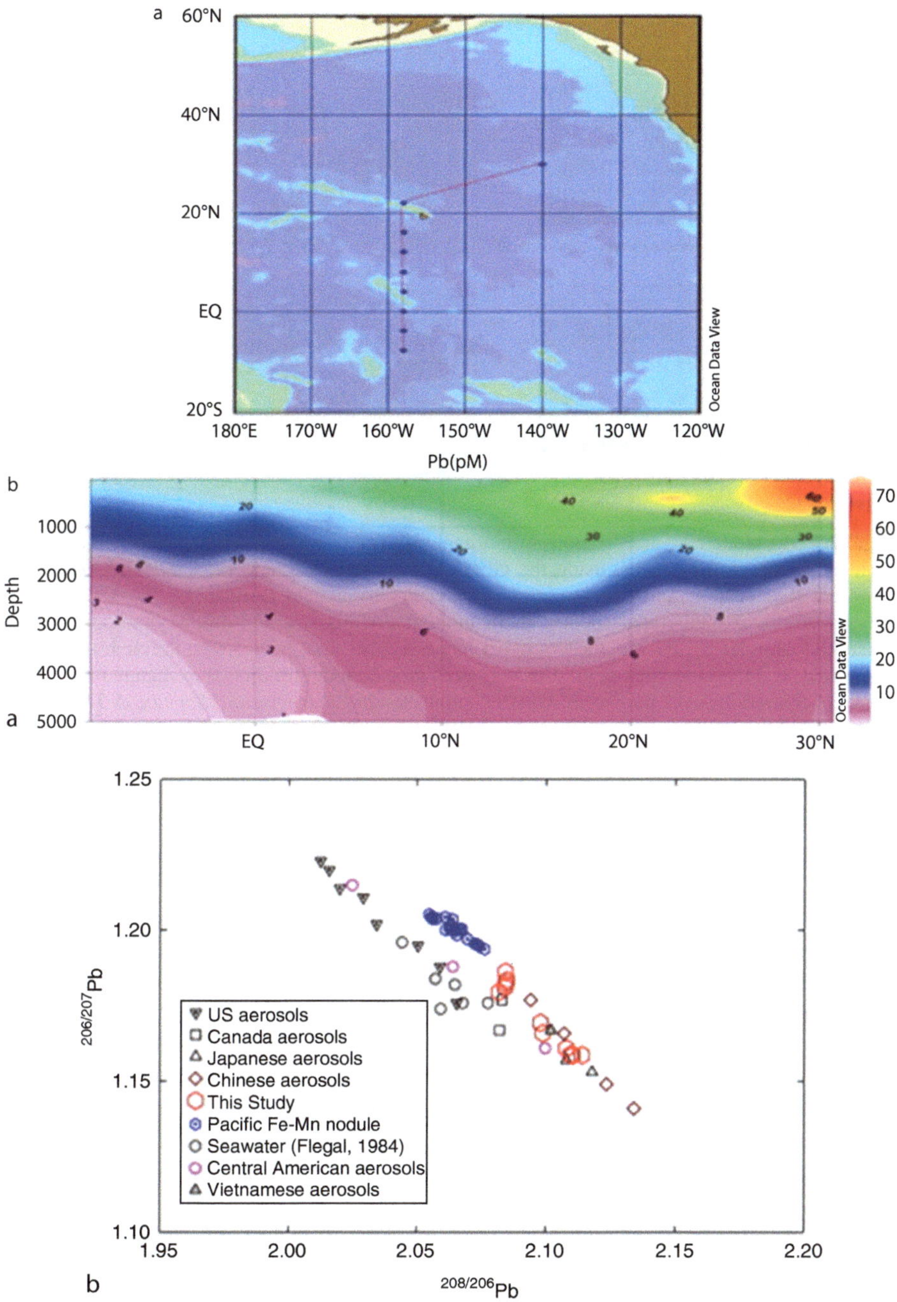

Fig. 12.8 (**a**) Cruise track and horizontal and vertical distributions of Pb collected during the cruise; (**b**) Relationship between the various isotope ratios of Pb for the samples collected during the cruise with comparison to other materials for which isotopes have been measured, and which could be sources for Pb in this region. Taken from J.F. Wu, R. Rember et al. "Isotopic evidence for the source of lead in the North Pacific abyssal water" published in *Geochimica et Cosmochimica Acta* volume 74(16), pages 4629–4638. Copyright (2010), and published with permission from Elsevier

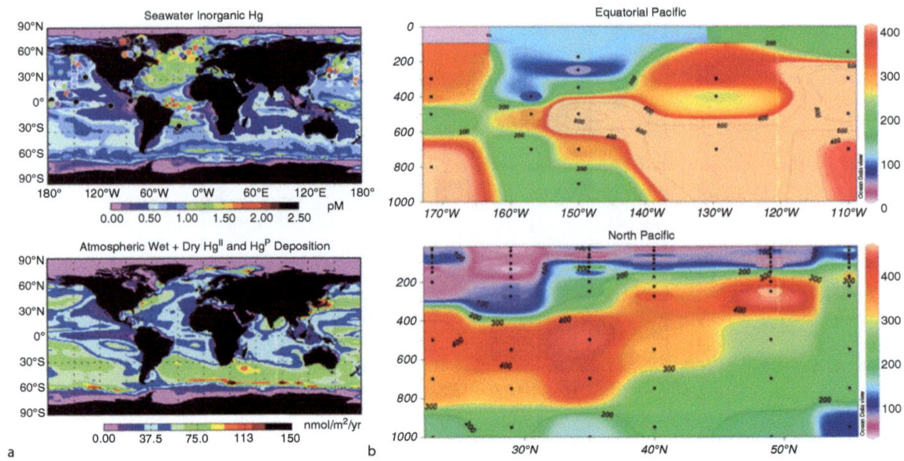

Fig. 12.9 (**a**) Modeled concentrations of inorganic Hg and of wet and dry deposition to the ocean surface for the global oceans, using the GEOS-Chem model framework. Circles in the figures represent the location and concentrations obtained during field campaigns; (**b**) Data for the concentrations of methylated species (both monomethylmercury and dimethylmercury) in the equatorial and North Pacific Ocean upper waters. The first figure was derived from A.L. Soerensen, E. M. Sunderland, et al. "An improved global model for air-sea exchange of mercury: High concentrations over the North Atlantic" published in *Environmental Science & Technology* volume 44(22), pp. 8574–8580. Copyright (2010) and reprinted with permission from the American Chemical Society. The second figure was compiled by Elsie Sunderland of Harvard University using information in Mason and Fitzgerald [68] and Sunderland et al. [101]

signal reflective of inputs via particle settling and from the anthropogenic enrichment of atmospheric deposition.

Most of the input of Pb to the ocean is atmospheric (Table 12.1) and the distribution of Pb in the ocean typically has a maximum concentration in the surface ocean and lower concentrations at depth (Fig. 12.8) [107], as well as higher concentrations in the Atlantic Ocean compared to the Pacific. The concentrations in the upper waters of the North Atlantic in the vicinity of Bermuda have been decreasing in the last 30 years (see Fig. 12.2) [106] from values above 150 pM in the late 1970s to concentrations around 50 pM today. It can be concluded that while the surface waters near Bermuda have decreased in concentration they are still substantially elevated above background and reflect the continual input of Pb from combustion and other industrial sources.

For Hg, both box and numerical models of the global cycle support the notion that anthropogenic releases of mercury (Hg) to the environment have impacted the biosphere substantially, enhancing deposition by a factor of 3–5, and influencing ocean concentrations ([39, 74, 70, 96, 102]). Besides atmospheric inputs, local and regional-scale contamination of the coastal zone by Hg has occurred due to runoff from the terrestrial environment and from point source inputs [103, 117]. Mercury distributions in surface waters reflect the magnitude of the atmospheric deposition source and the strength of local removal process (scavenging and gas evasion) superimposed on water circulation [63, 101, 102]. In some regions, concentrations change seasonally depending on the variability of atmospheric deposition, evasion, and removal of Hg by particulate sinking. Recent model output shows the potential variability that could exist for various Hg species in the ocean (Fig. 12.9a shows modeled inorganic Hg distributions). Concentrations vary between ocean basins due to differences in the relative impact of anthropogenic Hg in deposition (Fig. 12.9a), and these are changing over time – recently, more anthropogenic inputs from Asia and less from North America and Europe. Surface water concentrations are higher in the Atlantic Ocean and Mediterranean Sea than in the Pacific Ocean [25–27, 68, 69, 71, 73, 101, 118], and there is evidence for recent decreases in concentration in the North Atlantic and the Mediterranean Sea and these are consistent with model predictions [97, 98, 102].

The upper ocean Hg concentration can be transient, changing on short timescales due to changes in atmospheric inputs and seasonal water mixing [63]. For example, the input of Hg from the atmosphere was demonstrated through an increase in the concentration of Hg in the waters of the North Pacific Ocean seasonal mixed layer during the summer. However, this signal is eliminated as a result of deep water mixing in the fall. Seasonal mixing and latitudinal transport of sinking surface water within the permanent thermocline is a mechanism for the transport of Hg deposited at higher latitudes, which may have been deposited to the ocean a decade or more previously [85] to tropical and other regions ([72, 102]). The higher concentrations in the more recent data suggest the input of Hg to these mid-depth waters, and this coincides with an increasing anthropogenic signal from Asia, and such changes are not evident in the deep ocean waters [63, 101]. Such upper ocean cycling confounds the understanding of how the Hg concentration in ocean surface waters has changed as a result of increased anthropogenic inputs, and model predictions suggest that the response time of the ocean to changes in atmospheric Hg concentrations is decadal or longer [97].

The distribution of Hg in ocean waters reflects the sources and cycling as well as the internal cycling of mercury, methylation, demethylation, oxidation, and reduction. The biological production and destruction of methylated Hg species, primarily CH_3Hg and $(CH_3)_2Hg$ in the ocean, is important as the methylated forms are more toxic. In numerous profiles there appears to be an enhancement in the concentration of methylated Hg (CH_3Hg and $(CH_3)_2Hg$) at mid-depth [52, 101, 119, 122] (Fig. 12.9b). The analysis of the data and the relationships to environmental parameters suggest that the profiles can be best explained in terms of production of methylated Hg during the decomposition and remineralization of organic matter [52, 101]. The more recent studies examining the production of methylated Hg suggests that the presence of methylated Hg in low oxygen waters is more due to the same factors that cause the oxygen depletion – heightened bacterial activity, and slow vertical mixing, lack of water ventilation, and particulate scavenging – than to the activity of particular microorganisms [67, 101]. In freshwater and coastal environments and sediments, sulfate and iron-reducing bacteria have been demonstrated to be the most important methylating organisms [8, 120]. While little is known about the microorganisms or processes whereby Hg is methylated in the ocean, the fact that methylation has been demonstrated in relatively oxygenated waters, and the dominance in many instances of $(CH_3)_2Hg$, suggests that the pathways may be very different from those for freshwater and coastal environments, and in sediments. The net concentration of methylated Hg reflects the complex interactions that occur throughout the water column: production and destruction of methylated Hg by microbial processes and abiotic mechanisms; scavenging and release from particles; and uptake into the food chain.

Overall, there are strong similarities between the three major heavy metals in terms of their strong association with organic matter and the particulate phase. Mercury and Ag have similarities in that they are both Class B metals and form strong associations with reduced sulfide, and exist in seawater as chloride complexes in the absence of NOM. Lead can also be associated with sulfides and forms strong complexes, but to a lesser degree. Also, Pb does not form strong chloride complexes and is found as the neutral $PbCO_3$ complex in the absence of NOM (Table 12.1). Atmospheric sources are all important for these metals to the open ocean, and this is primarily related to the fact that their global cycles have been substantially impacted by anthropogenic sources. Besides Hg, the accumulation and fate of these elements in ocean microorganisms and in the oceanic food chain has been little examined and this could be an important area for future research.

Metalloids and Other Important Elements

The main metalloids of interest in marine systems (e.g., As, Se) exist as oxyanions, and in a number of oxidation states (Table 12.1), although they are also found as methylated compounds, or even as larger metalloid-containing species, such as arsenobetaine and selenoproteins. These organic species and their formation mechanisms are not detailed here but their biochemical formation is relatively well known. Arsenic can be found as either As^{III} or As^{V} and as mono-, di-, and tri-methyl arsenic in marine waters, and similar forms and speciation are found for Se and the other metalloids (Sb, Ge). It is thought that the methylation of As is a detoxification and elimination mechanism for As from phytoplankton as As^{V} can be taken up inadvertently by microorganisms in low phosphate waters (both exist as polyprotic acids with similar pKa's). Methylation involves pre-reduction of the As^{V} to As^{III} before being methylated. This is a different mechanism to Hg methylation by sulfate-reducing bacteria. There is also often As^{III} present in conjunction with phytoplankton in surface waters due to this reduction pathway, which is contrary to what is expected based on thermodynamic equilibrium calculations. However, As^{III} is a small fraction of the total, as are the methylated species [28, 29]. In estuarine environments and freshwaters, As^{III} and the methylated forms can be a larger fraction of the total dissolved As [53, 80].

The two main inorganic redox states of Se appear to cover a similar range in concentration, with deep waters having a ratio of Se^{IV}/Se^{VI} of >0.5, but <1. Both inorganic species appear to be depleted in the surface waters, likely due to their uptake and incorporation into biota. Selenium is an essential element although it is only required at low concentrations. Most of the bioorganic Se compounds are proteins and are therefore being continually produced through microbial processes.

Overall, there has been little recent study of the inorganic and organic speciation of the metalloids in the ocean water column. It is probable that new insights and understanding could be gained from the examination of the various fractions in more detail, especially the "organic fraction." It is not clear whether these compounds are derived directly from microorganisms and other biota, or are primarily produced during organic matter remineralization. It has however been shown that microorganisms contain small and large molecular weight metalloid-containing molecules.

Cycling of Organic Chemicals in the Ocean

There have been few detailed studies of the ocean distribution of organic chemicals and this relates both to their difficulty of measurement and the lack of research focus. However, the cycling and fate of other organic chemicals can be estimated based on the knowledge of these chemicals if their general chemical characteristics are known. The physiochemical parameters needed to predict their fate and transport include: (1) their solubility and volatility; (2) their Henry's Law Constant (K_H), which defines their relative solubility in water versus air; (3) their dissolved-particulate partition coefficient, typically normalized to particulate organic content (K_{OM}); and (4) their octanol-water partition coefficient (K_{OW}) which is a measure of the hydrophobicity of the compounds.

The solubility of organic compounds ranges widely but most compounds of environmental interest are relatively insoluble in seawater. Additionally, their environmental concentrations do not approach their saturated solubilities (from 10^{-2} M for small chain compounds to 10^{-11} M for some PCB congeners; [94]). Solubility is a strong function of the composition of the organic chemical as compounds that can be ionized in water due to the presence of acidic or basic groups (e.g., carboxylic acids, amine, and other N-containing groups, phenols and alcohols and thiols) will be more soluble that a similar compound without an ionizable group. The pH of seawater is around 8.2, it has high ionic strength and water molecules are polar and therefore ionizable compounds will tend to dissociate or be protonated to some degree as a result. Nonpolar compounds are the most hydrophobic as there is little interaction between the compounds and water.

Related to solubility is the air-water partition coefficient (K_H = partial pressure (P_i)/dissolved water concentration (C_W)), which for environmental science can be considered as a measure of the potential for a compound to be lost from solution via gas exchange. Compounds with high values are volatile, but the degree to which a compound would be lost from water to the air is a function of K_H and the relative concentration of the compound in each phase. Aliphatic hydrocarbons have

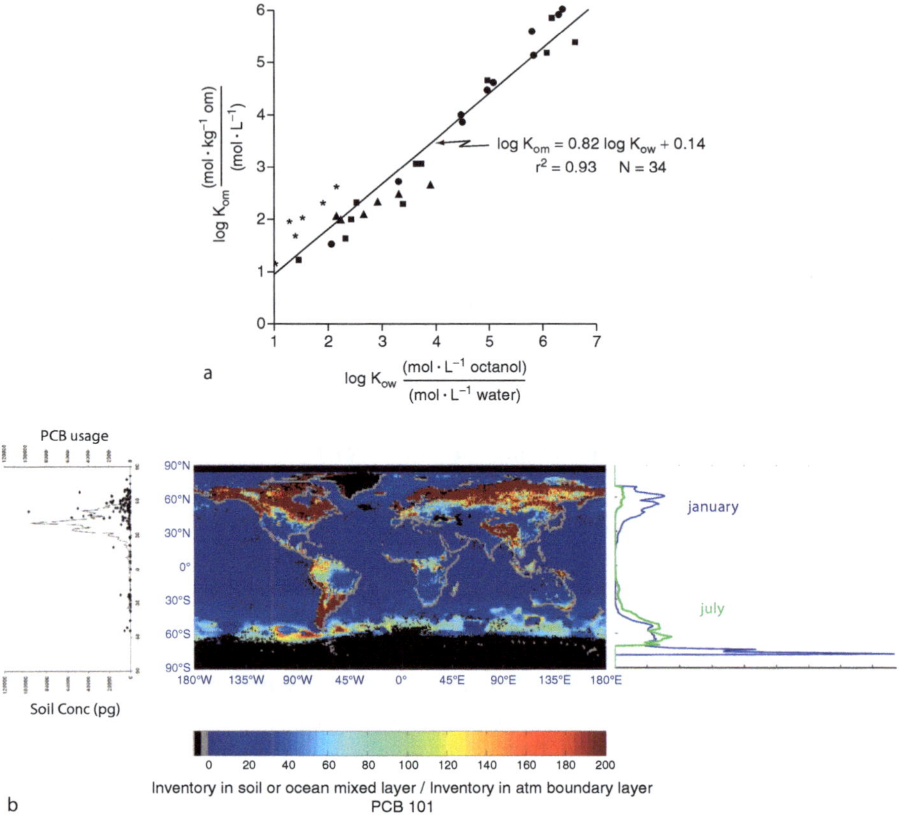

Fig. 12.10 (**a**) Relationship between the octanol-water partition coefficient and the dissolved-solid partition coefficient (normalized to organic content) for a variety of organic chemicals. Taken from R.P. Schwarzenbach, P. M. Gschwend, et al. "Environmental Organic Chemistry" published by John Wiley and Sons. Copyright (1993) and reproduced with permission; (**b**) the distribution of PCB in various reservoirs of the biosphere: the historical usage and soil concentrations are shown in the left panel; the model predictions of the maximum reservoir capacity in the center figure, with the right figure showing the particular distribution for the Atlantic Ocean. Taken from R. Lohmann, K. Breivik, J. Dachs and D. Muir "Global fate of POPs: Current and future research directions" published in Environmental Pollution, volume 150, pp. 150–165. Copyright (2007) and published with permission from Elsevier; (**c**) the global spatial variability for PCB 52 in the atmosphere (gas phase) (*top*), dissolved surface ocean (*middle*) and predicted air-sea and sinking fluxes (*bottom*). Taken from J. Dachs, R. Lohmann, W.A. Ockenden, L. Mejanelle, S.J. Eisenreich and K.C. Jones "Oceanic biogeochemical controls on global dynamics of persistent organic pollutants" published in Environmental Science and Technology, volume 36, pp. 4229–4237. Copyright (2002) and reprinted with permission from the American Chemical Society

relatively high K_H (~10^2–10^4 atm/M), PCBs intermediate (~0.01–1 atm/M), and PAHs lower values (~0.001–0.1 atm/M) [94].

The degree to which organic chemicals will remain dissolved in an environmental solution or partition into particulate phases, or become associated with high

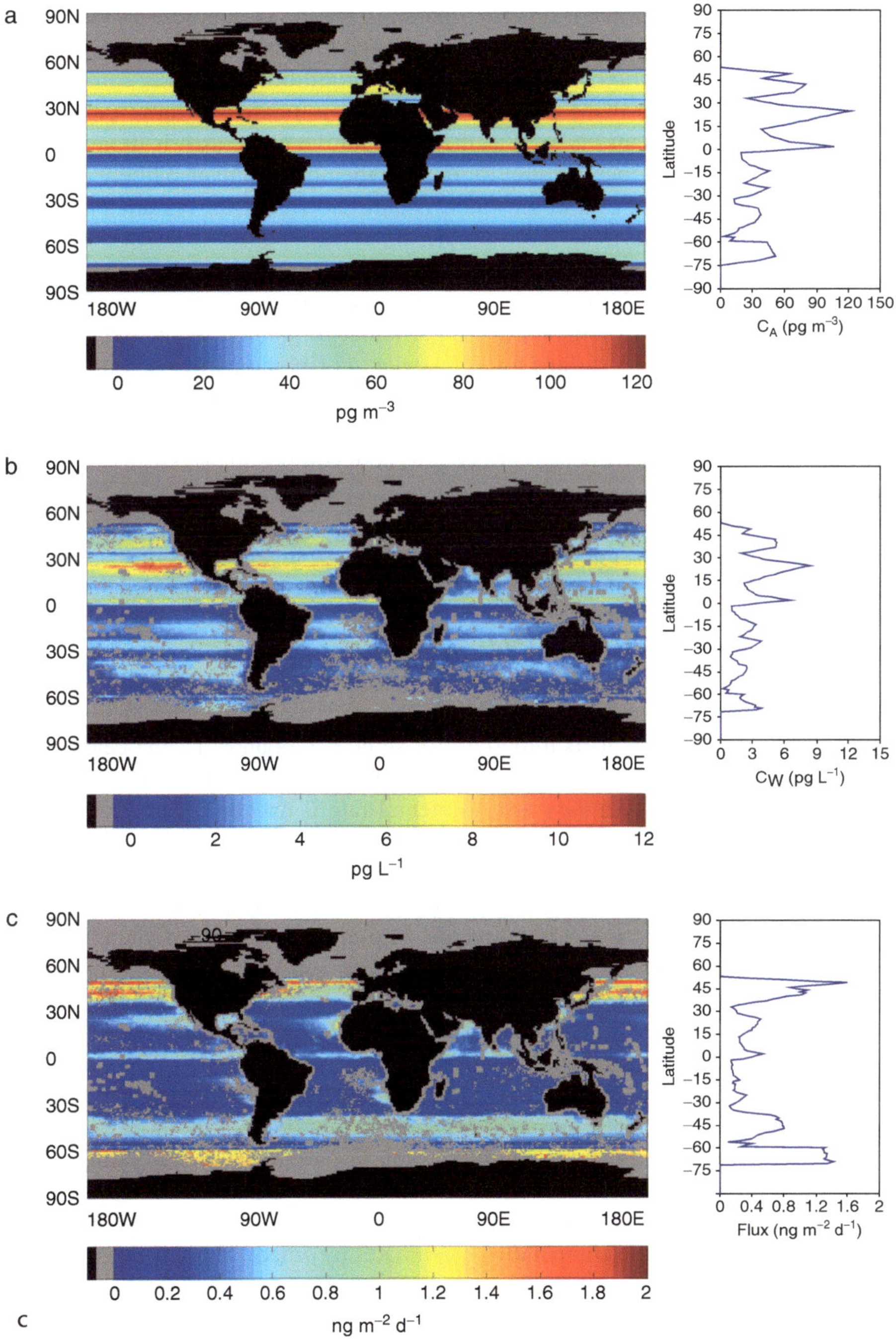

Fig. 12.10 (continued)

molecular weight dissolved organic material, such as humic substances, is a function of their hydrophobicity which is related to their solubility and is related to K_{OW}, which is the ratio of the concentration at equilibrium in water compared to octanol. The dissolved-POM partition coefficient is related to these parameters (Fig. 12.10a) [94]. Octanol is used as the comparative phase as it is thought to be a reasonable representative medium for the cell membrane and also likely provides a reasonable proxy for POM in seawater. Again, the values for organic chemicals range over a wide scale. The more soluble the compound, the smaller is the K_{OW} and K_{OM}.

As POPs are hydrophobic, they also attach to particles in the water, and the degree of association is a function of the particulate organic content, and therefore can be removed from the surface to deep ocean via sinking particles, especially in regions of high primary productivity. Thus, the factors controlling the overall distribution and fate of the more volatile POPs are similar to that of Hg. The degree to which the deep ocean is being enriched in these compounds is related to the rate of additional from above compared to their rate of removal to the deep sediment. As most of the organic matter sinking into the deeper ocean is remineralized, it is likely that the degree of incorporation of POPs into deep ocean sediments is less than that for metals and other compounds that form relatively strong associations with mineral phases.

The oceans play an important role in controlling the environmental transport, fate, and sinks of many POPs at regional and global scales [50, 58]. While the atmosphere is the most important global transport mechanism for many POPs, transport via ocean currents is important for some compounds. Deep ocean waters are usually considered the final sink for POPs, although an evaluation of the importance of their role in their environmental fate is uncertain, and modeling of their fate in the deep ocean has not been considered in a detailed fashion. POPs bound to sinking POM are exported from the mixed layer and will either accumulate in the deep ocean waters, as do the nutrients and some elements (Ag, Al), or will be finally removed to deep sediments. Marine sediments, and potentially the continental shelf could represent an important reservoir for POPs [59]. Gustafsson et al. [51] examined the settling fluxes of PCBs in the coastal and offshore North Atlantic Ocean by measuring their concentrations on particles and using ^{234}Th to estimate particle settling velocity. As may be suspected, the ratio of the flux to the sediment surface compared to that of the export flux from the mixed layer decreased with increasing water depth, and with the aqueous solubility of the PCB congeners, suggesting overall release from the dissolution of particles with increasing depth. Depending on the individual congener, the flux rates ranged from <1 to 10 pmol m^{-2} day^{-1} at the offshore stations to almost 100 pmol m^{-2} day^{-1} at the nearshore locations. Such fluxes are comparable to estimates by others. These authors inferred from the PCB distributions that atmospheric input was dominant source to the open ocean locations, and that particle scavenging and removal from the mixed layer was a more important sink for these compounds than photochemical decomposition in the marine boundary layer. Given the larger surface area of the open ocean environment, the flux in this region was a bigger fraction of the total input to deeper waters than the more nearshore waters, even though the fluxes in the shallow regions were higher [51].

Deep ocean sediments are likely important for many POPs. The relative importance of removal to the deep ocean and sediments versus reemission to the atmosphere will be different for different chemicals. Records of POP concentrations in deep ocean sediments are sparse and therefore it is difficult to evaluate for many compounds their degree of removal to the deep sediment. Estimates for $\sum$DDT are that 4–6% of these compounds have been removed to the deep ocean sediments and that most of the amount added to the ocean during their use is either still resident in the ocean or returned to the atmosphere.

Water column distributions and knowledge of the inter-basin variability in concentration of organic contaminants is much less well known that that of nutrients and metals but there are some similarities in the chemistry and partitioning that allows inferences to be made concerning their fate and transport in the ocean and the main processes controlling their internal cycling. Many of the POPs are sufficiently volatile and insoluble that air-sea gas exchange is an important part of their global cycle and determines to a large degree the extent of their accumulation and persistence in the ocean. For compounds that are relatively volatile, their upper ocean concentration is determined by the rate of input relative to the rate of re-emission to the atmosphere. For example, for PCBs, their upper ocean concentration is determined primarily by the gas exchange component with wet deposition and dry deposition of PCBs being relatively minor components [30]. The impact of the historical extent of emissions and atmospheric concentration on the ocean distribution of PCBs is illustrated with the model data shown in Fig. 12.10b. This further illustrates the importance of gas exchange on controlling concentrations as for the open ocean atmospheric particulate concentrations are low and these aerosols are dominated by large sea salt particles which are not effective is trapping POPs given their low organic content. It has been estimated that approximately 10% of the global releases of PCBs resides in the upper ocean [16], and similarly the upper ocean is a major reservoir for other compounds such as $\sum$DDT [99] and PFOS and their precursors.

Modeling studies suggest that the ocean, particularly in the Northern Hemisphere, was a net sink for $\sum$DDT during the period of its heightened use but that it is now a source as the banning and limited use of DDT has lead to a decrease in the atmospheric burden and as a result a degassing of $\sum$DDT from the ocean surface to the atmosphere [99]. The change from sink to source was slowest for the Northern Hemisphere as the air concentration in the region was buffered to a larger degree by re-emission of these chemicals from the terrestrial environment. This is because of both the relatively higher inputs of these chemicals in the Northern Hemisphere as well as the higher relative land mass in the north. The model results suggest that the ocean surface in the Southern Hemisphere became a source to the atmosphere in the late 1980s/early 1990s while the situation was delayed for about a decade in the Northern Hemisphere. The ocean gyre regions became a source more rapidly that regions of higher atmospheric inputs due to enhanced wet deposition, or regions with higher productivity as the enhanced removal of these compounds from the surface ocean to deep in association with particulate sinking delayed the buildup in surface concentrations relative to those in the atmosphere and resulted in a longer persistence of ocean uptake. This example again illustrates the importance of ocean circulation and primary productivity as factors that have a large influence over the overall fate and cycling of POPs.

Given the lack of data for many chemicals in the ocean, there have been several efforts over the years to model the systems to understand the fate and transport of these chemicals. Wania and Mackay [105] developed a zonal climatically differentiated nonequilibrium fugacity mass balance model for POPs including air, fresh and ocean water, and sediments/soils with advective connections to examine the main processes involved in the portioning and distribution of these chemicals across environmental media. Not surprisingly, temperature and chemical reactivity were highly important variables, as were the transport rates and the chemical composition, such as the media organic content and other physicochemical properties (e.g., solubility, K_{OW}, vapor pressure). Dachs et al. [30] further examined the biogeochemical controls over the cycling of POPs in the ocean and demonstrated the importance of air-sea exchange, and as a result, temperature in the cycling of the volatile components across this interface. Additionally, given the importance of particulate removal in the cycling of POPs in the upper ocean as a sink, and air-sea exchange as a potential source or sink, they determined that influx of gaseous POPs to the ocean was actually higher in productive regions as the particulate removal depleted the surface layers and resulted in net deposition of gaseous species from the atmosphere. This is shown in Fig. 12.10c, which shows the model estimated parameters for gaseous concentrations over the ocean, predicted dissolved surface ocean concentrations, and the extent of air-sea exchange for a particular PCB congener. These results show that fluxes are into the ocean overall and are driven by the atmospheric concentration.

Overall, the results of the limited studies of POPs in the ocean indicate that their cycling is controlled by their degree to which gas exchange at the sea surface is important in their geochemical cycling. The more soluble compounds will be less impacted by gas exchange and the more hydrophobic compounds will attach to particles and be removed from the surface ocean more efficiently, driving further gas exchange input. Overall, while many of these chemicals have been banned in many parts of the world, or have now more restricted usage, there is a strong legacy remaining in the biosphere because of their buildup in the terrestrial soils, and because of their overall volatility, they can be reemitted to the atmosphere and continually contribute to the contamination of the ocean. Finally, it is clear that there are many similarities in the cycling and ocean biogeochemistry of elements and organic chemicals and that productivity and carbon export from the surface ocean are main drivers on upper ocean concentrations and the extent to which the ocean is a net sink. For many chemicals, buildup is occurring in the deep ocean as a result of human activity and this legacy will remain for an extended period given the slow deep cycling of the ocean waters.

Future Directions

The above discussions on the cycling of pollutant elements and chemicals in the ocean indicates that there has been both an increased awareness of the impact of human activities on the ocean and on its environmental health, as well as on that of the organisms that reside in the ocean, and on the consumers of ocean seafood, including and particularly humans. There has been a notable increase in the amount of study and regulation globally concerning the ocean and many nations are involved currently in assessments that will lead to further understanding and regulation. One such activity is the international GEOTRACERS program which is designed to measure the concentrations and distributions of a whole suite of metals and isotopes in all the major ocean basins, and this activity in its initial few years has produced many interesting and exciting results that are mentioned above. Additionally, there is a continual improvement in the analytical and sampling technology, and the development of continuous or in situ samplers for a variety of chemicals that are increasing enormously the amount of information gathered on chemicals in the ocean. Previously, ship time and resources have been a serious limiting factor but the developments of these new technologies are increasing the databases on chemicals in an exponential fashion, and there is now a need to develop extensive methods to compare and contrast and validate the measurements being made by a suite of countries around the globe.

The increasing information and knowledge that is occurring in terms of the elements is not happening to the same extent for the organic chemicals. This is the result of the fact that current analytical techniques are still time consuming and their detection limits are often comparable to the concentrations that are being measured. Further improvement in analytical methods is needed to further enhance the databases on these chemicals, which is still very sparse, especially for open ocean and deep waters. Furthermore, there is a constant development of new chemicals by industry and therefore while the understanding of past chemicals and their fate and transport are becoming better understood, there is little information on the newer chemicals released into the environment. To be able to keep pace with such developments and changes, it is necessary to develop an understanding of the underlying fundamental processes that control organic chemical fate and transport and transformation in the ocean. Only with such understanding can reliable predictions be made; currently most understanding is developed in an empirical fashion.

Model development and application can provide insights where there is a lack of data but the models themselves are only as good as their extent of validation, and so while it is possible to a degree to use models to fill gaps in data and knowledge, there is also a need to collect data and to constantly improve the models through validation and comparison with the data. Only through such interactions can our understanding of the complex cycling of chemicals in the ocean be further improved.

Bibliography

Primary Literature

1. Akeredolu FA et al (1994) The flux of anthropogenic trace-metals into the arctic from the mid-latitudes in 1979/80. Atmos Environ 28:1557–1572
2. Anderson RF (2004) Chemical tracers of particle transport. In: Enderfield H (ed) The oceans and marine geochemistry, vol 6. Elsevier, Amsterdam, pp 247–273
3. Arimoto R et al (1995) Trace-elements in the atmosphere over the North Atlantic. J Geophys Res Atmos 100(D1):1199–1213
4. Arimoto R, Duce RA, Ray BJ, Tomza U (2003) Dry deposition of trace elements to the western North Atlantic. Global Biogeochem Cycles 17(1)
5. Arimoto R, Gao Y, Zhou M-Y, Soo D, Chen L, Gu D, Wang Z, Zhang X (1997) Atmospheric deposition of trace elements to the western Pacific Basin. In: Baker J (ed) Atmospheric deposition of contaminants to the great lakes and to coastal waters. SETAC Press, Pensacola, pp 195–208
6. Baker J, Poster DL, Clark CA, Church TM, Scudlark JR, Ondov JM, Dickhut RM, Cutter G (1997) Loadings of atmospheric trace elements and organic contaminants to the Chesapeake Bay. In: Baker J (ed) Atmospheric deposition of contaminants to the great lakes and coastal waters. SETAC Press, Pensacola, pp 171–194
7. Barbante C et al (2004) Historical record of European emissions of heavy metals to the atmosphere since the 1650s from alpine snow/ice cores drilled near Monte Rosa. Environ Sci Technol 38(15):4085–4090
8. Benoit JM, Gilmour CC et al (2003) Geochemical and biological controls over methylmercury production and degradation in aquatic ecosystems. Biogeochem Environ Important Trace Elem 835:262–297
9. Berg T, Steinnes E (2005) Atmospheric transport of metals. In: Sigel A, Sigel H, Sigel RKO (eds) Biogeochemistry, availability and transport of metals in the environment, vol 44, Metal ions in biological systems. Taylor & Francis, Boca Raton
10. Bergan T, Gallardo L, Rodhe H (1999) Mercury in the global troposphere: a three-dimensional model study. Atmos Environ 33(10):1575–1585
11. Billen G, Garnier J (2007) River basin nutrient delivery to the coastal sea: assessing its potential to sustain new production of non-siliceous algae. Mar Chem 106(1–2):148–160
12. Boesch DF, Burroughs RH, Baker JE, Mason RP, Rowe CL, Siefert RL (2001) Marine pollution in the United States. Pew Oceans Commission Report, Washington, DC
13. Boutron C et al (2004) Anthropogenic lead in polar snow and ice archives. C R Geosci 336(10):847–867
14. Boyle EA, Sherrell RM et al (1994) Lead variability in the western North Atlantic Ocean and central Greenland ice – implications for the search for decadal trends in anthropogenic emissions. Geochim Cosmochim Acta 58(15):3227–3238
15. Boyle EA (2010) In: Steele JH, Thorpe SA, Turekian KK (eds) Marine chemistry and geochemistry. Elsevier, Amsterdam, pp 273–280
16. Breivik K, Wania F (2002) Mass budgets, pathways, and equilibrium states of two hexachlorocyclohexanes in the Baltic Sea environment. Environ Sci Technol 36(5):1024–1032
17. Broecker WS, Peng T-H (1982) Tracers in the sea. Eldigio Press, New York
18. Bruland K, Lohan MC (2004) Controls of trace metals in seawater. In: Elderfield H (ed), Holland HD, Turekian KK (Exec eds) The oceans and marine geochemistry, vol 6. Amsterdam, Elsevier, pp 23–47
19. Bruland KW, Coale KH (1986) Surface water 234Th/238U disequilibria: spatial and temporal variations of scavenging rates within the Pacific Ocean. In: Burton JD, Brewer PG, Chester R (eds) Dynamic processes in the chemistry of the upper ocean. Plenum Press, New York, pp 159–172

20. Callender E (2005) Heavy metals in the environment – historical trends. In: Lollar B, Holland HD, Turekian KK (eds) Environmental geochemistry, vol 5, Treatise on geochemistry. Elsevier, Amsterdam, pp 67–105
21. Chester R (2003) Marine geochemistry. Blackwell, Malden
22. Clegg S, Whitfield M (1990) A generalized model for the scavenging of trace metals in the open ocean–I. Particle cycling. Deep Sea Res 37:809–832
23. Conley DJ, Schelske CL et al (1993) Modification of the biogeochemical cycle of silica with eutrophication. Mar Ecol Prog Ser 81:121–128
24. Cossa D, Averty B et al (2009) The origin of methylmercury in open Mediterranean waters. Limnol Oceanogr 54(3):837–844
25. Cossa D, Cotte-Krief MH et al (2004) Total mercury in the water column near the shelf edge of the European continental margin. Mar Chem 90(1–4):21–29
26. Cossa D, Martin J-M, Takayanagi K, Sanjuan J (1997) The distribution and cycling of mercury species in the western Mediterranean. Deep Sea Res 44:721–740
27. Cossa D, Michel P et al (1992) Vertical mercury profile in relation to arsenic, cadmium and copper at the eastern North Atlantic ICES reference station. Oceanol Acta 15:603–608
28. Cutter GA (2010) Metalloids and oxyanions. In: Steele JH, Thorpe SA, Turekian KK (eds) Marine chemistry and geochemistry. Elsevier, Amsterdam, pp 64–71
29. Cutter GA, Cutter LS (2006) The biogeochemistry of arsenic and antimony in the North Pacific Ocean. Geochem Geophys Geosyst 7, doi:2007JD009750/2005GC001159
30. Dachs J, Lohmann R, Ockenden WA, Mejanelle L, Eisenreich SJ, Jones KC (2002) Oceanic biogeochemical controls on global dynamics of persistent organic pollutants. Environ Sci Technol 36:4229–4237
31. Donat J, Dryden C (2010) Transition metals and heavy metal speciation. In: Steele JH, Thorpe SA, Turekian KK (eds) Marine chemistry and geochemistry. Elsevier, Amsterdam, pp 72–80
32. Donat JR, Bruland KW (1995) Trace elements in the ocean. In: Salbu B, Steinnes E (eds) Trace metals in natural waters. CRC Press, Boca Raton, pp 247–281
33. Duce RA (2010) In: Steele JH, Thorpe SA, Turekian KK (eds) Marine chemistry and geochemistry. Elsevier, Amsterdam, pp 281–290
34. Duce R, Liss PS, Merrill JT et al (1991) The atmospheric input of trace species to the world ocean. Global Biogeochem Cycles 5:193–259
35. Dupont CL, Nelson RK, Bashir S, Moffett JW, Ahner BA (2004) Novel copper-binding and nitrogen-rich thiols produced and exuded by *Emiliania huxleyi*. Limnol Oceanogr 49:1754–1762
36. Erel Y, Patterson CC (1994) Leakage of industrial lead into the hydrocycle. Geochim Cosmochim Acta 58(15):3289–3296
37. Farrington JW (2010) Chlorinated hydrocarbons. In: Steele JH, Thorpe SA, Turekian KK (eds) Marine chemistry and geochemistry. Elsevier, Amsterdam, pp 163–174
38. Fine RA (2010) CFCs in the ocean. In: Steele JH, Thorpe SA, Turekian KK (eds) Marine chemistry and geochemistry. Elsevier, Amsterdam, pp 155–162
39. Fitzgerald WF, Lamborg CH, Hammerschmidt CR (2007) Marine biogeochemical cycling of mercury. Chem Rev 107:641–662
40. Fitzgerald WF, Mason RP (1996) The global mercury cycle: oceanic and anthropogenic aspects. In: Baeyens W et al (eds) Global and regional mercury cycles: sources, fluxes and mass balances. Academic, Netherlands, pp 85–108
41. Gabrielli P et al (2005) Variations in atmospheric trace elements in Dome C (East Antarctica) ice over the last two climatic cycles. Atmos Environ 39(34):6420–6429
42. Gaillardet J, Viers J, Dupre B (2004) Trace elements in river waters. In: Holland H, Turekian KK (eds) Treatise on geochemistry, vol 5. Elsevier, Amsterdam, pp 225–272
43. German CA, Von Damm KL (2004) Hydrothermal processes. In: Elderfield H (ed) The oceans and marine geochemistry, vol 6, Treatise on geochemistry. Elsevier, Amsterdam, pp 181–222

44. GESAMP, Bowmer T, Kershaw PT (eds) (2010) Proceedings of the GESAMP international workshop on microplastic particles as a vector in transporting persistent, bioaccumualtive and toxic substances in the ocean, Paris. GESAMP Report # 82, 68pp
45. GESAMP (2001) A sea of troubles. GESAMP, The Hague. GESAMP Report # 70, 40pp
46. GESAMP (1993) Impact of oil and related chemicals on the marine environment. GESAMP, London. GESAMP Report # 50, 180pp
47. GESAMP (1990) Review of potential harmful substances – choosing priority organochlorines for marine hazard assessment. Food and Agriculture Organization of the United Nations, Rome. GESAMP Report # 42, 24pp
48. GESAMP (1989) The atmospheric input of trace species to the World's ocean. World Meteorological Organization, Geneva. GESAMP Report # 38, 124pp
49. Gioia R et al (2008a) Polychlorinated biphenyls in air and water of the North Atlantic and Arctic Ocean. J Geophys Res 113(D19302). D19302 doi:10.1029/2007JD009750
50. Gioia R et al (2008) Polychlorinated biphenyls (PCBs) in air and seawater of the Atlantic Ocean: sources, trends and processes. Environ Sci Technol 42:1416–1422
51. Gustafsson O, Gschwend PM, Buesseler KO (1997) Settling removal rates of PCBs into the northwestern Atlantic derived from ^{238}U–^{234}Th disequilibria. Environ Sci Technol 31:3544–3550
52. Heimburger LE, Cossa D et al (2010) Methyl mercury distributions in relation to the presence of nano- and picophytoplankton in an oceanic water column (Ligurian Sea, North-western Mediterranean). Geochim Cosmochim Acta 74(19):5549–5559
53. Hellweger FL (2005) Dynamics of arsenic speciation in surface waters: As(III) production by algae. Appl Organomet Chem 19(6):727–735
54. Herzig PM, Hannington MD (2000) Input from the deep: hot vents and cold seeps. In: Schultz H, Zabel M (eds) Marine geochemistry. Springer, Heidelberg, pp 397–416
55. Hong SM, Candelone JP, Soutif M, Boutron CF (1996) A reconstruction of changes in copper production and copper emissions to the atmosphere during the past 7000 years. Sci Total Environ 188(2–3):183–193
56. Hopkinson BM, Roe KL et al (2008) Heme uptake by *Microscilla marina* and evidence for heme uptake systems in the genomes of diverse marine bacteria. Appl Environ Microbiol 74(20):6263–6270
57. HTAP, Dutchak S, Zuber A (eds) (2010) Hemispheric transport of air pollutants part c: persistent organic pollutants. ECE, Geneva. Air Pollution Studies #19
58. Iwata H et al (1994) Geographical distribution of persistent organochlorines in air, water and sediments from Asia and Oceania, and their implications for global redistribution from lower latitudes. Environ Pollut 85(1):15–33
59. Jonsson A, Gustafsson O, Axelman J, Sunberg H (2003) Global accounting of PCBs in the continental shelf sediments. Environ Sci Technol 37:245–255
60. Kim G, Alleman LY, Church TM (1999) Atmospheric depositional fluxes of trace elements, Pb-210, and Be-7 to the Sargasso Sea. Global Biogeochem Cycles 13(4):1183–1192
61. Krishnaswami S (2005) Uranium-thorium series isotopes in ocean profiles. In: Steele JH, Thorpe SA, Turekian KK (eds) Marine chemistry and geochemistry. Elsevier, Amsterdam, pp 214–224
62. Landing W, Bruland KH (1987) The contrasting biogeochemistry of iron and manganese in the Pacific Ocean. Geochim Cosmochim Acta 51:29–43
63. Laurier FJG, Mason RP et al (2004) Mercury distributions in the North Pacific Ocean – 20 years of observations. Mar Chem 90(1–4):3–19
64. Lohmann R, Breivik K, Dachs J, Muir D (2007) Global fate of POPs: current and future research directions. Environ Pollut 150:150–165
65. Luther GW (2005) Manganese(II) oxidation and Mn(IV) reduction in the environment – two one-electron transfer steps versus a single two-electron step. Geomicrobiol J 22(3–4):195–203

66. Luther GW (2010) The role of one- and two-electron transfer reactions in forming thermodynamically unstable intermediates as barriers in multi-electron redox reactions. Aquat Geochem 16(3):395–420
67. Malcolm EG, Schaefer JK et al (2010) Mercury methylation in oxygen deficient zones of the oceans: no evidence for the predominance of anaerobes. Mar Chem 122:11–19
68. Mason RP, Fitzgerald WF (1993) The distribution and biogeochemical cycling of mercury in the equatorial Pacific Ocean. Deep Sea Res 40(9):1897–1924
69. Mason RP, Fitzgerald WF (1996) Sources, sinks and biogeochemical cycling of mercury in the ocean. In: Baeyens W et al (eds) Global and regional mercury cycles: sources, fluxes and mass balances. Academic, Netherlands, pp 249–272
70. Mason RP, Fitzgerald WF et al (1994) The biogeochemical cycling of elemental mercury: anthropogenic influences. Geochim Cosmochim Acta 58(15):3191–3198
71. Mason RP, Lawson NM et al (2001) Mercury in the Atlantic Ocean: factors controlling air-sea exchange of mercury and its distribution in the upper waters. Deep Sea Res II 48:2829–2853
72. Mason RP, O'Donnell J et al (1994) Elemental mercury cycling within the mixed layer of the equatorial Pacific Ocean. In: Watras CJ, Huckabee JW (eds) Mercury as a global pollutant: towards integration and synthesis. Lewis, Boca Raton, pp 83–97
73. Mason RP, Sullivan KA (1999) The distribution and speciation of mercury in the south and equatorial Atlantic. Deep Sea Res Part II 46(5):937–956
74. Mason RP, Sheu GR (2002) Role of the ocean in the global mercury cycle. Global Biogeochem Cycles 16(4): Art. No. 1093
75. Maybeck M (2004) Global occurence of major elements in rivers. In: Drever JI (ed) Surface and ground water, weathering and soils, treatise on geochemistry, vol 5. Elesevier, Amsterdam, pp 207–223
76. Morel F, Milligan AJ, Saito MA (2004) Marine bioinorganic chemistry: the role of trace metals in the oceanic cycles of major nutrients. In: Elderfield H (ed) The oceans and marine geochemistry. In: Holland HD, Turekian KK (Exec eds) Treatise on geochemistry, vol 6. Elsevier/Pergamon, Amsterdam
77. Morel FMM, Hering JG (1993) Principals and applications of aquatic chemistry. Wiley, New York
78. Murnane R, Sarmiento J et al (1990) Thorium isotopes, particle cycling models, and inverse calculations of model rate constants. J Geophys Res 95:16,195–16,206
79. Ndung'u K, Thomas MA et al (2001) Silver in the western equatorial and South Atlantic Ocean. Deep Sea Res Part II 48(13):2933–2945
80. Nice AJ, Lung WS et al (2008) Modeling arsenic in the Patuxent estuary. Environ Sci Technol 42(13):4804–4810
81. Nozaki Y (1997) A fresh look at element distributions in the North Pacific. http://www.agu.org/eos_elec/97025e.html
82. Nriagu J (1978) Properties and the biogeochemical cycle of lead. In: Nriagu J (ed) The biogeochemistry of lead in the environment. Elsevier, Amsterdam, pp 1–14
83. Okbamichael M, Sanudo-Wilhelmy SA (2005) Direct determination of vitamin B-1 in seawater by solid-phase extraction and high-performance liquid chromatography quantification. Limnol Oceanogr Methods 3:241–246
84. Orians KJ, Merrin CL (2010) Refractory metals. In: Steele JH, Thorpe SA, Turekian KK (eds) Marine chemistry and geochemistry. Elsevier, Amsterdam, pp 52–63
85. Pickard GL, Emery WJ (1982) Descriptive physical oceanography. Pergamon Press, Oxford, 249pp
86. Pirrone N, Costa P, Pacyna JM, Ferrara R (2001) Mercury emissions to the atmosphere from natural and anthropogenic sources in the Mediterranean region. Atmos Environ 35(17):2997–3006
87. Pohl C, Loffler A, Schmidt M, Seifert T (2006) A trace metal (Pb, Cd, Zn, Cu) balance for surface waters in the eastern Gotland Basin, Baltic Sea. J Mar Syst 60(3–4):381–395

88. Ranville MA, Cutter GA et al (2010) Aeloian contamination of Se and Ag in the North Pacific Ocean from Asian fossil fuel combustion. Environ Sci Technol 44:1587–1593
89. Ranville MA, Flegal AR (2005) Silver in the North Pacific Ocean. Geochem Geophys Geosyst 6: Art. No. Q03M01
90. Ryaboshapko II, Gusev A, Afinogenova O, Berg T, Hjellbrekke A-G (1999) Monitoring and modeling of lead, cadmium and mercury transboundary transport in the atmosphere of Europe. EMEP, Meteorological Synthesizing Centre – East, Moscow
91. Sanudo-Wilhelmy SA, Flegal AR (1992) Anthropogenic silver in the southern California bight – a new tracer of sewage in coastal waters. Environ Sci Technol 26:2147–2151
92. Schaule B, Patterson CC (1983) Perturbations of the natural lead depth profile in the Sargasso Sea by industrial lead. In: Wong C, Boyle E, Bruland KW, Burton JD, Goldberg ED (eds) Trace metals in seawater. Plenum Press, New York, pp 487–503
93. Schlitzer R (2010) Inverse modeling of tracers and nutrients. In: Steele JH, Thorpe SA, Turekian KK (eds) Marine chemistry and geochemistry. Elsevier, Amsterdam, pp 188–199
94. Schwarzenbach RP, Gschwend PM et al (1993) Environmental organic chemistry. Wiley, New York
95. Scudlark J, Church TM (1997) Atmospheric deposition of trace metals to the mid-Atlantic bight. In: Baker J (ed) Atmospheric deposition of contaminants to the great lakes and coastal waters. SETAC Press, Pensacola, pp 195–208
96. Selin NE, Jacob DJ et al (2008) Global 3-D land-ocean–atmosphere model for mercury: present-day vs. preindustrial cycles and anthropogenic enhancement factors for deposition. Global Biogeochem Cycles (Accepted)
97. Selin NE, Sunderland EM et al (2010) Sources of mercury exposure for US seafood consumers: implications for policy. Environ Health Perspect 118(1):137–143
98. Soerensen AL, Sunderland EM et al (2010) An improved global model for air–sea exchange of mercury: high concentrations over the north Atlantic. Environ Sci Technol 44(22): 8574–8580
99. Stemmler I, Lammel G (2009) Cycling of DDT in the global oceans 1950–2002: world ocean returns the pollutant. Geophys Res Lett 36(L24602)
100. Stumm W, Morgan JJ (1996) Aquatic chemistry. Wiley, New York
101. Sunderland EM, Krabbenhoft DP et al (2009) Mercury sources, distribution, and bioavailability in the North Pacific Ocean: Insights from data and models. Global Biogeochem Cycles 23
102. Sunderland EM, Mason RP (2007) Human impacts on open ocean mercury concentrations. Global Biogeochem Cycles 21:GB4022. doi:10.1029/2006GB002876
103. USEPA (1997) The incidence and severity of sediment contamination in surface water of the United States. Office of Science and Technology, U.S. Environmental Protection Agency, Washington, DC, (Three volumes) http://water.epa.gov/polwaster/sediments/cs/upload/nsqszed-complete.pdf
104. Von Damm KL (2010) Hydrothermal vent fluids, chemistry of. In: Steele JH, Thorpe SA, Turekian KK (eds) Marine chemistry and geochemistry. Elsevier, Amsterdam, pp 81–88
105. Wania F, Mackay D (1995) A global distribution model for persistent organic chemicals. Sci Total Environ 160/161:211–232
106. Wu JF, Boyle EA (1997) Lead in the western North Atlantic Ocean: completed response to leaded gasoline phaseout. Geochim Cosmochim Acta 61(15):3279–3283
107. Wu JF, Rember R et al (2010) Isotopic evidence for the source of lead in the North Pacific abyssal water. Geochim Cosmochim Acta 74(16):4629–4638
108. Zhang Y, Amakawa H et al (2001) Oceanic profiles of dissolved silver: precise measurements in the basins of the western North Pacific, Sea of Okhotsk and the Japan Sea. Mar Chem 75:151–163
109. Zhang Y, Obata H et al (2004) Silver in the Pacific Ocean and the Bering Sea. Geochem J 38:623–633
110. Libes S (1992) Introduction to marine biogeochemistry. Elsevier, San Diego, 909 pp

111. Iwata H, Tanabe S, Sakal N, Tatsukawa R (1993) Distribution of persistent organochlorines in the oceanic air and surface seawater and the role of ocean on their global transport and fate. Environ Sci Technol 27:1080–1098
112. van der Loeff MMR (2010) Uranium-thorium decay series in the ocean: overview, Chapter 23. In: Steele JH, Thorpe SA, Turekian KK (eds) Marine chemistry and geochemistry. Elsevier, Amsterdam, pp 203–213
113. Lal D (2010) Cosmogenic isotopes, Chapter 25. In: Steele JH, Thorpe SA, Turekian KK (eds) Marine chemistry and geochemistry. Elsevier, Amsterdam, pp 225–234
114. Key RM (2010) Radiocarbon, Chapter 26. In: Steele JH, Thorpe SA, Turekian KK (eds) Marine chemistry and geochemistry. Elsevier, Amsterdam, pp 235–250
115. Rabalais NN (2010) Hypoxia, Chapter 34. In: Steele JH, Thorpe SA, Turekian KK (eds) Marine chemistry and geochemistry. Elsevier, Amsterdam, pp 306–314
116. Sunda WG (2010) Trace metal nutrients, Chapter 4. In: Steele JH, Thorpe SA, Turekian KK (eds) Marine chemistry and geochemistry. Elsevier, Amsterdam, pp 17–28
117. Balcom PH et al (2004) Mercury sources and cycling in the Connecticut River and Long Island Sound. Marine Chem 90(1–4):53–74
118. Gill GA, Fitzgerald WF (1988) Vertical mercury distributions in the oceans. Geochim Cosmochim Acta 52:1719–1728
119. Cossa D et al (2011) Mercury in the Southern Ocean. Geochim Cosmochim Acta 75 (14):4037–4052
120. Hollweg T, Gilmour C, Mason R (2009) Factors controlling the methylation of mercury in sediments of the Chesapeake Bay and mid-Atlantic continental shelf. Marine Chem
121. Mason RP (2005) Air-sea exchange and marine boundary layer atmospheric transformations of mercury and their importance in the global mercury cycle. In: Pirrone N, Mahaffey KR (eds) Dynamics of mercury pollution on regional and global scales. Springer, New York, pp 213–239
122. Mason RP, Fitzgerald WF (1990) Mercury speciation in the equatorial Pacific Ocean. EOS 71 (43):1413
123. Cutter GA (1993) Metalloids in wet deposition on Bermuda – concentrations, sources and fluxes. JGR-Atmos 98(D9):16777–16786

Chapter 13
Subsurface Fate and Transport of Chemicals

Frank T. Barranco Jr., Samantha L. Saalfield, Frederick J. Tenbus, and Brian P. Shedd

Glossary

Abiotic	Not relating to life, as in abiotic chemical reactions that occur independent of living organisms.
Absorption	Retention of a chemical within a solid material.
Adsorption	Adhesion of a chemical to the surface of a solid.
Advection	Transport of a solute within a fluid in the direction of the bulk fluid's flow.
Aerobic	Requiring oxygen.
Air sparging	Injecting air or oxygen into an aquifer.
Aliphatic	Organic compounds not containing an aromatic ring.
Anaerobic	Without oxygen.
Best management practice (BMP)	Techniques generally accepted as effective for achieving a particular goal, for example minimizing the environmental impact of remediation.
Biodegradation	Use of living organisms to clean up contaminated environmental media.
Biotic	Relating to life, as in a biotic reaction mediated by living organisms.
Bioventing	The addition of air (or oxygen) under, at times, an induced lowering of water table to promote aerobic biodegradation of subsurface contaminants in the unsaturated zone.

This chapter was originally published as part of the Encyclopedia of Sustainability Science and Technology edited by Robert A. Meyers. DOI:10.1007/978-1-4419-0851-3

F.T. Barranco Jr. (✉) • S.L. Saalfield • F.J. Tenbus • B.P. Shedd
EA Engineering, Science, and Technology, Inc., 15 Loveton Circle, Sparks, MD 21152, USA
e-mail: fbarranco@eaest.com; ssaalfield@eaest.com; ftenbus@eaest.com; bshedd@eaest.com

J.S. Gulliver (ed.), *Transport and Fate of Chemicals in the Environment: Selected Entries from the Encyclopedia of Sustainability Science and Technology*, DOI 10.1007/978-1-4614-5731-2_13, © Springer Science+Business Media New York 2012

Catalytic oxidizer	Remediation technology equipment that uses a catalyst to accelerate the chemical oxidation of hydrocarbons with oxygen in a vapor effluent stream.
Chemisorption	Adhesion of a chemical to the surface of a solid, specifically through a chemical reaction occurring at the surface.
Confined aquifer	A water-bearing geologic strata that is situated between impermeable layers (clays and silt layers), leading to higher pressure of the groundwater in this unit.
Extraction well	A well used to remove liquid or gas from the subsurface.
Fate and transport	Encompasses how contaminants move through environmental media and how long the contaminants persist or how fast they are degraded.
Feasibility study	A document that describes and analyzes potential cleanup alternatives for a site and recommends selection of an effective and efficient alternative.
Green and sustainable remediation	Environmental cleanup that is designed and performed with consideration of the environmental impacts of the technologies used.
Half-life	Time required for half of the molecules of a chemical to decay or be degraded.
Henry's law	Law that is used to describe the volatility of a chemical, by describing the equilibrium between the vapor phase and dissolved forms of the chemical.
Hydrocarbons	Chemical compounds that consist of carbon and hydrogen.
Hydrodynamic dispersion	Transport within a fluid in directions other than the primary direction of fluid flow. This process decreases contaminant concentrations while increasing the total volume of fluid contaminated.
Hydrodynamics	The process of the motion of groundwater.
Hydrogeology	Discipline dealing with the properties and characteristics of groundwater.
Hydrolysis	Reaction that splits a chemical into two parts by adding a water molecule, through addition of a hydrogen ion to one fragment of the chemical and addition of a hydroxyl group to the other fragment.
Hydrophobic	Having an aversion to water. Typically describes a contaminant that associates with nonpolar substances (such as oils and organic matter) rather than polar substances like water.
Injection well	A well used for injection of fluids, gases, and/or chemicals for remediation.

Inorganic	Describes chemicals that are not organic, including metals and common anions (sulfate, nitrate, etc.).
Interfacial tension	Tension at the interface between a liquid of one chemical and a solid, liquid, or gas of another chemical. One of the primary determinants of NAPL mobility in the subsurface.
Interim remedial action	A remedial action taken to address immediate risks to human health or the environment before long-term remedial goals are achieved.
Leaching	Dissolution of relatively soluble chemicals and removal by water transport.
Life cycle analysis (LCA)	Evaluation of the environmental impacts of all stages of a product or process.
Light nonaqueous phase liquid (LNAPL)	A nonaqueous phase liquid that is less dense than water and therefore floats on the water table, including petroleum hydrocarbon fuels and lubricating oils.
Liquid density	Mass per unit volume of a liquid.
Liquid viscosity	Resistance of a liquid to being deformed. Higher viscosity is associated with more resistance to flow, or less fluidity.
Lower explosive limit	The concentration of a compound in air below which it will not ignite.
Microaerophilic	Requiring only small amounts of oxygen.
Monoaromatic hydrocarbons (MAHs)	Organic chemicals containing one aromatic ring, which are common petroleum derivatives.
Nonaqueous phase liquid (NAPL)	A liquid, such as oil, that remains in a separate phase in the groundwater and can act as a source of organic contaminants to groundwater and soil.
Organic	Describes a category of chemicals that typically contain carbon with hydrogen, oxygen, and/or nitrogen. Most organic compounds can be degraded to carbon dioxide, water, and other simple components.
Oxidation	Chemical reaction in which a chemical of interest loses electrons. The chemical that takes the electrons is known as the *oxidant.* Includes "rusting" of metals and processes that degrade organic matter to carbon dioxide.
Partitioning	Distribution of a chemical between the solid, fluid, and/or gas phases, in proportions reflecting its affinity for each phase, as described by the *partition coefficient*.
Permeability	Tendency of a material to allow fluids to flow through it.

Persistent organic pollutants (POPs)	Chemicals that do not readily degrade under environmental conditions and, therefore, persist in environmental media.
Polychlorinated biphenyls (PCBs)	Organic chemicals with chlorine atoms attached to two benzene (aromatic) rings, which were widely used as dielectric and coolant fluids, for example in transformers.
Polycyclic aromatic hydrocarbons (PAHs)	Organic chemicals containing more than one aromatic ring, which are common by-products of coal combustion.
Porosity	Fraction of a material that is void space. Can be primary (original, from when the geological material was formed) or secondary (formed later, by selective dissolution or fracturing).
Precipitation	Formation of a solid from dissolved chemicals in a solution.
Preferential flow	Faster movement of groundwater through certain, more porous or permeable, portions of the subsurface, which can result in localized, rapid contaminant transport.
Redox	Term used to describe the related processes of reduction and oxidation.
Reduction	Chemical reaction in which electrons are gained by a chemical of interest. The chemical donating the electrons is known as the *reductant*. Includes the reduction of oxygen gas during aerobic respiration and the reduction of other chemicals (nitrate, iron, carbon dioxide) during anaerobic respiration.
Remedial action	Action taken to remove or contain a hazardous substance in the environment.
Remediation	Cleanup or other methods used to remove or contain hazardous materials.
Risk assessment	Qualitative and quantitative evaluation of the risk posed to human health and/or the environment by contaminants.
Saturated zone	The portion of the subsurface below the water table, where the pressure of water within the pores is at a pressure equal to or greater than atmospheric pressure.
Solubility	Ability of a chemical to dissolve into (i.e., mix with and become incorporated into) another substance. Unless otherwise specified, in an environmental context, solubility is typically used to refer to solubility of a chemical in water.

Sorption (verb: to sorb)	Attachment of a chemical to a solid, which removes the chemical from the dissolved phase. See also adsorption, absorption, and chemisorptions.
Speciation	The chemical form (phase, redox state, molecular structure) in which an element exists. Important determinant of metal mobility in the environment.
Subsurface	The zone beneath the surface of the earth, including geologic strata and groundwater.
Transport mechanisms	Processes by which contaminants move through the environment.
Unsaturated/vadose zone	The subsurface zone between land surface and the water table where the moisture content is less than atmospheric pressure (i.e., soil pores are not completely filled with water).
Vapor pressure	Pressure of the vapor of a chemical that exists in equilibrium with the chemical's solid or liquid phase.
Volatility	Tendency of a chemical to vaporize, or go into the gaseous phase.

Definition of the Subject and Its Importance

Since the onset of subsurface remediation in the 1970s, there has been a need for a more appropriate balance between the protectiveness of environmental cleanup technologies and the concept of environmental sustainability. This entry explores the implementation of innovative green and sustainable practices deemed appropriate for the remedial technologies that address the most common classes of persistent and toxic subsurface contaminants.

Introduction

Preceding and during the early timeframe of environmental remediation, there was a lack of concern and knowledge about the fate and persistence of chemicals released to the subsurface. There was a prevailing but unfounded assumption that the subsurface environment would sorb or attenuate almost unlimited amounts of contaminants. Much to our dismay, this has been shown to be false through major advancements in analytical chemistry techniques over the last 2 decades. Once subsurface transport mechanisms were understood, it became obvious that contaminants released at or near the surface may make their way deep into subsurface environments, including confined aquifers and bedrock settings. The fate and transport of environmental contaminants in subsurface environments are

Table 13.1 Summary of properties that affect fate and transport of organic and inorganic contaminants in the subsurface

Contaminant properties	
Organic contaminants	*Inorganic contaminants*
Solubility	Solubility
Fluid density	Redox (reduction-oxidation)
Viscosity	Speciation
Interfacial tension	Adsorption
Carbon partition coefficient	Reactivity
Henry's law constant	Vapor pressure
Biological degradation	
Vapor pressure	
Subsurface/hydrogeologic properties	
Rock type and characteristics	
Hydraulic conductivity	
Aquifer redox chemistry	
Specific surface area of minerals	

significantly affected by two categories of properties (Table 13.1): (1) the geologic characteristics of the subsurface environment, which defines the intrinsic properties of the soil (or rock) and imparts the characteristics of groundwater flow through that media, and (2) the properties of the contaminants, which define the physicochemical and biological processes that affect their fate and persistence.

The prevailing group of processes affecting subsurface fate and contaminant transport are hydrodynamic processes, partitioning, biotic reactions, and abiotic reactions. Hydrodynamic processes impact contaminant transport through groundwater advection, hydrodynamic dispersion, and potential preferential flow. Partitioning affects contaminant distribution and dispersal by allowing interchange of contaminant from one subsurface medium to another (e.g., soil, groundwater, soil gas) through means of adsorption, absorption, or chemisorption. Biotic reactions can affect contaminant transport by degradation (or immobilization) of the contaminant in oxidation or reduction reactions. More specifically, biotic processes, which occur under aerobic, microaerophilic, or anaerobic conditions, can lead to degradation or immobilization (by reaction or precipitation), depending on the type of contaminant. Abiotic reactions affect contaminant transport in the subsurface by promoting interactions between the contaminant and groundwater or stationary media (e.g., soil, bedrock), causing the contaminant to degrade or change in form (e.g., hydrolysis, redox reactions). These processes take place within the saturated zone and the unsaturated zone of the subsurface. Of these two subsurface zones, heightened concern is often paid to the saturated zone because contaminants are in direct contact with groundwater, which may be used as a potable or irrigational resource or for other purposes. The unsaturated zone, or vadose zone, overlies the saturated zone (i.e., above the water table) and is an important contributor to contaminant fate and transport through processes such as leaching and migration to the saturated groundwater zone.

Sources of chemicals released to the subsurface are varied, but generally include (1) underground and above ground storage tanks, (2) septic tanks, (3) agricultural activities, (4) municipal and industrial landfills and dumps, (5) regulated and abandoned hazardous waste sites, (6) injection wells, and (7) other industrial sites [1]. The types of chemicals that have been historically released to the subsurface and have the potential to cause adverse effects to human health or the environment are generally divided into organic compounds and inorganic compounds (or inorganic elements). Several of the more common classes of organic and inorganic contaminants found in soil and groundwater are as follows:

- Petroleum hydrocarbons and derivatives used as fuels such as gasoline, diesel fuel, jet fuel, and heating oil. These fuels consist of many organic chemical components, including monoaromatic hydrocarbons (MAHs) like benzene, toluene, ethylbenzene, and xylene (BTEX), and low-molecular-weight polycyclic aromatic hydrocarbons (PAHs), such as naphthalene, and are generally less dense in hydrocarbon fluid form than water, contain components that are highly volatile, and are sparingly to moderately soluble in water. When present as an immiscible phase (in water), they tend to persist over reasonably long time frames as light non-aqueous phase liquids (LNAPLs). However, LNAPL constituents will dissolve in water, sorb onto soils, and/or partition into the vapor phase.
- Chlorinated aliphatic hydrocarbons used as solvents, degreasers, and dry cleaning fluids, such as tetrachloroethylene (PCE), trichloroethylene (TCE), and carbon tetrachloride (CT). These compounds generally are distilled for use in industrial settings in a pure form or as relatively simple mixtures. When present as an immiscible phase, they generally are denser than water, highly volatile, and also sparingly soluble in water. Hence, they can exist as dense non-aqueous phase liquids (DNAPLs), dissolved in water, sorbed onto soils, and/or present in the vapor phase.
- Polychlorinated biphenyls (PCBs), pesticides, wood-preserving chemicals (such as creosote), and fossil fuel combustion/gas manufacturing gas by-products, such as coal tars and high-molecular-weight PAHs. These contaminants tend to be dense, highly viscous fluids (when present an immiscible phase) with constituents that sorb strongly onto soils, are nearly insoluble in water, exhibit low volatility, and are persistent in the subsurface environment because they chemically and biologically degrade very slowly. When present as an immiscible phase, they generally are found as DNAPL or are sorbed onto soils.
- Explosives and energetic compounds such as solid-rocket fuels and propellants, including trinitrotoluene (TNT), plastic explosives, perchlorate, and munitions components. These constituents are often found as solids on or near the land surface, although some (such as perchlorate) can be found dissolved in groundwater.
- Metals such as arsenic, lead, chromium (VI), mercury, cadmium, and others. These generally are found either dissolved in groundwater, present as elements in the solid phase, or present within the rock matrices as mineral components.

Each type of contaminant has a distinct set of physicochemical characteristics that define its behavior and migration within subsurface environments. Hydrogeology and rock/mineral geochemistry also have a significant influence on defining the fate and migration of contaminants. As a function of the contaminant's physicochemical properties and geologic characteristics, the above listed classes of organic chemicals exist in the subsurface as one or more of four phases: (1) mobile or residually entrapped nonaqueous phase liquid (NAPL), (2) dissolved phase in groundwater, (3) sorbed phase to solid aquifer media, and (4) vapor phase in soil gas. Both the properties of the chemical and that of the subsurface control the dynamic evolution of phase transfer, including the duration of time that these organic chemical remains within these phases following a spill or release. From the point of release, organic chemicals generally exist in the NAPL phase, with eventual partitioning to one or more of the other phases with time. Subsurface NAPLs can exist as a pure chemical or as a bulk mixture of chemicals.

Physical and chemical properties that have a major effect on the fate, transport, and persistence of the classes of typical organic contaminants are shown in Table 13.2. Properties such as solubility (in water) determine the degree to which a contaminant persists in the subsurface as an immiscible fluid or solid. If an immiscible fluid phase persists in the subsurface, fluid density governs whether the fluid acts as an LNAPL or DNAPL (e.g., petroleum hydrocarbons, with a fluid density less than 1 g/mL, form LNAPLs, whereas chlorinated hydrocarbons form DNAPLs). Properties like organic carbon partition coefficient (K_{oc}) and Henry's law constant effect the tendency of a chemical to partition under equilibrium conditions from a source release (e.g., contaminant liquid or solid) to subsurface media (e.g., soil, groundwater, and/or soil gas). The relatively high Henry's law constants and solubilities of hydrocarbons, for example, indicate that hydrocarbons are more likely to partition into air and groundwater than are PAHs, PCBs, and pesticides.

The half-lives for biological degradation of the contaminants, also shown in Table 13.2, provide an indication (using first order kinetics) of how quickly a chemical will biologically degrade in the subsurface. Petroleum hydrocarbons and chlorinated hydrocarbons have been observed to degrade under aerobic and anaerobic conditions, respectively, at reaction rates determined to be moderate to fast for both natural and engineered remedial systems. Conversely, PCBs and certain PAHs have been shown to undergo very slow to negligible rates of aerobic and/or anaerobic biodegradation in natural subsurface settings. Explosives and energetic compounds have been shown on a constituent by constituent basis to biologically or chemically degrade (in presence of reductants and oxidants, respectively) under subsurface conditions attainable with the aid of remedial technologies.

The most common inorganic compounds identified in the subsurface include metal contaminants. Although metals are natural constituents of soils, anthropogenic metals enter the soil through a variety of means including (1) leaching of municipal or industrial solid wastes, (2) storm water runoff and infiltration, (3) industrial by-products, (4) dredged materials, (5) mining and smelting operations, (6) atmospheric emissions from coal or oil combustion, (7) ash and

Table 13.2 Physicochemical properties of organic contaminants that affect contaminant fate, transport, and persistence (Refs. [2–5])

Class of contaminant	Henry's law constant (atm. m^3/mol)	Solubility (g/L)	Fluid density (g/cm^3)	log K_{oc}	Biodegradation half-life: Aerobic	Biodegradation half-life: Anaerobic
Chlorinated hydrocarbons	0.00674–0.015	0.15–6.3	1.2–1.6	0.4–2.4	4 weeks–1 year	11 weeks–4.5 years
Petroleum hydrocarbons	0.00046–0.015	0.03–200	0.86–0.88	1.5–3.6	2 days–6 months	8 days–2 years
Heavy PAHs	Minimal–0.0003	0.0000003–0.004	NA (solid)	1.3–7.5	12 days–5.2 years	50 days–21 years
PCBs	0.000003–750	0.00001–0.002	1.182–1.566	2.4–6.4	1.5–9 years	–
Pesticides	0.0000002–0.002	0.0000004–0.5	NA (solid)	2.1–6.3	2 days–16 years	1–294 days
Explosives/energetics	Minimal–0.00000002	0.13–250	NA (solid)	NA	4 weeks–6 months (TNT)	4 weeks–6 months (TNT)

NA not applicable

slag from coal or oil combustion, (8) and sludge residues from wastewater treatment. Typical metals identified as contaminants in subsurface environments include arsenic, chromium, cadmium, copper, lead, mercury, nickel, silver, and zinc. Radionuclides are a separate class of contaminants that are inorganic and due to their specificity are not discussed in this entry.

The physicochemical properties of inorganic chemicals that govern their fate, partitioning, and migration within subsurface environments include solubility, reduction–oxidation (redox) speciation, reactivity, and vapor pressure. Properties of subsurface media that play a defining role in the fate of inorganic contaminants within subsurface environments include rock type and characteristics (primary and secondary porosity, mineral composition, fracture density), hydraulic conductivity, redox chemistry, and specific surface area of minerals present. Metals are not degraded by biological or chemical reactions, though they can conveniently be rendered unavailable through precipitation reactions or transformed via oxidation/reduction reactions to less toxic species.

Properties of subsurface media that play a defining role in the fate, partitioning, and transport of organic contaminants within subsurface environments include rock type (i.e., porous or fracture flow), primary and secondary porosity, hydraulic conductivity, and total organic carbon.

Background

Beginning in the 1970s with the advent of subsurface remediation, the environmental industry invested heavily in remediation systems without an adequate understanding of the degree of contaminant cleanup (or the duration of the cleanup). Although these early efforts were well intended, the actions taken with the technologies utilized were not justified based on the costs of the cleanups, the overutilization of resources, the intensive amount of energy consumed, and the insufficient contaminant removal. Past presumptive remedies such as soil "excavation and off-site disposal" and groundwater "pump and treat" are prime examples of these generally wasteful remedies considered unsustainable onto the future.

Beyond this early misunderstanding about remedial technology selection to achieve efficient and sustained removal of subsurface contaminants, there has been a long-standing misperception that cleanup shall continue to pristine conditions, a goal that as turns out is largely unattainable in most circumstances. At the expense of nearly 2 decades of numerous examples of unsuccessful and costly cleanups, we have learned considerably from these early mistakes. Today, in many cases, there is a robust process and a wealth of regulatory guidance to determine more pragmatic yet adequate remedial action objectives and cleanup performance criteria. These cleanup goals and objectives are most often defined by human health risk assessment [6], ecological risk assessment [7], risk-based

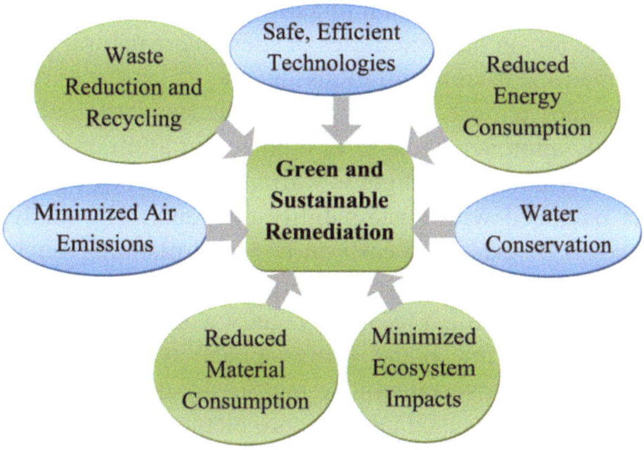

Fig. 13.1 Key elements of green and sustainable remediation

corrective action [8, 9], or by what is reasonably achievable based on the best practicable remedial technology options (generally referred to as best available technology [BAT]).

In addition to these well-considered approaches that have provided a practicable risk-related basis for cleanup end goals, there is growing interest in the last 5 years to incorporate green and sustainable remediation (GSR) concepts throughout the remedial action process, while continuing to provide acceptable long-term protection of human health and the environment. As suggested by the Sustainable Remediation Forum (SURF), sustainable remediation is defined as a "remedy or combination of remedies whose net benefit on human health and the environment is maximized through the judicious use of limited resources" [10]. To this end, sustainable remediation employs solutions that minimize the environmental footprint while providing maximum net environmental benefit over the remedial lifecycle. To realize the benefits of sustainable remediation requires the use of green and/or renewable energy sources, conservation of water and energy, decreasing waste, and formulating integrated sustainability policies. With sustainable remediation, the goal should be to (1) develop and implement safe remedial solutions that are minimally disruptive to the environment, (2) realize energy savings through creative design and value engineering, (3) embrace waste minimization and recycling concepts, and (4) emit the least amount of pollutants and greenhouse gases to the atmosphere. To the extent practical, the following sustainability elements should be applied to remedial solutions incorporating GSR practices (Fig. 13.1): (1) short-and long-term energy and water consumption, (2) greenhouse gas emissions and air pollutants, (3) ecosystem impacts, (4) material consumption, and (5) waste minimization and/or recycling.

The recent recognition of the balance between adequacy of cleanup and sustainability concepts has not, in some cases, dispelled the notion of initiating cleanups with little thought to the protective character (human health and environmental), resource utilization, or safety of the action. As evidenced by the continued use of pump and treat systems, there are still regulatory mandates for cleanup with little decrease in contaminant mass (and therefore little decrease in overall risk). There are, in fact, examples of sites where natural attenuation is actively decreasing contaminant concentrations in the subsurface, yet the overseeing regulatory agencies require a more traditional, less sustainably oriented, significantly more costly approach because of outdated or arbitrary remediation goals set for the site. Clearly a new paradigm is warranted, whereby cleanup protectiveness of a remedy is balanced with sustainability elements. The analysis to establish this balance should be evaluated early in remedial planning, such as during a feasibility study of remedial alternatives. Incorporation of sustainability into the balancing criteria evaluated during a feasibility study would help to assure that this process happens. In the interim, many professionals in the environmental industry have been documenting sustainability metrics by qualitatively and/or quantitatively scoring the degree of sustainability core elements to be utilized on cleanup actions. Such evaluations have ranged from simple qualitative review of the available list of best management practices (BMPs), utilizing those that are applicable, to the performance of quantitative and complex life cycle analysis (LCAs) for optimization of the cleanup over the remedial lifetime. The following section provides valuable information on industry lessons learned from implemented LCAs and BMPs for sustainable remediation practices, as applied to the typical classes of subsurface contaminants described in this document.

Sustainable Remediation Practices for Classes of Typical Subsurface Contaminants

Remediation sites comprise a range of sizes, proximity to human and/or ecological receptors, proximity to man-made infrastructure, site accessibility, environmental complexity, type of contaminants and their chemical, physical, and toxicity characteristics, complexity of the circumstances surrounding the release(s), and so on. All of these factors affect the feasibility of corrective actions, and even the ability to implement any corrective action. Because of this, no generalized discussion of sustainable remediation practices can be complete, because the topic is simply too broad. However, the information described below, is comprised of remedial measures applied recently that can be characterized as sustainable practices.

With this in mind, this chapter will focus on sustainable practices honed over time for the general classes of subsurface contaminants discussed above. The discussions are premised on the following assumptions:

- Releases to the subsurface have stopped; i.e., that the pipelines have been repaired, the underground storage tanks have been removed, the uncontrolled landfill is no longer receiving toxic materials and has been capped, etc.

- Any discussion of "sources," with regard to groundwater, refers to concentrated areas of contaminants such as NAPL or highly contaminated soils that are present but relatively stable.
- A risk assessment has been completed (if necessary), and remedial action objectives have been defined before remedial implementation.
- Emergency or interim measures (e.g., protection or replacement of domestic water supplies affected by the release) are in place as needed.
- Site characterization has progressed to a point where remediation methods can be considered within a feasibility study (FS) or a focused feasibility study (FFS).
- Remedial systems are designed to destroy (or remove) contaminants with the knowledge and benefit of exploiting the physical and/or chemical characteristics of the contaminants.
- Sustainable measures for the contaminants discussed have been optimized over time as a result of trial and error evaluation with full-scale remedial systems and technology innovation or breakthroughs.

These discussions readily address the lessons learned and resulting sustainable optimizations (or BMPs) of various technologies for typical contaminants.

Petroleum Hydrocarbons

Petroleum hydrocarbons, including MAHs and low-molecular-weight PAHs, often occur in persistent LNAPLs in the subsurface. Therefore, the primary method of remediation historically has relied on product recovery within a cone of depression produced by water table drawdown from groundwater pumping.

As a result of energy inefficiencies and high cost of groundwater extraction (and treatment), a more sustainable approach evolved involving product recovery with skimmer pumps. Skimmer pumps are designed to remove LNAPL from the water table surface. The skimmer floats on the water table and has an interval with a hydrophobic (water-rejecting) screen that is open to the LNAPL layer within a monitoring well. LNAPL is drawn into the skimmer and flows through a flexible tube to a reservoir where it is pumped to the surface. The following practices are often employed with product recovery systems to reduce energy consumption, minimize site impacts, and improve the overall sustainability of the treatment [11]:

- Power product recovery components or auxiliary equipment with low energy demand, such as renewable energy off-grid wind turbines or photovoltaic (PV) systems.
- Such systems relying on off-grid energy should be equipped with deep-cycle batteries to provide steady power.
- Eliminate the long-distance transport of incoming materials and equipment or outgoing remedial-derived wastes (i.e., recovered product). To that end, consolidate deliveries/pickups to avoid deploying partially filled vehicles.

- Recycle separated product (LNAPL) through local fuel or waste recyclers.
- Optimize product recovery through proper equipment sizing and frequent reassessment based on treatment performance.
- Establish operating or cleanup performance criteria that could trigger use of less intensive polishing technologies as cleanup progresses and LNAPL recovery rates decline.

Skimmer pumps are capable of removing LNAPL down to a sheen, but do not reduce dissolved contaminant concentrations or the mass of contaminants sorbed onto soil. In addition, skimmer pumps, unless supplemented with vacuum enhancements, generally are not capable of providing complete capture of a mobile accumulation of free-phase LNAPL that is migrating under natural subsurface conditions. In such cases, groundwater extraction is sparingly used to capture and contain the mobile, free-phase LNAPL for recovery. The following sustainable practices are often employed to minimize groundwater extraction while optimizing LNAPL capture:

- Perform groundwater capture zone analyses using empirical calculations or numerical groundwater modeling as basis of design for groundwater capture and, therefore, LNAPL containment.
- Calibrate and refine groundwater extraction network and flow rates with calculations or modeling after system startup with observed drawdown conditions from corrected groundwater elevations.
- Monitor and periodically optimize the groundwater extraction network and flow rates to maximize LNAPL recovery rates as a result of higher producing wells.
- Monitor extraction well change in head at a given rate over time, or specific capacity, to ensure continued efficient well operation. Rehabilitate wells periodically, if decreased specific capacity is observed, to maximize extraction well longevity.
- Design system to minimize the total amount of piping including length, surface area, bends, and elbows to maximize transport efficiency.
- Engineer extraction wells to maximize efficiency through decreased head loss and optimized flow rates by designing the well gravel pack to match the formation and using the largest feasible slot size to maximize open cross-sectional area. This decreases velocity and prevents migration of fines by allowing a less-turbulent flow into the well.

Sustainable features of skimmer pumps are that they require little energy to operate, and the energy can be obtained from sustainable sources such as solar panels (Fig. 13.2). Recovered LNAPL can be recycled to minimize waste. The method is relatively slow and inefficient for LNAPL removal, but it can be effective at small sites. Skimmer systems can be enhanced by applying a vacuum, which can speed up LNAPL recovery for a relatively minor additional energy input that could also be obtained from sustainable power sources. The use of skimmer pumps

Fig. 13.2 Example of solar powered system with photovoltaic panels for powering product skimming

generally does not exacerbate existing subsurface conditions or make additional remediation efforts more difficult or less effective.

Soil vapor extraction (SVE) can be effective at removal of volatile organic compounds (VOCs) of petroleum hydrocarbons within the unsaturated zone of the subsurface present as LNAPL, sorbed to contaminated soil, or in the vapor phase. One or more SVE wells (screened in the unsaturated zone above the water table) are constructed, and a vacuum system is installed and manifolded to the wells. The vacuum draws the soil air (which is contaminated with VOCs) out of the subsurface through a treatment system (described below). The contaminated vapors are replaced by fresh air from vent wells or from other parts of the subsurface. This allows the volatile fraction of the VOC to evaporate at a rate that is proportional to the Henry's law coefficient. These volatiles are then carried by the air into the SVE system, allowing for further evaporation into the fresher soil air, and so on.

SVE systems are efficient for VOC removal in the unsaturated zone primarily because volatiles preferentially partition into the vapor phase, and air is relatively easy to move in the subsurface compared to liquids. SVE systems tend to lose efficiency with age as the volatile fractions of the contaminants are removed and the less volatile fractions remain. Because not all of the components of refined petroleum hydrocarbons are volatile, SVE by itself cannot achieve complete subsurface remediation. In addition, SVE does not affect contaminants below the water table, because they are not in contact with the soil air. However, certain adaptations can be made to an SVE system to facilitate contaminant removal from the subsurface.

Examples of these adaptations include artificially lowering the water table to expose more soil contaminants to the air or forcing air into the aquifer below the water table to strip out the dissolved or sorbed volatile compounds in the saturated zone through a process known as air sparging.

Nearly all SVE systems require treatment of the vapors removed from the subsurface soil. The most efficient treatment depends on the contaminant concentrations within the vapor stream. Essentially, low concentrations are most cost effectively removed using a sorption medium such as granular activated carbon (GAC), which is a specific type of charcoal. GAC requires no energy input to remove organic compounds from soil vapors while deployed on an SVE system. Because GAC is an absorbent material, however, it does not destroy contaminants, but transfers them to another medium. Once most of the sorption sites are filled and "breakthrough" (i.e., contaminants are no longer captured by the absorbent material) occurs, GAC must be regenerated off-site (requiring energy) or replaced with virgin material (also requiring energy). The virgin material commonly used to manufacture GAC is coconut shells, which are a sustainable resource.

If vapor concentrations are too high, breakthrough will occur quickly and GAC change-out costs can become prohibitive. Above certain concentration threshold, however, a treatment method that uses a catalytic oxidizer may become the most appropriate choice. Catalytic oxidizers are energy intensive because their optimal performance occurs at a relatively high temperature (300–600°C). However, the oxidation of fuel components at sufficient concentrations releases heat that is used to maintain the optimal temperature, greatly increasing efficiency and decreasing or eliminating the need for external energy inputs. As a result, the sustainability of the remediation system is enhanced.

For very high concentrations of VOCs that approach or exceed the lower explosive limit of the vapors, high temperature oxidation of the fuel components can be used to actually run the SVE system in whole or in part. The technology uses an internal combustion engine powered by the extracted vapors and an auxiliary fuel source such as propane if needed. Under the right conditions, such a system requires no external power supply, and if a generator module is added, the system can supply power that can be used to operate lights or other electrical equipment onsite. Off-gas emissions from these units are equivalent to those from operating an automobile engine.

If properly used, SVE does not generally exacerbate existing subsurface conditions or make additional remediation efforts more difficult or less effective. If improperly vented, however, the negative pressure generated by the SVE can cause mounding of the water table that creates a hydraulic gradient capable of spreading LNAPL away from the vapor extraction well(s), possibly increasing the lateral extent of the contaminated area.

Most SVE systems have esthetic impacts that reduce their green characteristics. One undesirable aspect is that they tend to be noisy. This can be mitigated by methods such as surrounding the SVE unit with fences lined with commercially available sound-insulating blankets. Another undesirable aspect is that the systems have a visual impact. They typically are not large, with a footprint on the order of

10 m^2 or less and a height less than 3 m, but they have an "industrial" look about them. This could be mitigated with appropriate landscaping that could include components like fencing with small trees, shrubs, or even large potted plants. Typically, SVE systems are operated almost constantly for a period of approximately 2–3 years, so the esthetic impacts can be a substantial nuisance to nearby workers, passers-by, or residents. The following BMPs are examples of ways to promote a more sustainable SVE approach:

- Optimize extraction configuration and rates by manifolding several vapor points and periodically manipulating valves to minimize "dead zones" and reduce remediation timeframe, energy consumption, and noise.
- Utilize appropriate vapor treatment with GAC by vapor concentration (low concentrations – energy efficient; low concentrations result in infrequent change out; cat-ox for intermediate to high concentrations – appropriate concentrations improve oxidation efficiency; thermal oxidation for very high concentrations – can utilize contaminant vapors as a fuel source).

Multiphase extraction combines a pump system to remove LNAPL and contaminated groundwater along with SVE to remove soil vapors. The pump system has the effect of lowering the water table, creating a gradient for the LNAPL to migrate toward the extraction well while it removes contaminated water and LNAPL. This effect also exposes more of the subsurface to the soil air, enabling SVE to act on a larger volume of subsurface material.

If the water table is relatively shallow (less than about 8 m below land surface), removal of all three contaminant phases can be accomplished from the surface by applying SVE within the unsaturated zone along with suction to a downhole drop tube with its end located at or slightly below the oil–water interface. At the surface, the oil and water are separated. The oil can then be recycled (or used to power the internal combustion engine if appropriate), and the water gets treated and released back into the environment (Fig. 13.3).

Multiphase extraction can be energy intensive, but as with SVE, green and sustainable options exist. For example, the power to operate the system can sometimes be obtained from the petroleum hydrocarbon vapors and/or the LNAPL, considerably reducing or possibly eliminating the need for external power sources. Multiphase extraction generally is fast and effective, reducing overall energy use along with reducing long-term esthetic impacts from noise and infrastructure. It can be deployed quickly as all or part of an interim remedial action to mitigate urgent cleanup requirements and reduce the need for additional remedial actions that may be less sustainable. Even if it is not effective at a site, multiphase extraction generally does not exacerbate existing conditions or make additional remediation efforts more difficult or less effective. As with SVE, visual and noise impacts from the extraction system can represent an esthetic nuisance, but can be mitigated in a similar fashion with appropriate landscaping and noise dampening components. Although groundwater pumping (and treatment) is included in multiphase extraction for the purpose of exposing subsurface soil to unsaturated conditions, many of the pitfalls (e.g., energy-intensive systems, low dissolved

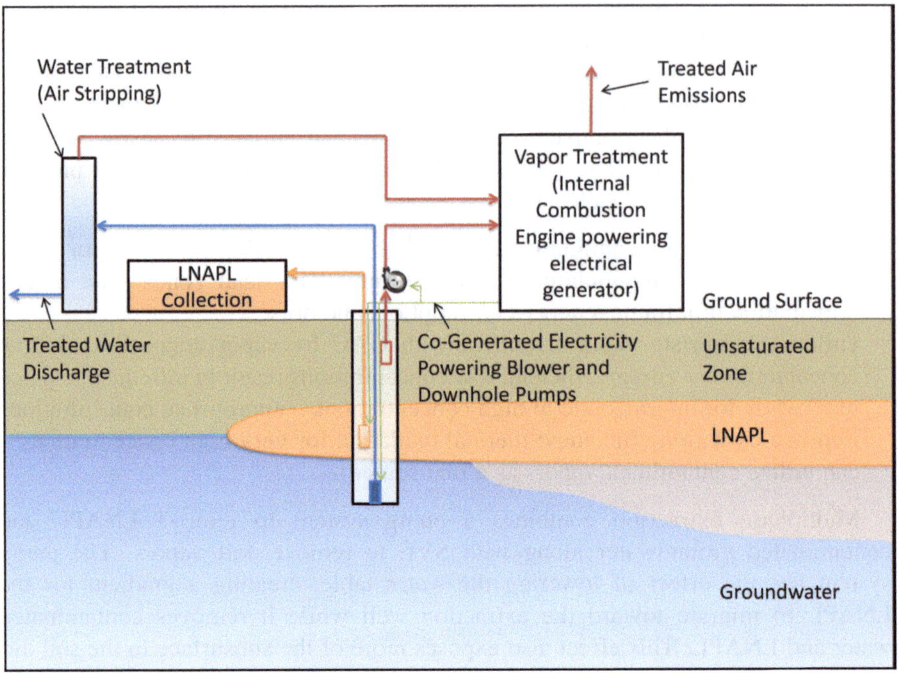

Fig. 13.3 Concept-level remedial strategy for green and sustainable multiphase NAPL recovery

phase contaminant removal, and high treatment cost) described earlier in this entry for pump and treat systems would also apply here. Therefore, caution is warranted when employing groundwater extraction and treatment for multiphase recovery or for product recovery. The following BMPs are offered to provide a sustainable approach to groundwater extraction and treatment with multiphase extraction systems:

- System design should consider modular treatment components that can be removed or added as needed.
- Variable frequency drive pumps can be used to optimize performance and reduce energy usage.
- Use of gravity flow where feasible to reduce the need for transfer pumps.
- System operation should have optimized extraction network to minimize pumping of clean water.
- Use of energy efficient equipment and green energy from alternative energy providers.
- Regenerate GAC onsite; recycle process residuals.

Monitored natural attenuation (MNA) is a highly sustainable remediation method that can be used for releases of petroleum hydrocarbon fuels, distillates, or contamination from other organic compounds. The method utilizes naturally occurring processes such as biodegradation to reduce contaminant mass and concentrations that are dissolved in groundwater, present as LNAPL, or sorbed to the soil. Biodegradation generally destroys most of the components of refined petroleum hydrocarbons, producing innocuous by-products such as carbon dioxide and water. As a result, MNA requires little to no external energy input and generates minimal waste.

Biodegradation of petroleum hydrocarbons readily occurs in many environments. Microbes have undergone natural selection for millennia, resulting in microbial communities that have evolved remarkable capabilities to utilize every bit of energy that can be extracted from oil constituents [12]. Biodegradation occurs even in extreme conditions. Fouling of fuel system components in aircraft due to microbial growth in kerosene-based jet fuel was recognized as early as 1956 [13]. For practical purposes, this means that in many (if not most) cases, degradation of refined petroleum hydrocarbons will proceed under natural conditions without external energy inputs.

MNA as the sole remediation method has certain limitations that must be addressed prior to and during implementation. First, it must be demonstrated that adverse impacts to human and ecological receptors are not occurring and are unlikely to occur in the future. Second, sufficient evidence for effective natural attenuation must be shown to exist at a given site. Such evidence can include among other things a demonstrably shrinking or stable dissolved plume, a reduction of dissolved concentrations in groundwater over time, and the use of stable isotopes to determine degradation rates. Third, long-term monitoring is required to ensure that conditions do not develop to change the efficacy of natural attenuation as a remedial action. MNA, in a practical sense, does not diverge from the activities associated with site investigation and monitoring. For this reason, the following BMPs utilized for site investigation are introduced with MNA activities [14]:

- Perform fewer field mobilizations through the use of flexible work plans and real-time field measurements as well as onsite mobile laboratory analyses to determine the next course of action during a single sampling event.
- Utilize small-scale direct push technology drilling equipment for invasive work or monitoring well installation to reduce fuel consumption, reduce drilling time, lower air emissions, lower water consumption, produce less noise, and minimize site clearing and physical impact.
- Use groundwater low-flow sampling equipment to minimize sampling purge volumes, reduce energy consumption, and reduced derived waste.
- Onsite treatment and recycling of MNA-derived wastes, including site clearing by onsite composting or landscaping and treatment/reuse of extracted groundwater for equipment decontamination.
- Collect the meteorological information (e.g., sun duration, wind direction and velocity) necessary to support the design and installation of off-grid alternative energy for auxiliary power for MNA monitoring.
- Use of solar or wind-powered telemetry systems to remotely transmit logging data directly to project offices.

In a variety of circumstances, MNA is used in conjunction with other methods that may not be sufficient to provide complete remediation of a site by themselves. There is a saying within the remediation industry that "90% of the remedial effort goes into cleaning up the last 10% of contamination." While this is not always true, monitored natural attenuation can provide a means in which the "last 10%" of the contamination does not have to be actively remediated because the final cleanup can be accomplished naturally and sustainably.

Enhanced bioremediation is designed to stimulate contaminant biodegradation by indigenous microbial populations [15]. Petroleum hydrocarbons degrade most rapidly under aerobic (i.e., oxygenated) conditions, although other electron acceptors may be utilized as reactants in biodegradation. Enhanced bioremediation of these compounds involves the injection of amendments to the contaminated subsurface to increase oxygen/electron acceptor concentration in the unsaturated and/or saturated zones. Enhanced bioremediation can supplement monitored natural attenuation in many cases to improve performance and reduce the time needed for complete remediation.

Several sustainable methods for promoting bioremediation by adding oxygen to the subsurface exist. One method used for soil remediation is known as bioventing. Bioventing works by injecting atmospheric air at low rates into the unsaturated subsurface zone to displace oxygen-depleted air, thereby stimulating the growth of aerobic microorganisms and improving biodegradation rates. Generally, the air is injected into the ground through well points screened in the contaminated unsaturated zone using a blower similar to that found on many SVE systems operated at low injection rates. Low air injection rates are important, particularly in populated areas, due to the potential for soil vapor migration into basements or other inhabited spaces. The low injection rates are advantageous for sustainable remediation, as external energy inputs are relatively low and could be supplied by renewable sources. A novel approach for air injection using mechanical windmill power (US Patent No. 6,109,358) represents a potential application of green bioventing technology.

Oxygen addition below the water table for bioremediation purposes can be accomplished through chemical or mechanical means. Proprietary slow-release oxygenating compounds for enhancing aerobic bioremediation are available in several forms, including powder (designed to be mixed with water), solids, and filter bags. Compounds in powder form generally are mixed with water and injected into the subsurface using direct push technology. Compounds in solid form or filter bags are typically suspended within the screened or open interval of monitoring-type wells.

Mechanical means of adding oxygen (for bioremediation or physical air stripping) below the water table usually takes the form of air sparging. Air sparging works by injecting air under pressure through one or more wells completely screened in the saturated zone for the purpose of providing oxygen to the groundwater. The air travels upward through the porous medium, along preferred flow paths that form a dendritic pattern (similar to the branches on a tree), until it reaches the unsaturated zone and becomes exposed to the soil.

Introduction of air in this manner will have two desirable outcomes if it is working well. First, the air will strip out volatile organic compounds from the water and saturated soil, transferring these compounds into the vapor phase where they can be removed through SVE (as needed). Second, the groundwater will pick up oxygen and become more aerobic, stimulating biodegradation. Air flow rate and injection pressure generally define whether physical air stripping (of VOCs) or in-place bioremediation will be the dominant process. If aerobic bioremediation is promoted through air introduction, the process is referred to as biosparging. The following sustainable BMPs can be applied to various types of bioremediation systems for treatment of petroleum hydrocarbons [15]:

- Maximize use of existing or new wells (to avoid resource overutilization) for addition of reagents that will act as electron acceptors in the biodegradation of petroleum hydrocarbons.
- Design and use of bioremediation recirculation cells allowing multiple passes of oxygenated groundwater through fewer wells.
- If oxygen additive is in liquid form, add to the subsurface via trickling gravity-feed system if high-pressure injections are unnecessary to assure proper distribution.
- If pressurized injection of air is required, evaluate the feasibility of pulsing rather than continuous injection to increase the efficiency of delivery.
- Employ modular, portable units that can be modified or incrementally reduced as needed.
- Employ photovoltaic panels or wind turbines to generate auxiliary power for trailer or for equipment, such as air blowers.

Physical air sparging (for the purposes of stripping and removing VOCs from the saturated zone) does not work at every site. It will not succeed if the dendritic patterns of the upwardly traveling sparge zone form with only a few branches, as contact between the air and the contaminants is essential. The efficacy of air sparging can be determined through relatively simple pilot tests, however.

Air sparging will require energy inputs to run the air pumps for a considerable period of time, depending on the size of the impacted area. The systems generally are run in conjunction with SVE, which is needed for vapor control, so 2–3 years of operation is not uncommon. If run with an SVE system, visual and noise esthetic impacts of air sparging are not substantially increased from an SVE system alone.

Phytoremediation is a remedial technology that uses plants to extract, destroy, and/or contain contaminants in environmental media. Petroleum hydrocarbons can be degraded within the plants or by plant by-products that are excreted into the soil, or volatilized into the atmosphere through transpiration.

To date, one of the most commonly implemented types of phytoremediation of petroleum hydrocarbons in shallow soils is rhizodegradation, or the destruction of contaminants by microbes whose activity is promoted by plant roots. Plants used for rhizodegradation include mulberry, hybrid poplars, grasses, cattails, and rice [16]. Phytodegradation, or destruction of the contaminants within the plant, and phytovolatilization of the contaminants can also be performed by a variety of trees, scrubs, and herbaceous plants.

Trees including poplars, cottonwoods, and willows can also be used to achieve hydraulic control to contain groundwater plumes. Often the same plants are used for both contaminant remediation and containment through hydraulic control. BMPs for phytoremediation include techniques for minimizing the impact of generic site operations, including energy conservation, waste minimization, and use of onsite resources.

The sustainability benefits of phytoremediation technologies include minimal site disturbance, leading to operational and esthetic benefits, and minimal energy inputs. Current technologies are limited in their applications by such factors as plants' limited contaminant tolerance, limited depth of influence of plant roots, and often lengthy remediation timeframes. Ongoing research and pilot studies are focused on developing new and improved methodologies for phytoremediation.

In situ chemical oxidation (also known as ISCO or Chem-Ox) is a method designed to oxidize contaminants using reactions that break apart chemical bonds, completely destroying the petroleum hydrocarbon compounds. The primary delivery method for the chemical oxidants is injecting them in liquid form into the aquifer using direct push methods or permanent injection wells. The method can work quite well at many sites, but like any remediation method, it is not universally applicable. Typical problems include daylighting (chemical oxidants flowing out onto the land surface or other inappropriate places), preferential oxidation of ambient organic carbon not related to contamination, desorbing of petroleum hydrocarbons from the soil into LNAPL form, and poor contact between the oxidant and the contaminants.

The sustainability advantages of ISCO is that it is fast acting, that it destroys contaminants rather than transferring them to another medium, and that it is injected using small equipment with a resultant small carbon footprint and minimal site disturbance. Direct push methods usually leave a boring with a diameter of 5 cm or less, which is filled in accordance with regulations and finished at the surface with a sod plug, asphalt patch, or other material such that little to no visual evidence of the hole exists. Remedial actions using ISCO are most applicable to small sites and generally are completed after two to six injection events, leaving no long-term esthetic impacts. Some or all of the following BMPs are examples of ways to promote a more sustainable ISCO approach:

- Minimize operational impact by constructing little to no long-term or permanent infrastructure.
- Use of extracted groundwater for onsite mixing of ISCO reagents.
- Use of direct push injections over injection wells if minimal injections are required.
- Reuse existing injection wells if multiple ISCO injection events are necessary.

Excavation and treatment or disposal is a remediation method not commonly associated with a green and sustainable approach. However, in many cases, it can be a direct and verifiable way of removing grossly contaminated material, which can then allow for more sustainable remedial methods to be implemented. Excavation can have many undesirable aspects during and after implementation, including things such as noise, dust, heavy equipment, large trucks, scarred landscapes, and so on. While these things may be necessary, they can be mitigated with proper planning and implementation of the remedial action. Some or all of the following BMPs are examples of ways that can help improve the green characteristics of excavation actions [17]:

- Devote time and resources to conducting low-impact site characterization (using direct push methods, field screening, etc.) both before and during the remedial action. Excavation footprints can sometimes be reduced if efforts are made to minimize the removal of clean material.
- Use appropriately sized equipment for excavation and hauling to minimize noise, dust, erosion, and fuel efficiency. Utilize energy efficient operating procedures such as minimizing idling and performing routine maintenance to improve fuel efficiency.
- Consider onsite treatment or using the closest treatment and disposal facility to minimize hauling. Purchase supplies that are produced locally and use local contractors when possible. Scout for local or onsite backfill material in the planning stages and during implementation.
- Salvage uncontaminated vegetation and organic debris for use as mulch or compost.
- Salvage uncontaminated objects and materials (such as steel, storage containers, etc.) for recycling, reuse, resale, or donation if possible.
- Revegetate and restore excavated areas as quickly as possible. Use native rather than imported plants. Judicious use of diverse plants and prior planning of final landforms can be utilized to create valuable habitat.

Chlorinated Hydrocarbons

Chlorinated aliphatic hydrocarbons, which include chlorinated VOCs, are important industrial chemicals frequently used as solvents for dry cleaning, degreasing, metal cleaning, paint stripping, and electronics. They tend to be highly volatile, toxic to humans, and denser than water. Common chlorinated solvents include chlorinated ethenes such as tetrachloroethylene (also known as perchloroethylene or PCE), trichloroethene (TCE), 1,2-dichloroethene (1,2-DCE), and vinyl chloride (VC); chlorinated ethanes such as 1,1,1-trichloroethane and 1,2-dichloroethane; and chlorinated methanes such as carbon tetrachloride, chloroform, and methylene chloride.

Like refined petroleum hydrocarbons, chlorinated VOCs can be found in the vapor phase, dissolved in groundwater, sorbed to soil, or as a NAPL. However, because chlorinated solvents are denser than water, NAPL will be present as DNAPL. Because of its higher density and other properties such as high viscosity, a DNAPL tends to be considerably harder to remove, destroy, or otherwise remediate than LNAPL.

Remediation methods for chlorinated VOCs include some that are the same as or similar to those discussed in the section on petroleum hydrocarbon remediation. These and other remediation methods are discussed below in the context of chlorinated hydrocarbons cleanup. However, the BMPs offered as sustainable practices discussed previously by technology for petroleum hydrocarbons have not been repeated in this section.

Soil vapor extraction (SVE) can be effective at removal of chlorinated VOCs present in the unsaturated zone of the subsurface that is sorbed to contaminated soil or present in the vapor phase. As described earlier, one or more SVE wells (screened in the unsaturated zone above the water table) are constructed and a vacuum system is installed with a manifold to the wells. The vacuum draws the air from the soil (which is contaminated with VOCs) out of the subsurface through a treatment system. The contaminated vapors are replaced by fresh air from vent wells or from other parts of the subsurface, allowing the VOCs to evaporate and drawn to the SVE for removal. Representative BMPs for SVE systems discussed in the previous section on petroleum hydrocarbons would be applicable for chlorinated VOCs with the exception of those related to vapor control technology.

Treatment of extracted soil vapor for chlorinated VOCs is more complicated than it is for fuel hydrocarbons. Chlorinated VOCs are not combustible, and when oxidized, one of the by-products is hydrochloric acid, which is highly corrosive, and can severely damage or ruin catalytic or thermal oxidizers. The vapors can be treated with GAC, but this is not green or sustainable and could result in considerable change-out expense if concentrations are high.

For higher concentrations, it is sometimes possible to recover the chlorinated solvents from the vapor stream for recycling. One method uses refrigerated condensation to recover VOCs from the vapor phase as liquid [18]. This type of technology is energy intensive, but can be considered a green remediation method based on the high efficiency of product recovery from the vapor stream.

Multiphase extraction is sometimes used to recover DNAPL, contaminated water, and soil vapor from the subsurface. DNAPL, however, is usually more viscous than LNAPL and is more prone to adhering to the soil matrix, making it difficult to draw the liquid into a well for product recovery. In most cases, the recoverable volume is small and the recovered solvent has been degraded to less than commercial quality, so recycling or reuse of the DNAPL generally would not be cost effective. As a result, most of the material would be disposed of off-site. As with the use of multiphase extraction for the recovery of petroleum hydrocarbons, caution is warranted when employing groundwater recovery and treatment. The BMPs offered for groundwater recovery with multiphase extraction of petroleum hydrocarbons would also apply to the recovery of chlorinated VOCs.

Monitored natural attenuation is a highly sustainable remediation method that can be very effective for the destruction of certain chlorinated hydrocarbons, particularly the chlorinated ethenes. The method exploits naturally occurring processes such as abiotic and biotic degradation to reduce contaminant mass and concentrations that are dissolved in groundwater, present as DNAPL, or sorbed to the soil. The process known as reductive dechlorination sequentially strips chlorine atoms off of each compound, starting with highly chlorinated compounds such as PCE and TCE through 1,2-DCE and VC, ultimately producing innocuous by-products such as carbon dioxide, chloride ions, and water. As with petroleum hydrocarbons, monitored natural attenuation of chlorinated ethenes requires little to no external energy input and generates minimal waste.

Biodegradation of chlorinated hydrocarbons through reductive dechlorination does not occur in all subsurface environments. Reductive dechlorination proceeds best under anaerobic (i.e., oxygen-depleted) conditions as opposed to the aerobic conditions that are most favorable for degradation of petroleum hydrocarbons. Additionally, reductive dechlorination requires both electron acceptors (i.e., the chlorinated ethenes) and electron donors (generally organic carbon that can come from anthropogenic sources such as landfill leachate or natural sources such as decaying vegetation) [19]. Third, microbial populations capable of facilitating reductive dechlorination are not always present, even if anaerobic conditions exist and adequate organic carbon is available. Therefore, even though MNA can be a green and sustainable remediation method, it is not always applicable or effective. In such cases, other methods need to be implemented.

Enhanced bioremediation, which is a method designed to stimulate contaminant biodegradation by indigenous microbial populations [15], can be a green and sustainable alternative similar to MNA. Enhanced bioremediation of chlorinated compounds involves the injection of amendments to the contaminated subsurface to promote bacterial processes that consume oxygen in the unsaturated and/or saturated zones, thereby initiating or enhancing reductive dechlorination. In cases where natural attenuation is actively occurring, enhanced bioremediation can supplement monitored natural attenuation to improve performance and reduce the time needed for complete remediation. If biodegradation is not occurring under natural conditions, enhanced bioremediation can initiate the process by creating conditions that are amenable to reductive dechlorination.

Enhanced biostimulation of chlorinated compounds typically uses fast-acting and/or slow-acting, food-grade carbon sources such as lactate, emulsified oils, molasses, or proprietary compounds developed specifically for this purpose. In many cases, injection of these carbon sources can be done with low energy input (such as controlled gravity feed), and they can be left in the subsurface because they are nontoxic and will eventually be completely consumed by the microbial populations. Enhanced bioremediation also destroys contaminants rather than transferring them to another medium, decreasing undesirable environmental impacts that can occur with some other remedial actions.

Like MNA, enhanced bioremediation can be used as a green and sustainable component of a larger scale remedial action involving media other than groundwater. Specific mixtures of carbon substrate can be used as needed to quickly induce anaerobic conditions or to react slowly and last longer. Other parameters, such as viscosity, density, emulsification, and so on, can be modified based on site conditions and remediation goals or requirements.

In some cases, the natural microbial populations are insufficient to degrade chlorinated hydrocarbons such as PCE and TCE to nontoxic endpoints, which can cause an undesirable buildup of 1,2-DCE and/or VC. Because VC is more toxic than TCE, this result is not acceptable from a regulatory or human health perspective. In such cases, bioaugmentation can be used, in which microbial consortia capable of complete degradation of chlorinated ethenes are grown in a laboratory, acclimated to site conditions, and injected into the subsurface with an appropriate carbon substrate. At present (2011), the only commercially available microbial consortia for this purpose are of the *Dehalococcoides* genus.

In situ chemical oxidation (also known as ISCO or Chem-Ox) is used to directly oxidize chlorinated hydrocarbons, typically by injection of oxidants in liquid form into the aquifer using direct push methods or permanent injection wells. The limitations to using ISCO for chlorinated hydrocarbons are similar to those discussed above for petroleum hydrocarbons and include daylighting, preferential oxidation of ambient organic carbon not related to contamination, and poor contact between the oxidant and the contaminants.

In situ chemical reduction can be used in place of or in conjunction with enhanced bioremediation. Its purpose is to chemically induce strongly reducing conditions capable of abiotically destroying chlorinated hydrocarbons. A typical reducing agent is zero valent iron (ZVI), which is a highly reactive metal that can be deployed as a powder in a permeable reactive barrier or injected into the subsurface in the form of a liquid suspension. The method is considered to be green and sustainable because it is long lasting, not harmful to the environment, and chemically destroys contaminants. The method also can have a residual effect of creating sustained reducing conditions in the subsurface that serve to enhance bioremediation, increasing its utility as a remedial tool.

Phytoremediation can be an effective technology for remediating chlorinated hydrocarbons in soil or groundwater. Plants used for phytodegradation of PCE, TCE, and DCE include poplars and cottonwoods. As discussed above, these trees can also be used to achieve hydraulic control to contain groundwater plumes. Rhizodegradation and phytovolatilization of chlorinated hydrocarbons have also been observed [16].

PCBs, Pesticides, and High-Molecular-Weight PAHs

PCBs, pesticides, and high-molecular weight PAHs are three of the most common classes of persistent organic pollutants (POPs), toxic chemicals that resist chemical and/or biological degradation and therefore persist in the environment. Although many of these compounds are now banned or highly regulated in the United States, their environmental persistence necessitates ongoing remediation efforts of historically contaminated sites.

Many POPs are polychlorinated or polycyclic aromatic compounds. The incorporation of multiple halogen atoms (chlorine, bromine, etc.) or multiple aromatic rings into organic compounds tends to make the compounds more difficult to degrade both chemically and biologically. Because POPs are difficult to degrade under environmental conditions, containment or ex situ treatments are typically required.

POPs are typically characterized by low solubility and, therefore, are most often found contaminating solid materials including soils and sediments. Incineration is often the presumed remedy for solid wastes contaminated with POPs, which are destroyed at high temperatures. Benefits of incineration include near-complete destruction of a wide variety of POPs and the ability to treat large volumes of contaminated material on relatively short timeframes. However, incineration can create toxic by-products (e.g., dioxins and furans) or volatilize heavy metals and, therefore, must be carefully engineered to prevent adverse health effects. Incineration also requires large quantities of energy to excavate the contaminated soil and to reach temperatures that often exceed 1,200°C.

Specialized technologies developed as alternatives to incineration include a variety of techniques for ex situ chemical degradation of POPs [20]. However, these technologies typically require high temperatures and caustic solutions or solvents, due to the innate stability of the chemicals. Onsite reuse of treated soil for backfilling can make these technologies somewhat greener. Vitrification can also be used to destroy POPs at high temperature, while containing any residual contaminants in the resulting glass, though this technology is not considered sustainable, due to excessive energy consumption. The large environmental impacts of most technologies for POP treatment highlight both the difficulty of treating these classes of contaminants and the need for additional innovation in this area of remediation.

In response to the need for innovation, various specialized technologies have been developed for bioremediation of PCBs and specific pesticides, typically via dechlorination. These technologies use specialized organic amendments, nutrients, or proteins to stimulate microbial activity that drives dechlorination of the POPs. Bioremediation is useful only for wastes with low POP concentrations, due to the much slower rate of degradation relative to more energy-intensive processes. Because of the specific nature of microbial processes and the difficulty of either dechlorinating or breaking the polyaromatic structure of POPs, technologies are often specific to the compounds for which they were developed and require testing for applicability to other POPs. These technologies are significantly more sustainable than incineration and chemical degradation processes, due to significantly lower energy requirements and waste production, and because some of the technologies can be utilized in situ at sites where conditions allow.

Another emerging technology uses mechanical agitation of POP-contaminated waste with a chemical reactant in a ball mill or similar machine to drive dechlorination, such as for PCBs. Although this technology requires excavation of the contaminated material, it has relatively low energy requirements, because mechanical energy is substituted for thermal energy in promoting degradation of the POPs. The sustainability of variations on this method is further affected by the nature of the chemical reactant used and waste products created.

Ongoing research into application of phytoremediation to POPs has revealed plants capable of phytoextraction and rhizodegradation of PCBs, pesticides, and PAHs. Phytoremediation is only possible where POP concentrations are low enough so that they are not toxic to the plants, and it is particularly promising for treatment of low-level pesticide contamination of surface soils. Where feasible, this technology offers a highly green and sustainable alternative to the high energy intensity and site disturbance required by most POP remediation technologies.

Explosives and Energetics

Explosive and energetic compounds that are in widespread use in military and industrial applications are common contaminants of concern, on account of their relative solubility and persistence in the environment. Common examples of these contaminants include perchlorate, 2,4,6-trinitrotoluene (TNT), hexahydro-1,3,5-trinitro-1,3,5-triazine (RDX), and octahydro-1,3,5,7-tetranitro-1,3,5,7-tetrazocine (HMX).

As with POPs, costly and energy-intensive remedial technologies such as incineration are often used to treat energetic compounds in soil. However, these compounds are less recalcitrant than POPs and can be degraded biologically under appropriate conditions.

Although microbial degradation of these compounds is typically slow, enhanced bioremediation of perchlorate, TNT, RMX, and HMX can be achieved by addition of a variety of carbon and/or nutrient amendments to groundwater, either in situ or ex situ. Investigations have shown that bacteria capable of degrading these energetic compounds, under both oxic and anoxic conditions, are widespread in the environment, but specific bacteria can also be added to promote degradation [21, 22]. Ex situ technologies that have been proven effective include a variety of bioreactors and composting operations. Organic energetic compounds (TNT, RMX, HMX) can be degraded to other organic by-products, or mineralized all the way to carbon dioxide. Perchlorate is reduced to chloride and oxygen gas.

Phytoremediation technologies for energetic compounds include rhizodegradation and phytodegradation of TNT by a variety of plants including hybrid poplars. Studies have also provided evidence for phytoremediation of perchlorate, RDX, and HMX. Constructed wetlands have also been found to be effective for remediating a variety of energetic compounds, through a combination of bioremediation and phytoremediation.

The BMPs and sustainability benefits of microbial degradation and phytoremediation of explosive and energetic compounds are similar to those discussed above for other classes of organic contaminants.

Metals

The diversity of metals and sources of metal contamination in the environment necessitates diverse contaminant- and site-specific remedial approaches. Common metal contaminants identified for remediation include arsenic, chromium, lead, and mercury. Modern and historical sources of metal contamination include mining and smelting operations (including acid mine drainage), steel production, landfills, firing ranges, battery recycling and disposal operations, metal plating facilities, wood treatment facilities, coal combustion by-products (e.g., coal ash), dyes and paints, lead arsenical pesticides, and leaded fuels.

Unlike organic contaminants, metals cannot be biologically or chemically degraded to innocuous by-products. Therefore, remediation of metals typically relies either on removal or on sequestration in solids, which limits mobility and potential exposure. Various methods exist for increasing the sustainability of technologies for both removal and sequestration.

In situ stabilization and geochemical fixation of metals uses chemical or biological processes to transform metals in soils or aquifers from soluble and/or toxic to insoluble and/or nontoxic forms. The chemistry of stabilization is dependent on the specific metal(s) of concern. Different metals have distinct properties that affect their mobility and bioavailability under different environmental conditions, and remedial technologies must be designed to address the specific site conditions relative to the properties of the metals of concern.

Redox state can be an important determinant of solubility and toxicity for redox-active metals, such as chromium, arsenic, iron, and manganese. For example, hexavalent (oxidized) chromium is much more soluble and toxic than trivalent (reduced) chromium, whereas iron and manganese are typically less soluble and toxic under oxidized conditions. Reducing conditions can also lead to sulfide production, and sequester metals through the formation of metal sulfide minerals. Therefore, addition of chemicals that stimulate reduction or oxygenation of the contaminated soils or aquifers can be designed to cause the metal(s) of concern to precipitate and become less bioavailable.

Various properties of soil and water geochemistry also affect metal behavior. Many metals are more soluble and bioavailable under acidic (low pH) conditions, and carbonate, in the form of lime or limestone, can be added to increase pH. The presence of other geochemical species (e.g., phosphates or organic matter) can cause metals such as lead, zinc, copper, cadmium, nickel, and uranium to be retained in the solid phase. Phosphate causes metals to precipitate, whereas addition of organic matter tends to increase metal sorption to soils. Addition of solids that tend to adsorb metals, such as iron oxyhydroxides, can also effectively sequester metals.

Implementation of in situ stabilization typically requires that an aqueous or solid chemical reactant be injected into the contaminated aquifer or mixed with impacted soils. Application of this technology in aquifers therefore requires the installation of injection wells, while remediation of soil requires equipment sufficient to mix to the depth of desired treatment. The BMPs for in situ stabilization include some discussed above for enhanced biodegradation and ISCO, technologies which require similar injection of reactants, and for excavation, which requires site disturbance similar to soil mixing.

As an in situ technology, stabilization offers relatively low energy requirements, as compared to extractive technologies. The environmental impacts of the stabilization as applied to groundwater are dependent on the frequency of injection events and level of automation of the system, which can minimize the number trips to the site for operations and maintenance of the injection system. Metals stabilization in soils has minimal energy requirements after the initial mixing. Use of recycled materials, such as composts, wood or coal fly ash, or red mud from the alumina industry, as treatment additives can further increase the overall sustainability of the remedial effort.

Limitations of this technology include the difficulty of achieving thorough mixing with the contaminated medium, possible limitations to future site use, and the necessity of treatability studies to demonstrate that metal solubilities will meet criteria after treatment. However, in sites where the contaminant and physical characteristics are well suited for in situ stabilization, it can be a highly effective and environmentally sustainable remedial option.

Phytoremediation of metals uses plants to extract and/or contain contaminants in environmental media. Whereas remediation of organic compounds by plants often focuses on contaminant degradation, phytoremediation of metals focuses on accumulating the metals within plant tissues, volatilizing them through the plants, stabilizing them in soils, or otherwise limiting metal mobility and bioavailability.

A wide variety of metals can be removed from solid materials or water through phytoextraction, which results in accumulation of metals in plant tissues. Plants used for phytoextraction are often chosen for their ability to tolerate and/or "hyperaccumulate" a specific metal of concern. For example, Chinese brake ferns are used to remove arsenic from shallow soils (Fig. 13.4). Because the metals are not degraded, plants typically must be harvested and disposed of as waste following metal uptake. Plants are also used for phytovolatilization of certain metals (e.g., selenium, mercury, and arsenic) through transpiration.

Plant-based containment technologies for metals include phytostabilization within the soil, achieved through chemical or biological processes that decrease metal solubility, and hydraulic control, which uses water uptake by trees to limit the mobility of dissolved metals.

Monitored natural attenuation, discussed in detail above in reference to organic contaminants, is also applicable to dissolved metals in aquifers, in cases where the natural groundwater chemistry results in immobilization of the contaminants. For example, hexavalent chromium may be naturally oxidized in aquifers with sufficient dissolved oxygen. Precipitation of metals at naturally high pH or adsorption of metals to solid materials present in the aquifer (e.g., minerals and organic matter) can also result in the attainment of remedial goals for dissolved metals.

As with remediation of organic contaminants, monitored natural attenuation of metals requires collection of evidence to indicate that natural attenuation is occurring and that no adverse impacts to humans or the environment are present. Where

Fig. 13.4 Use of Chinese brake ferns for remediation of arsenic in shallow soils

these conditions are met, this remedial strategy is a highly sustainable method for achieving remedial goals with minimal energy inputs.

Excavation is not typically considered a green or sustainable technology, due to its high energy demands and the need to contain excavated waste within a disposal facility. However, as discussed for petroleum hydrocarbons above, certain methods can be incorporated into excavation projects to improve their sustainability. For metals, one additional option is excavation followed by metal recycling, which can be an environmentally responsible remedial option for waste with high metals concentrations, such as slag from metal smelting operations and munitions from firing ranges. The economic feasibility of recycling is often dependent on the concentration of metals, the quantity of contaminated material, and proximity of recycling facilities.

Future Directions

This entry has provided practical concepts, value engineering principles, and BMPs for the current state of remedial technologies for the typical classes of environmental contaminants observed in the subsurface. These concepts and practices represent the qualitative principles currently utilized to improve or increase the sustainability of subsurface remediation.

Although these BMPs are helpful, is it not possible based on BMP application to evaluate their ultimate benefit to improving sustainability. In the near future, the environmental industry will move away from the general practice of subjectively reviewing and applying BMPs from a master list to the practice of applying project-specific, robust, quantitative analyses of sustainability benefit for a project. Green and sustainable remediation organizations like SURF have already created a compilation of comprehensive metrics (Metrics Toolbox) that can be used to evaluate, track, and forecast a remedy's ability to achieve certain outcomes in relation to sustainability goals [23]. In the future, metrics like those in SURF's Metrics Toolbox will be supplemented with a wider suite of metrics to perform analyses of sustainability key elements in remedial program decision making.

Although it is envisioned that greener remedies have a distinct place within regulatory programs, such as the US Environmental Protection Agency (EPA) Comprehensive Environmental Response, Compensation, and Liability Act (CERCLA) or Resource Conservation and Recovery Act (RCRA), program-specific regulatory criteria do not currently address the social and economic considerations of sustainability. The US EPA is presently attempting to clarify the role of green remediation within the CERCLA and RCRA programs; however, its ability to include the social and economic benefits in the remedy selection process may exceed the authority of these programs. The task of defining what is meant by the term "sustainable" in terms of remedial measures remains an ongoing effort [24] and will continue to evolve in coming decades through collaboration between researchers, regulators, and environmental remediation professionals.

Bibliography

Primary Literature

1. Office of Technology Assessment (OTA) (1984) Protecting the nation's groundwater from contamination, report
2. Howard PH, Boethling RS, Jarvis WM, Meylan WM, Michalenko EM (1991) Handbook of environmental degradation rates. Lewis, Chelsea, p 725
3. Suthersan SS (1997) Remediation engineering: design concepts. Lewis, Boca Raton, p 362
4. Kalderis D, Juhasz AL, Boopathy R, Comfort S (2011) Soils contaminated with explosives: environmental fate and evaluation of state-of-the-art remediation processes (IUPAC technical report). Pure Appl Chem 83:1407–1484
5. Liu L, Tindall JA, Friedel MJ (2007) Biodegradation of PAHs and PCBs in soils and sludges. Water Air Soil Pollut 181(1–4):281–296
6. U.S. Environmental Protection Agency (EPA) (1989) Risk assessment guidance for superfund. Volume I: Human health evaluation manual, Part A, EPA/540/1-89/002. U.S. Environmental Protection Agency, Office of Emergency and Remedial Response, Washington, DC
7. U.S. Environmental Protection Agency (EPA) (1989) Risk assessment guidance for superfund. Volume II: Environmental evaluation manual, EPA/540/1-89/001. U.S. Environmental Protection Agency, Office of Emergency and Remedial Response, Washington, DC
8. Office of Solid Waste and Emergency Response (1995) Use of Risk-based decision making in UST corrective action programs. OSWER Directive 9610.17, Mar 1995. Office of Solid Waste and Emergency Response, Washington, DC

9. Foster Wheeler Environmental Corporation (Foster Wheeler) (1998) RBCA fate and transport models: compendium and selection guidance, prepared for ASTM, Nov 1998, pp 1–26
10. Ellis DE, Hadley PW (2009) Sustainable remediation white paper – integrating sustainable principles, practices, and metrics into remediation projects, remediation. Wiley Interscience. doi:10.1002/rem.20210
11. U.S. Environmental Protection Agency (USEPA) (2011) Green remediation best management practices – sites with leaking underground storage tank systems, EPA 542-F-11-008, June 2011. Office of Solid Waste and Emergency Response, Office of Emergency and Remedial Response, Washington, DC
12. American Academy of Microbiology (2011) Microbes and oil spills (FAQ), p14
13. Crane CR, Sanders DC (1967) Evaluation of a biocidal turbine fuel additive, Aug 1967. Federal Aviation Administration AM 67-21, Washington, DC, p12
14. U.S. Environmental Protection Agency (USEPA) (2009) Green remediation best management practices – site investigation, EPA 542-F-09-004, Dec 2009. Office of Solid Waste and Emergency Response, Office of Emergency and Remedial Response, Washington, DC
15. U.S. Environmental Protection Agency (USEPA) (2010) Green remediation best management practices – bioremediation, EPA 542-F-10-006, Mar 2010. Office of Solid Waste and Emergency Response, Office of Emergency and Remedial Response, Washington, DC
16. U.S. Environmental Protection Agency (USEPA) (2000) Introduction to phytoremediation, EPA/600/R-99/107, Feb 2000. Office of Research and Development, Cincinnati
17. U.S. Environmental Protection Agency (USEPA) (2008) Green remediation – best management practices for excavation and surface restoration, EPA 542-F-08-012, Dec 2008. Office of Solid Waste and Emergency Response, Office of Emergency and Remedial Response, Washington, DC
18. Kessel L, Squire J, Holland K (2008) Sustainable soil remediation by refrigerated condensation at sites with "high-concentration" recalcitrant compounds and NAPL – two case studies. Remediation 19:53–72
19. U.S. Environmental Protection Agency (USEPA) (1998) Technical protocol for evaluating natural attenuation of chlorinated solvents in ground water, EPA/600/R-98/128, Sept 1998. Office of Research and Development, Cincinnati
20. U.S. Environmental Protection Agency (USEPA) (2010) Reference guide to non-combustion technologies for remediation of persistent organic pollutants in soil, 2nd edn. EPA 542-R-09-007, Sept 2010. Office of Solid Waste and Emergency Response, Office of Emergency and Remedial Response, Washington, DC
21. Interstate Technology and Regulatory Council (ITRC) (2005) Perchlorate: overview of issues, status, and remedial options, Sept 2005. ITRC, Washington, DC
22. Esteve-Nunez A, Caballero A, Ramos JL (2001) Biological degradation of 2, 4, 6-trinitrotoluene. Microbiol Mol Biol Rev 63:335–352
23. Butler PB, Larsen-Hallock L, Lewis R, Glenn C, Armstead R (2011) Metrics for integrating sustainability evaluations into remediation projects. Remediation 21(3):81–87
24. Interstate Technology and Regulatory Council (ITRC) (2011) Green and sustainable remediation: state of the science and practice. ITRC, Washington, DC, 43 pp

Books and Reviews

Fetter CW (1999) Contaminant hydrogeology, 2nd edn. Waveland, Long Grove, p500
Holland KS et al (2011) Framework for integrating sustainability into remediation projects. Remediation 21(3):7–38

Illaszewicz J, Gibson K (2009) Green and sustainable remediation: creating a framework for environmentally friendly site cleanup. Environ Qual Manag 18(4):1–8

Pankow JF, Cherry JA (1996) Dense chlorinated solvents and other DNAPLs in groundwater. Waterloo, Portland, 522 pp

Schwarzenback RP, Gschwend PM, Imboden DM (1993) Environmental organic chemistry. Wiley, New York, 681 pp

Stumm WS, Morgan JJ (1981) Aquatic chemistry: an introduction emphasizing chemical equilibria in natural waters, 2nd edn. Wiley, New York, 780 pp

U.S. Environmental Protection Agency (USEPA) (2008a) Green remediation: incorporating sustainable environmental practices into remediation of contaminated sites, EPA 542-R-08-002, Apr 2008. Office of Solid Waste and Emergency Response, Washington, DC

Index

J.S. Gulliver (ed.), *Transport and Fate of Chemicals in the Environment: Selected Entries from the Encyclopedia of Sustainability Science and Technology*, DOI 10.1007/978-1-4614-5731-2,

MIX
Papier aus verantwortungsvollen Quellen
Paper from responsible sources
FSC® C105338

If you have any concerns about our products,
you can contact us on
ProductSafety@springernature.com

In case Publisher is established outside the EU,
the EU authorized representative is:
Springer Nature Customer Service Center GmbH
Europaplatz 3, 69115 Heidelberg, Germany

Printed by Libri Plureos GmbH
in Hamburg, Germany